Green Building and Energy Efficiency

Green Building and Energy Efficiency

Edited by **Matilda Schmidt**

SYRAWOOD
PUBLISHING HOUSE
New York

Published by Syrawood Publishing House,
750 Third Avenue, 9th Floor,
New York, NY 10017, USA
www.syrawoodpublishinghouse.com

Green Building and Energy Efficiency
Edited by Matilda Schmidt

International Standard Book Number: 978-1-68286-323-7 (Hardback)

Printed in China.

Contents

Preface

Green buildings and energy efficiency have become a very prominent part of civil engineering and architecture. The main objective behind these innovations is to encourage the design and construction of buildings that comply with environmental sustainability and energy efficiency measures. This book discusses the fundamental as well as modern approaches for energy efficiency, environmental quality enhancement, use of renewable energy technologies, and sustainable architecture and designs, etc. From theories to research to practical applications, case studies related to all contemporary topics of relevance to this field have been included in this book. As this field is emerging at a fast pace, this book will help the readers to better understand the concepts of green building and energy efficiency.

Significant researches are present in this book. Intensive efforts have been employed by authors to make this book an outstanding discourse. This book contains the enlightening chapters which have been written on the basis of significant researches done by the experts.

Finally, I would also like to thank all the members involved in this book for being a team and meeting all the deadlines for the submission of their respective works. I would also like to thank my friends and family for being supportive in my efforts.

Editor

1

Zero-Energy and Beyond: A Paradigm Shift in Assessment

Ronald Rovers [1,2]

[1] Sustainable Building Support and Consultancy, Waalre 5581 HH, The Netherlands; E-Mail: r.rovers@sustainablebuilding.info

[2] Research Institute Built Environment of Tomorrow (RiBuilT), Zuyd University of Applied Sciences, Heerlen 6419 DJ, The Netherlands

Academic Editor: Alice Moncaster

Abstract: The world is on the eve of major changes in the way energy and material are used, and the building and construction sector is at the forefront. One of the revolutionary changes is that for 0-energy houses and buildings. Many countries already have some projects established, and legislation is following, first requiring *near* 0-energy, but undoubtedly this will evolve into 0-energy as basic requirement. Buildings will generate all required energy from within their building lot, from incoming free and renewable energy sources: solar radiation and earth core heat mainly. In other words, there are no polluting or depleting issues anymore related to energy consumed to operate a building. This will change the whole approach in evaluation and optimization of the environmental performance of buildings: while the energy-driven measures for buildings become obsolete, it will be materials needed for this transition that have to become the main focus, as argued in this paper.

Keywords: 0-energy; buildings assessment; exergy; solar radiation; materials; closed cycle; paradigm shift

---

1. Introduction

Europe is on the eve of major changes in building and construction, and one of the revolutionary changes is in the direction of 0-energy houses and buildings. Many countries already have exemplary pilots established, and in a country like Germany they are on regular sale. Legislation is following fast, at first demanding *near* 0-energy buildings by 2020 [1], but undoubtedly this will evolve into 0-energy as a basic requirement for new constructions.

A 0-energy building can be described as a building that generates all required energy from within its building lot, based on incoming free and renewable energy sources, mainly solar radiation and earth core heat. The operational energy consumption will be free of pollution and fossil fuel depletion. This will change the whole approach in the evaluation and optimization of the environmental performance of buildings. The energy-driven measures for buildings will become obsolete and materials will take over as the main concern to address, as we will argue in this paper.

2. Focus on Energy Efficiency

For some 40 years now the building sector has targeted energy "reduction" measures. At first packages of separate measures were introduced, such as making buildings and houses airtight, introducing double glazing and insulating floors and roofs. This was followed by a more integral approach, supported and stimulated by government regulation and legislation about methodology, setting mandatory targets in energy demand for houses and offices.

More recently a race has started to maximize the reduction in energy demand. Reduction has been brought to the extreme with the introduction of the so-called Passive House standard, requiring less than 15 kWh/m^2 for heating.

In recent years an additional option has become widely available: energy generation by renewable sources. Off and on-site solar collectors for heating experienced a fast introduction in sunny countries like Greece and Israel, followed by a breakthrough for solar electricity panels, "Photovoltaic (PV) panels", mainly because the Chinese were able to break open the protected industrialized world market for Photovoltaic panels with a rapid decrease of consumer prices.

Thus, a "double" approach to tackle fossil fuels became possible: reducing operational energy demand on one side and producing renewable energy to supply that reduced demand on site, on the other side, resulting in the quest for 0-energy houses and buildings. This combination will change the playing field drastically, and, as this paper will show, optimizing building designs for operational energy only has become sub-optimizing.

The exact definition of a 0-energy house is not the subject of this paper. It has been studied by many, among others by the International Energy Agency (IEA) annex 52 work, and an in depth analysis of the term was performed earlier by the author [2,3]. The definition that is used in this paper is that a 0-energy building produces as much renewable energy on site (*i.e.*, below or behind the building) as it requires operationally, with the electricity grid to balance overproduction and shortages due to seasonal and temporal differences.

The main reason for the initial race towards extreme energy reduction in buildings was, and still is, (the *side* effects of) the use of fossil fuels. The environmental impact on a global scale of fossil fuels consumption on a small scale and with low population levels might not be significant, but with a global population of 7 billion people (and growing) in combination with a fossil fuel based society, the notion has evolved that we are significantly disrupting our own appreciated environment, climate change being the most visible and threatening effect on a global scale. Other side effects include air pollution, by sulfur and nitrogen oxides, smog in the world's cities, and air pollution by fine-dust, creating health problems. There are a range of other, often more local, negative side effects, such as large scale oil spills, tar sand extraction, mountain top removal, and virgin forest removal. These side effects ruin nature on a regular basis.

This is not to speak of causing wars, with oil and gas stocks as the major cause of clashes [4]. In the end, we even face depletion of these fossil fuel resources, which would not only eliminate an energy resource, but create a problem in the production of other valuable products, for instance medicines and some plastics.

Despite the focus on energy efficiency, mankind has not been very successful in reducing fossil fuel use: up to today the demand is still on the rise. When assessing buildings it can be concluded that houses are heated more efficiently, but at the same time have largely increased in number and size eliminating the achievements while at the same time raising in required comfort standards. An Amsterdam canal-house in 1900 required the same amount of energy for heating as a well insulated Dutch reference house of the year 2000 [5]. The difference is that the family in the Amsterdam canal house only heated one room, and only in the evening when all the family was sitting together. In the year 2000 all (and larger) rooms are heated for a comfortable 20 degrees all day long. The increase in efficiency is lost in rebound effects such as more heated space, higher temperatures and for longer periods [6].

2.1. Energy and Materials Use

The race for ever more energy efficiency led to the development of the "Passive-house concept". This original German approach, aims to reduce energy demand for heating as rigorously as possible, setting targets for new housing at less than 15 kWh/m^2 and at less than 25 kWh/m^2 for existing, renovated dwellings. In order to reach these levels of efficiency, thick insulation packages are applied, of up to 40–50 cm with traditional materials, as well as extreme airtightness and cold bridge interruption (Figure 1).

Consequently, extreme reduction measures require enormous amounts of materials investments. The (fossil) energy impacts are reduced, but at the cost of increasing material impacts.

It seems a logical approach, when the only aim is to reduce fossil fuel consumption, and the material consumption is left out of the evaluation.

To bridge the gap between this low energy demand and 0(-fossil fuels)-energy buildings, building connected renewable energy production is introduced, requiring additional installations and conversion devices like PV panels, inverters, solar heat collectors, heat storage tank, or aquifer devices, compared to centralized energy production.

Reaching 0-energy, therefore, implies two things change: firstly, fossil fuels (related impacts) have no relevance anymore in the evaluation of operational energy decisions, since there are none involved anymore; and secondly, the step from low (fossil demand) to 0 (fossil demand) requires a drastic increase in materials use. This is where the paradigm shift in approach starts: for 0-energy buildings, operational energy, in general, is not an issue of relevance anymore. Evaluating energy demand has become obsolete: the sole question resulting is that of optimizing materials related impacts.

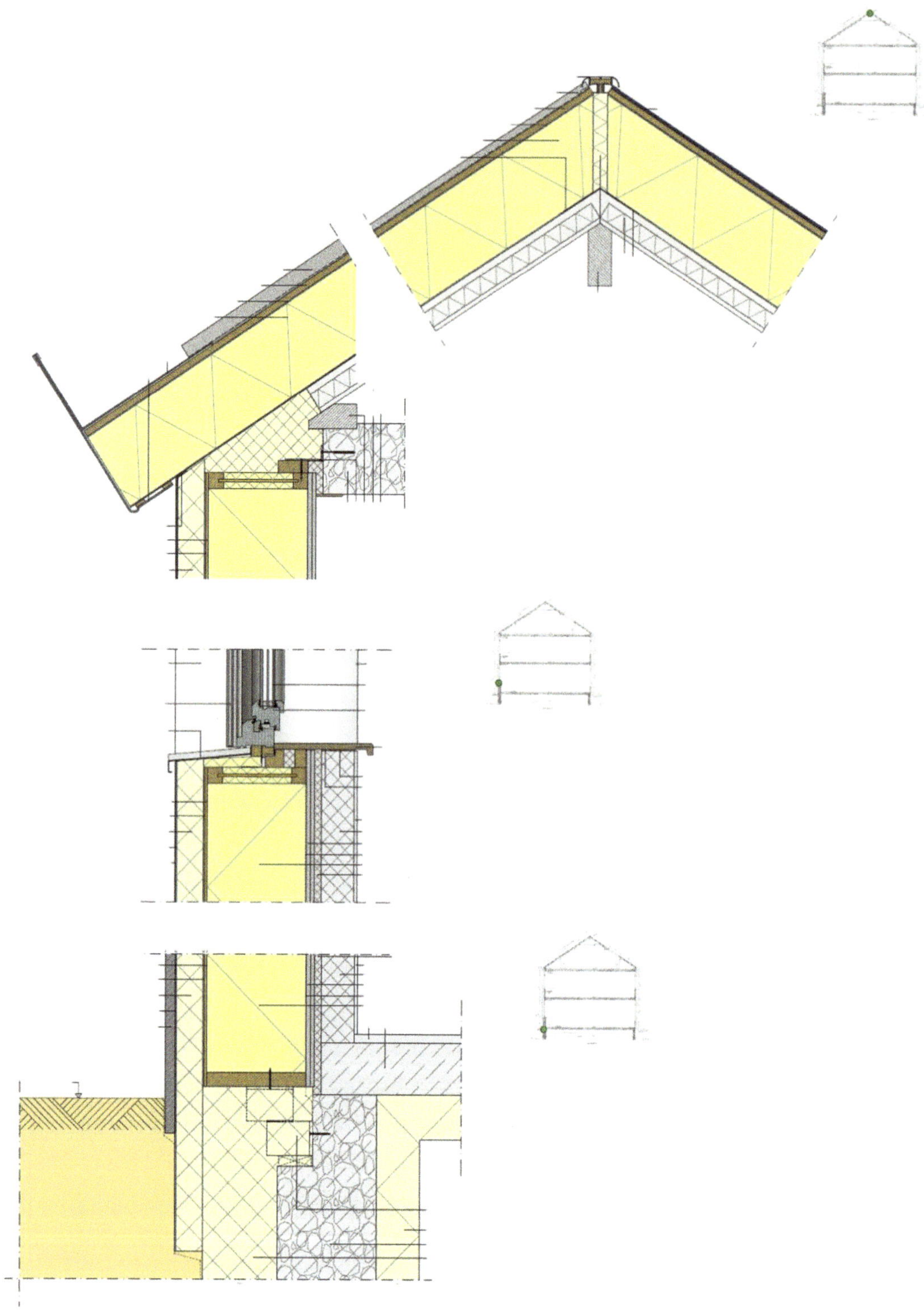

Figure 1. Passive house level renovation in a local project showing the large additional layer of materials needed for energy reduction only (yellow) [7].

This requires a complete shift in focus for environmental optimization of our buildings, geared towards a minimal balance between demand and production, in materials impact.

In fact, energy demand for a 0-energy house does not imply in any way a low energy level. The energy balance can exist at any level, low, average or high demand: The balance between demand and supply is determined by the lowest impact of the combined material investments in demand reduction and supply devices and materials.

The situation is best illustrated with imagining the design of a new building: Why in the conceptual design phase, would you start with insulation levels and wall thickness, when no fossil fuel based energy is planned to be used? Reducing (operational energy) emissions is not a target. They (the emissions) are absent by definition. As a designer you will have to start finding ways to maximize energy generation on site, and use the productive/conversion elements in the design. From there, the next step is to find out how much of the possible demand can be covered. From there, the balancing act begins: finding the optimum between materials to reduce energy consumption and materials to produce energy.

This paradigm shift has consequences for the way in which we assess buildings, the way we measure and value energy impacts, and ultimately how legislation is developed.

2.2. Energy Reduction versus Energy Production

A first observation is, that in case of 0-energy buildings, the energy quest, related to the environmental impacts of fossil fuels, is limited to fossil fuels used to harvest and produce materials and products for buildings.

Fossil energy embodied in materials and products will become the only contributor to climate change and fuel depletion from buildings. The focus therefore shifts to embodied energy (from still fossil fuels), and material impacts in the broad sense. To create/construct "m^2 of shelter" a basic level of materials consumption is unavoidable, however, in order to optimize the energy-impact performance of that building, by means of the input of materials to limit energy demand or increase energy production on site, has now become a materials balance question. How much materials, is invested for energy demand reduction, and how much for on-site energy production? More of one decreases the need for the other, and vice versa. The main question is: Where is the balance?

A Dutch national reference terraced house of the 1970s was chosen to explore this balance in retrofitting the house for a 0-energy performance [8]. To reduce energy demand four possible levels of insulation were defined: (1) no insulation; (2) cavity wall insulation plus floor and roof between beams; (3) one level of outside skin insulation added; (4) second extra layer of outdoor insulation, requiring a supporting framework. This leads to four levels of (reduced) energy demand. To reach the 0-energy performance status, a certain amount of m^2 PV has to be installed, depending on the level of insulation and remaining heating demand. This leads to four combined packages: the amount of materials for insulation going up, the amount of materials for PV panels going down. The resulting calculation showed that the optimal strategy is to only insulate the cavity wall (40 mm). The optimal balance, in terms of lowest embodied energy for the combined materials input (insulation for reduction and solar panels for production) turns out to be at the second highest level of operational energy.

The total required operational energy decreases with each step, unmistakably, but that is not relevant because it is solar radiation! This operational energy is included in the PV panels: using PV panels, the material/product comes *inclusive* with free non-polluting energy output. So it is no issue anymore in the evaluation, and only embodied energy (from fossil fuels) is the fair impact evaluation (energy wise).

Even if we include (materials for) on site storage of energy (instead of using the (free-riding) grid for balancing energy), and accept losses in conversion steps, the optimum remains level 2 (cavity wall insulation).

More calculations have to be made for different cases, but the trend is obvious: only optimizing for operational energy is sub-optimizing: the materials impacts have become the decisive factor.

With (energy embodied in) materials becoming a main material related decisive factor for fossil fuels related CO_2 emissions, a subsequent question could be, can we develop a "0-embodied energy building", in other words, a 0-fossil-fuels-embodied-energy building, as an addition to the 0-fossil-fuels-operational-energy building?

3. Embodied Energy and Fossil Fuels Strategies

Materials becoming the crucial factor, will change the playing field on the construction market rigorously, since lowering embodied energy will be the competing field for building parts and devices (and industry), in relation to the (0-) operational energy balance established with their product. In order to address "0-energy" (no fossil fuels) in the production chain of building parts and equipment, three main strategies can be identified [9]:

1. Further reduction of material demand (-impacts) by alternative materials and design;
2. Production of materials and products on site (construction site), using renewable energy produced on site as well;
3. Collect products, which have been produced as (net) 0-embodied energy products by the construction industry, and arrange for suitable transport.

3.1. Reduction

Many studies have addressed material demand reduction, but there are examples from practice as well. The wood construction company we contacted for delivering the main load bearing construction for a pilot 100% biobased house [9], came up with a design which connects the (wooden) poles and beams with wooden pins, which reduces the embodied energy compared to construction with iron angles and screws. They see this as a future market opportunity.

3.2. On Site Production

Production of materials on the building site itself is an option for a few construction methods, such as earth ships and adobe construction, but we will not further elaborate on this option within this paper.

3.3. The Products Chain

Building parts production in a factory with a 0-energy targeted approach, will require the same approach as for the building itself: all energy used to produce materials and products should be from on site renewable production: on site can be at the quarry, or the product-factory.

There are a few examples of industries that already run on renewable energy, produced on site. For a pilot renovation project [10] we worked with a company that makes prefab wooden facade elements, producing one third of their energy by on site generated renewable energy and the aim of the company is 100% within two years. Therefore, when compiling the materials and products list for a building, a selection could be made for products from such companies, and have a lower fossil related embodied energy target (partially) realized.

A next element in the production chain using fossil fuels is transport. The question can be posed: how to establish 0-energy-transport? New developments are underway. Ocean vessels have been studied for instance, and can use renewable energy by means of wind sails (parachutes) to relieve the fuel engines, and up to 50% reduction of oil is possible as reported from the first trips [11]. However, a more obvious approach would be to reduce transport distances and use local materials in the first place.

It is obvious that going to 0-energy and beyond puts the focus on materials, and will require a complete shift in the materials production chain, and change the market functioning, as well as design and detailing buildings.

4. Materials: A Broader Perspective

Besides the focus on embodied energy, (being the remaining fossil fuel based fraction), the use of materials resources itself has many impacts: depletion of (re-)sources, pollution by production processes , and not to forget the so called "rucksack" effects: to mine materials comes with side effects like biotic and abiotic resource investments, air and water use, land degradation, *etc.* [8]. The availability of materials has never been part of the main discussion in the building sector, or the side effects. In the light of the 0-energy paradigm shift they will, however, become the major issue to address. This coincides with current developments in global materials stocks, which will require changes in materials treatment.

There are many discussions amongst scientists whether we will run out of (some of our) resources through current consumption patterns [12], sometimes related to money, sometimes related to geopolitical situations, sometimes based on the assumptions of still to discover large quantities [13,14].

However, for many important metals and minerals the situation is actually getting critical, since ore grades are reducing steadily, all over the world [15], and have sometimes reached 0.5%, while coming from a 2%–5% grade. There are not much signs for improvement, as can be seen by the discovery rates of new mining areas [16]. With population and consumption growing, the situation will only get worse. Even more threatening is the fact that this seems a trend for all metal sources. These all show the same trend in decreasing ore grades. This implies that the amount of energy invested in mining and treating ores is increasing fast. The current ore grade of 0.5% requires ~100 MJ/ton copper, which is more or less double from that of ore grade of 1% [17]. From 0.5 to 0.25, this will double again. So the point is not only scarcity, but concentration levels, directly influencing the power input to make more and more diluted resources available.

This requires industry to increase energy input to make materials available, which implies that embodied energy per unit product is not a fixed amount, but over time progressively growing per unit of material due to ore grade dilution. Whether that will be fossil energy or renewable energy is not so relevant: the first has nasty side effects and depletes, the other requires—again—large material inputs for capture and conversion devices. Even if industry would succeed in producing more and more with renewable energy, this again will require lots of materials for renewable energy generation (PV panels, wind turbines, *etc.*). The materials consumption is to be seen as a runaway process.

There are a few strategies that provide a way out of this rebound effect from 0-energy building:

- Reorganization of functions and services;
- Reducing the amount of materials consumption;
- The use of renewable materials: These (can) have no depletion effects (if regenerated), come with a smaller rucksack effect and on average lead to a lower embodied energy (for providing the same function). In fact it comes down to the same strategy we applied for years for the (operational) energy quest, but now solely targeted for materials: reduction and use of renewables;
- Reuse and recycling.

The most important and effective strategy is changing the way services are provided. A case study in a regional neighborhood seemed to point to the impossibility of reaching a 0-target for an existing urban district, while maintaining all the functions and services that people were used to. Rethinking the way the district functioned, revealed that many functions could be provided with the same outcome, but using a different strategy, as, for instance, for cleaning clothes. The solution is not in more energy efficient laundry machines, individually in each house, to reduce energy demand, but in laundry shops at district level: making all laundry machines (and their materials and energy burden) obsolete, and replaced by one, more industrial sized type. It also adds to local employment (with a zero energy pick up and delivery service). It decreases the energy and materials impacts combined, while maintaining the desired service: cleaning cloth. More options have been identified in this research [18].

An example of a reduction strategy is the one that was investigated in the 1980s, when energy efficiency had just been launched as a topic: it was about compartmentalization of houses: during a cold winter period it was seen as wise not to heat the whole house, but only main living areas. It could become the main strategy again, when we start optimizing 0-energy approaches in the sense described above.

The third option is change to renewable resources, also known currently as “biobased building”. It has growing interest and formal and governmental policies already addressed this approach. Renewable and or biobased materials have in general a much lower embodied energy, and rucksack effects are much lower on average as well. Average contemporary housing is around 5 GJ/m^2 floor embodied energy [19,20] while a mainly renewable materials based house can go down to 3 GJ/m^2 and even lower [9,21]. The IEA-SBC annex 57 working group recently established is especially addressing the use of embodied energy data for practitioners. Part of the work is bringing together literature and cases with embodied energy calculations [22].

Recycling and re-use are well known options. They do, however, not fully replace the impacts from new virgin resources. More research will be required when it comes to deciding how to deal with the first time use impacts: recycled materials are in principle nothing else than prolonged use, decreasing the impact per unit of service in time, but not free of impact [23].

5. Consequences for Assessment

The trend for 0-energy buildings will influence how buildings, and more in detail energy and materials performances of buildings, need to be evaluated. A few of the consequences for assessment methods and common indicators are discussed here.

5.1. CO_2

CO_2 emissions are a side-effect from (fossil) energy use, and are currently used as a favorite structuring indicator. Focusing on CO_2 emissions as parameters for improvement is focusing on an end of pipe effect from a specific cause. A cause, fossil fuel use, that has more indirect effects: for instance resource depletion. CO_2 focused approaches, can for instance lead to carbon capture and storage (CCS): in that case you score on reduction of CO_2, (and limit climate change possibly) but this has no effect regarding all other effects of the same cause, like (fossil) fuel depletion. This is in two ways not very wise: CCS is an end of pipe approach neglecting cause (and not even a real solution, it is "sweeping the dirt under the carpet", one day you will have to confront yourself with it again—compare this with nuclear waste), and secondly resources are still depleted, (and wars fought over it remain).

Not to mention that more and more discussion is ongoing about the fact that CO_2 (reduction) is not the effective measure to fight climate change that it was thought to be [24].

As the example of the 0-energy buildings paradigm shift shows, it is however of no use even to continue to address the CO_2 measure as the guiding principle for building evaluation. It might reduce CO_2 emissions, but not fossil fuel use, and overlooks the fact that the materials balance becomes guiding instead of operational energy. It makes no sense to assess a 0-energy building in terms of CO_2 for operational energy.

5.2. Primary Energy

Primary energy has been developed as a way to calculate system optimization from an energy point of view, assuming a limited system (for instance a building) and assuming fossil fuels are available and referred to as the highest energy density input. The "impact history" of fossil fuels is left out of the evaluation: fossil fuels are renewable energy as well, since they originate from biomass. Primary energy may have worked in the past century, but since we will be running out of fossil fuels, and are in a global transition towards 100% renewable energy by 2050 (if all goes well), with buildings at the forefront of running only on renewable energy, it is counterproductive and outdated to continue referring to primary (fossil) energy: calculations should be based on final energy, and how this can be provided with renewable sources. We would even like to make the case here to reverse it and relate it to prime solar radiation, to make a scientific proper comparison. First steps in that direction have been made already by Odum in the 1980s, known as the "Emergy" approach [25,26], and recently by our research group, known as the "Maxergy" approach using Embodied Land© as indicator. In this "Maxergy" project, materials (depletion and compensation) are introduced into the equation, with land use (to make "exergy" available via solar radiation) as the common denominator [27]. More work has to be done in this field.

5.3. Towards a New Assessment Methodology

In summary: As we want to become independent from fossil fuels somewhere towards 2050, we always have to optimize from an energy and materials combined evaluation, in which energy is in all cases provided with renewable energy. Ergo, the energy as such is not the issue, but the materials invested to reduce demand and harvest the renewable energy, for producing conversion devices. The whole optimization comes down to a materials evaluation. Besides the consequences for construction and production as above described, this has significant implications for many established evaluation approaches.

With the transition towards renewable energy and soon accompanied by a focus at renewable/biobased materials, the old reference model of CO_2 and primary energy will be obsolete. Assessment has to go through a transition as well.

- For most prevailing assessment tools, the development towards 0-energy buildings and better, will require a re-focus on the actual parameters measured, and how they are integral valued without weighing factors.
- An even more drastic change would be the practice of using benchmark buildings: many assessment tools refer to a benchmark building, and measure relative improvement form this past. This, however, has been designed and optimized from a fossil fuel optimized design perspective. The playing field has changed, and therefore the benchmark has to change as well: assessments should shift from relating to an old benchmark, with outdated references, to measuring how far of the building's performance is from the 0-situation. This will require the calculation the materials balance, knowing that, even if the building is not 0-energy today, somewhere in future it will be. Focusing today on only fossil fuels or CO_2 (-benchmarks), leads to sub-optimization, with devastating effects for the materials impacts and stocks.
- Data currently in use to evaluate materials and products have to be adapted. As argued before, the data are mostly given in primary energy. This is hindering the shift in approach, since it makes it difficult to compare solar energy-based activities with fossil fueled activities. Since the transition toward renewable energy will count for all energy used, the data should become available in final or end use energy for any function, to be able to compare the required input from solar radiation (and related material impacts). If not, we will continue measuring with two different systems, and delay transition enormously. If the data are shown in end use energy, both ways of calculation are possible, with the primary energy approach being phased out gradually.

5.4. Policies and Legislation

The transition will also affect the way we enforce developments with legislation. The mandatory Energy Performance Coefficient (EPC) in the Netherlands is further strengthened by January 2015 from 0.6 to 0.4 (requiring another 50% increase in efficiency). However, focusing only on energy is shifting an enormous burden to the materials part. Its intended that in future this EPC goes to 0, but again, without addressing the materials impact. As we argued in this paper, reducing energy demand is not the issue anymore when fossil fuels are no longer involved in operational energy.

The EPC has done its work over the past years, but should be replaced with a materials indicator, in order to avoid increasing environmental impact overall instead of reducing only energy demand.

A similar issue is at stake with the European Union (EU) legislation, more specifically, the directive known as the Energy Performance of Buildings Directive (EPBD) recast, dating from 2010, which requires countries to take measures to be implemented by 2018 and beyond [1]. The directive requires that all public buildings from 2018 on, and all other buildings from 2020 on, should be (near) 0-energy. The article 2 in the EPBD states that [1]:

> *"Nearly zero-energy building" means a building that has a very high energy performance. The nearly zero or very low amount of energy required should be covered to a very significant extent by energy from renewable sources, including energy from renewable sources produced on-site or nearby.*

The term "nearly" will cause much discussion, as it has already in many EU countries preparing for this. However, that is not the main point, since in due time the near-0 will become real-0 anyhow. More worrying is where it states "very low amount of energy required" to be "covered significantly by renewables". In going to 0-energy and beyond, it has been shown that low energy demand is not the right focus anymore, and a guarantee for sub optimizing environmental impact approaches overall is more important. It is even more worrying knowing that the EU is developing strong policies targeted at materials, has a biobased materials road map, and is preparing a method to value materials with footprint style indicators. Separating the policies for energy and material creates friction and does not lead to the intended lower environmental impact, overall.

A practical solution for EU countries could be to require buildings (and renovations) to be designed as real 0-energy buildings (with the balanced materials approach), and in construction leave some room for partial implementation of the design, for instance due to limited investment possibilities. The building could be completed without PV panels, but prepared for it when additional funds become available in future, thereby avoiding sub optimization and disinvestments.

6. Conclusions

The sun, in the form of solar radiation, is the only source available on this planet that adds potential to the system. All use of already available stocks on the planet can only be downgraded, depleting quality to perform functions appreciated by mankind, only to be avoided when solar radiation is used to upgrade materials to a higher quality level. Solar energy itself comes without impact, but to convert the solar energy into the useful forms again requires materials.

In a time when we start realizing this, and in which the building sector has already made the first steps towards a transition to use that solar radiation, it is time to recognize that the real burden has shifted towards the use of materials acquired from earthen stocks and (solar based) energy to make these materials available in the right product forms.

It is very exciting to see that the building sector is at the forefront of this transition. However, the consecutive step still has to be made: a paradigm shift in the approach and evaluation of resource use for buildings, requiring a completely different mindset to design and optimizing functions and services in our built environment.

Conflicts of Interest

The authors declare no conflicts of interest.

References

1. Directive 2010/31/EU of the European Parliament and of the Council of 19 May 2010 on the energy performance of buildings. Available online: http://goo.gl/zKc0Nz (accessed on 19 December 2014).
2. IEA SHC Task 40 combined with ECBCS Annex 52: "Net Zero Energy Solar Buildings". Available online: http://task40.iea-shc.org/ (accessed on 19 December 2014).
3. Rovers, R.; Rovers, V. 0-Energy or Carbon Neutral? Discussion Paper on Systems and Definitions. 2008. Available online: http://goo.gl/WqAo0H (accessed on 19 December 2014).
4. Christie, D.J.; Klare, M.T. Resource wars: The new landscape of global conflict. *Peace Confl. J. Peace Psychol.* **2001**, *7*, 375–377.
5. Rovers, R. Material-Neutral Building: Closed Cycle Accounting for Building Construction. In Proceedings of Smart and Sustainable Built Environments (SASBE) conference, Delft, The Netherlands, 15–19 June 2009.
6. Kibert, C. Policy Instruments for a Sustainable Built Environment. Available online: www.law.fsu.edu/journals/landuse/vol17_2/kibert.pdf (accessed on 19 December 2014).
7. Rovers, R. Evaluation of 0-Materials House Design. In Proceedings of PLEA2012—28th Conference, Lima, Perú, 7–9 November 2012.
8. Rovers, R.; Ritzen, M.; Gommans, L. Reducing Energy Demand—Or Producing More Energy? The Role and Impact of Materials in Zero or Near-Zero Energy Building Renovation. Available online: http://www.dewijkvanmorgen.nl/files/5135/20130328+ribuilt+report+-+versie+b%C3%A8ta+communicatie+%281%29.pdf (accessed on 19 December 2014).
9. Schmidt-Bleek, F. The Fossil Makers. Available online: http://www.factor10-institute.org/files/the_fossil_makers/FossilMakers_Intro.pdf (accessed on 19 December 2014).
10. Rovers, R. New energy retrofit concept: Renovation trains for mass housing. *Build. Res. Inf.* **2014**, *42*, 757–767.
11. SkySails Homepage. Available online: http://www.skysails.info/english/ (accessed on 19 December 2014).
12. Gordon, R.B. On the sustainability of metal supplies: A response to Tilton and Lagos. *Resour. Policy* **2007**, *32*, 24–28.
13. Skinner, B.J. Exploring the Resource Base. In Proceedings of the Workshop on "The Long-Run Availability of Minerals", Resources for the Future (RFF), Washington, DC, USA, 22–23 April 2001.
14. Allwood, J.; Cullen, J. *Sustainable Materials—With Both Eyes Open: Future Buildings, Vehicles, Products and Equipment—Made Efficiently and Made with Less New Material Paperback*; UIT Cambridge: Cambridge, UK, 2011.
15. Mudd, G.M.; Diesendorf, M. Sustainability of uranium mining and milling: Towards quantifying resources and eco-efficiency. *Environ. Sci. Technol.* **2008**, *42*, 2624–2630.
16. New Discoveries Relative to World Copper Production, 1992–2004; International Copper Study Group (ICSG): Lisbon, Portugal, 2012. Available online: http://copperalliance.org/

wordpress/wp-content/uploads/2012/01/2013-World-Copper-Factbook.pdf (accessed on 19 December 2014).
17. Mudd, G.M. An analysis of historic production trends in Australian base metal mining. *Ore Geol. Rev.* **2007**, *32*, 227–261.
18. Rovers, R.; Rovers, V.; Leduc, W.; Gommans, L.; Sap, H.; van Kann, F. Urban harvest + approach for 0-impact built environments, case Kerkrade west. *Int. J. Sustain. Build. Technol. Urban Dev.* **2011**, *2*, 111–117.
19. Hammond, G.P.; Jones, C.I. Embodied energy and carbon in construction materials. *Proc. Inst. Civil Engin. Energy* **2008**, *161*, 87–98.
20. Dixit, M.K.; Fernandez-Solis, J.L.; Lavy, S.; Culp, C.H. Need for an embodied energy measurement protocol for buildings: A review paper. *Renew. Sustain. Energy Rev.* **2012**, *16*, 3730–3743.
21. Henry, A.F.; Elambo, N.G.; Tah, J.H.M.; Fabrice, O.E.; Blanche, M.M. Embodied energy and CO_2 analyses of mud-brick and cement-block houses. *AIMS's Energy* **2014**, *2*, 18–40.
22. Malmqvist, T.; Birgisdottir, H.; Houlihan Wiberg, A.; Moncaster, A.; Brown, N.; John, V.; Passer, A.; Potting, J.; Soulti, E. Design Strategies for Low Embodied Energy and Greenhouse Gases in Buildings: Analyses of the IEA Annex 57 Case Studies. In Proceedings of World Sustainable Building Conference WSB14, Barcelona, Spain, 28–30 October 2014.
23. Rovers, R. Valuing the Impact of Recycling: Its New, unless a Theoretical Exploration of the Impact of Re-Use and Recycling. In Proceedings of the World Resource Forum, Davos, Switzerland, 7–9 October 2013.
24. Schmidt, M. Global Climate Change: The Wrong Parameter. In Proceedings of Rio 9, World Energy and Climate Event, Rio de Janeiro, Brazil, 17–19 March 2009.
25. Odum, H.T. *Environmental Accounting: Emergy and Environmental Decision Making*; Wiley: New York, NY, USA, 1996.
26. Hau, J.L.; Bakshi, B.R. Promise and problems of emergy analysis. *Ecol. Model.* **2004**, *178*, 215–225.
27. Rovers, R. Maxergy, and Embodied Land Indicator; RiBuilT: Heerlen, The Netherlands. Available online: http://www.sustainablebuilding.info/maxergy/ (accessed on 19 December 2014).

2

Building Information Modelling for Smart Built Environments

Jianchao Zhang *, Boon-Chong Seet and Tek Tjing Lie

Department of Electrical and Electronic Engineering, Auckland University of Technology, Auckland 1010, New Zealand; E-Mails: bseet@aut.ac.nz (B.-C.S.); tlie@aut.ac.nz (T.T.L.)

* Author to whom correspondence should be addressed; E-Mail: jizhang@aut.ac.nz

Academic Editor: George Baird

Abstract: Building information modelling (BIM) provides architectural 3D visualization and a standardized way to share and exchange building information. Recently, there has been an increasing interest in using BIM, not only for design and construction, but also the post-construction management of the built facility. With the emergence of smart built environment (SBE) technology, which embeds most spaces with smart objects to enhance the building's efficiency, security and comfort of its occupants, there is a need to understand and address the challenges BIM faces in the design, construction and management of future smart buildings. In this paper, we investigate how BIM can contribute to the development of SBE. Since BIM is designed to host information of the building throughout its life cycle, our investigation has covered phases from architecture design to facility management. Firstly, we extend BIM for the design phase to provide material/device profiling and the information exchange interface for various smart objects. Next, we propose a three-layer verification framework to assist BIM users in identifying possible defects in their SBE design. For the post-construction phase, we have designed a facility management tool to provide advanced energy management of smart grid-connected SBEs, where smart objects, as well as distributed energy resources (DERs) are deployed.

Keywords: building information modelling (BIM); design verification; energy management; smart built environments; smart grid

---

1. Introduction

Computer-aided design (CAD) techniques have been in use by the construction industry since the early 1980s. The next paradigm shift was the introduction of building information modelling (BIM) in the field of architecture, engineering and construction (AEC) in the mid-1990s [1].

BIM is a methodology enabled by a set of software tools and processes for facilitating the creation and use of the digital representation of the physical and functional characteristics of a facility [2]. In terms of software, BIM introduces exchangeable information formats, *i.e.*, International Foundation Classes (IFC), for modelling and visualizing building entities in 3D. In terms of processes, BIM facilitates the conveyance of building information from the design phase throughout the building life cycle, supporting cost management, construction management and facility management [3].

Thus, research on BIM as a methodology has been focused on the following two areas: (i) development of software tools and techniques for creating and evaluating new BIM artefacts that arise as building designs and technologies evolve; and (ii) the application and usage of BIM processes across the life-cycle of a building from pre-construction design to post-construction facility management.

Recent rapid advances in information and communication technologies (ICT) have led to their pervasive use across industry sectors, including building construction. An upcoming and important aspect of ICT use that we anticipate to take a central stage is the construction and management of emerging smart built environments (SBEs). Here, "smart built environment" refers to a built environment that has been embedded with smart objects, such as sensors and actuators, with computing and communication capabilities, making the environment sufficiently 'smart' to interact intelligently with and support their human users in their day-to-day activities [4].

In a smart building, the human users refer not only to the building occupants, but also to the building's owners/managers. The former are typically concerned with how a smart building could improve personal safety, comfort and productivity, while the latter are more concerned with how a smart building could better support their operation and management of the building. In addition, as buildings are a major source of energy consumption, accounting for 40% of primary energy consumption in most countries [5], SBE will be expected to harness its new technological capabilities to achieve an unprecedented level of energy efficiency.

Therefore, constructing SBEs can have a set of requirements and procedures not defined or typically considered in traditional construction settings. Although there has been research conducted on various aspects of SBEs, very little attention has been focused on the role and application of BIM in the pre- and post-construction processes of SBEs.

Therefore, in this paper, we investigate how BIM can contribute to the pre-construction design and post-construction management phases of SBE. The rest of the paper is organized as follow: in Section 2, we review related works. In Section 3, we introduce the general challenges and the opportunities for BIM in SBE. In Section 4, BIM is first extended in the building design phase to provide material/device profiling and an information exchange interface for sensors, smart meters and DERs. In Section 5, a facility management tool is developed to provide advanced energy management functions based on BIM produced in the previous phase. Finally, Section 6 concludes the paper.

2. Literature Review

In this section, we review the state-of-the-art on the design and management of smart built environments, as well as related BIM applications.

2.1. Design and Management of SBE

The design and management of SBE generally focus on two aspects: service and sensing. The service refers to that provided by SBE to support users in their daily lives. The authors in [6] discuss the service life cycle and design principles for a management framework to effectively manage autonomous and adaptive services in smart spaces. In [7], a smart shadow system was proposed to provide home users with real-world services, defined as services that affect the real world by changing the user's environmental factors, such as air conditioning and lighting, and to dynamically detect and resolve service conflicts.

On the other hand, some design approaches consider sensing as one essential component of SBE. The authors in [8] utilize a logic-based modelling language TCOZ (Timed Communicating Object Z) to describe constraints of the sensor imposed by its environmental conditions or relations with other sensors and its sensing pattern, such as periodic or conditional sensing. The sensor-based model was then applied to the design of smart spaces. In [9], ECA (event, condition, action) rules are introduced for describing home-based sensor-driven services, which can often cause a chain reaction, *i.e.*, one service may generate an outcome that automatically triggers another service. Therefore, a method is proposed to detect such service chains and the possible conflicts among these services.

2.2. Related BIM Applications

In recent years, researchers have also explored how SBE design and post-construction management could benefit from harnessing BIM information and capabilities. In [10], an indoor wireless sensor network (WSN) is designed with BIM data that provides a detailed description of the building environment, which is required for accurate predictions of signal propagation and, therefore, link quality between sensor nodes in the building. In [11], a three-dimensional smart space design framework was proposed in which space (e.g., furniture, walls, floors, *etc.*) was viewed as one dimension along with technologies (for ubiquitous computing) and living (e.g., safety, health and sustainability requirements) as the other two dimensions. Bhatt *et al.* [12] propose an ontology-based spatial-terminological inference approach to validate work-in-progress designs of smart environments. The approach checks the design for compliance with the spatial and functional constraints of environment entities based on available architecture data in IFC format.

While not specifically for smart built environments, it is also shown in [13] that BIM can benefit post-construction facility management, such as locating components, checking maintainability and creating digital assets through its powerful visualization, analysis and control capabilities. There are a number of existing works on integrating BIM with real-time information. For example, the Autodesk research group integrated BIM with sensors and meters to provide 3D visualization of building performance and life-cycle operation [14]. The Virtual Real-time Information System (VRIS) combines an Onuma cloud-based BIM tool [15] with a real-time sensor engine called the Virtual Real-time

Operating Centre (vROC) to provide building management functions [16]. However, when it comes to energy management, to the best of our knowledge, there are no existing BIM-based solutions in smart built environments which are smart grid [17] ready.

3. Utilizing BIM in Smart Built Environments

As mentioned before, a variety of smart objects will be ubiquitously and transparently installed in SBE to perform actions, such as sensing and control. These smart objects may interact with each other and with the environment and the environment users. The communication can be carried by either wireless (e.g., ZigBee, 802.11a/b/g/n) or wired (e.g., Ethernet, power line) information networks.

3.1. General Challenges

A number of questions arise when considering the life cycle of a smart built environment. First is how the smart objects are embedded in the environment. From the aspect of sensors, the physical location and the surrounding settings can significantly affect their ability to carry out specific tasks, e.g., ambient light or occupancy sensing. From the aspect of the information network, whether a wireless sensor network (WSN) or wired Ethernet is deployed, it should be designed to offer the smart objects an excellent level of communication.

Second is how the smart objects interact with the environment. The smart objects are situated in a specific surrounding to carry out their tasks and often require the input of space data, such as building floor plans. Furthermore, in carrying their tasks, the confluence of the actions by different smart objects may involve affecting a common set of environment factors, the impact of which should be studied and understood within a spatial context. Last, but not least, when conducting a building performance analysis, such as an energy efficiency analysis of the SBE, not only the information from sensors/meters is vital, but the building architectural/geometry data are also indispensable.

Third is the maintenance of such smart objects in a building's post-construction phase. In building management, it is common for facility managers to complain about either incomplete or inaccurate (not up-to-date) documentation [18]. To maintain a smart built environment that is more complicated than traditional structures, more building design and construction data will need to be documented and conveyed to the facility managers. In the event that the company that designed and constructed the building is no longer in business, building management can continue to function properly provided that the documented information is both comprehensive and reliable.

3.2. Introducing BIM in Smart Built Environments

BIM hosts the collaborative architectural information and provides the semantic knowledge of the building. With the emergence of smart built environment technology, BIM should be further developed to be capable of seamlessly integrating smart objects in building design, verifying the SBE design and feeding smart objects with relevant building-related information.

Designing SBE with BIM is both advantageous and convenient. Firstly, designers of SBE can utilize the building knowledge of BIM for planning the layouts of sensors, tags, actuators and meters. The performance of smart objects can be verified against known constraints and their layouts optimized for the best functional performance. Secondly, BIM also serves as a data repository for the physical information of smart objects. For maintenance and asset tracking in the building post-construction phase, the hardware information of smart objects can be recorded, and their installation locations can be documented and visualized in 3D.

On the other hand, BIM provides a perfect ontology database for SBE. Smart objects may be designed and manufactured by different vendors. The data they provide may vary in structure, and they may communicate using different protocols. Middleware is a popular approach to addressing such issues with heterogeneous smart objects [19]. With the introduction of BIM, each smart object can be profiled through its information exchange interface. Because BIM is standard compliant, the middleware can extract data formats of smart objects and other building information for viewing as an ontology database.

3.3. Energy Management in Smart Built Environments

Energy management is an important part of facility management. In SBEs, energy management can be enhanced through smart objects, such as temperature, occupancy and ambient light sensors, that provide data for estimating the building's energy requirements, understanding the building's energy usage patterns and decision-making by building control systems to achieve a balance between a building's energy efficiency and the comfort level of its occupants.

As awareness of energy efficiency grows within the building industry, the trend of deploying various sensors in the buildings will only become more common in energy management practice. Furthermore, with the introduction of a smart grid, buildings have an active role in the power system, as they exchange electrical power information with the power grid via their smart meters [20]. Sub-metering systems are also important to achieve energy awareness for building management, as they can provide high-resolution monitoring data down to the individual appliance level.

Another significant aspect that challenges energy management today is the penetration of DERs, such as photovoltaic (PV) panels and micro wind turbines. With on-site DERs, the role of building changes from a pure energy consumer to both an energy producer and consumer (or "prosumer") [21]. Correspondingly, the role of SBE is also extended to include the tasks of performing energy generation forecast, load scheduling, storing and feeding energy back to the grid. The monitoring and control of DERs is therefore vital for energy management in SBEs.

In essence, smart meters and DERs can be viewed as real-time information objects, which are a type of smart object that can be embedded in the BIM design of SBE, and the energy management system can benefit from the profiling capability of BIM, as shown in Figure 1.

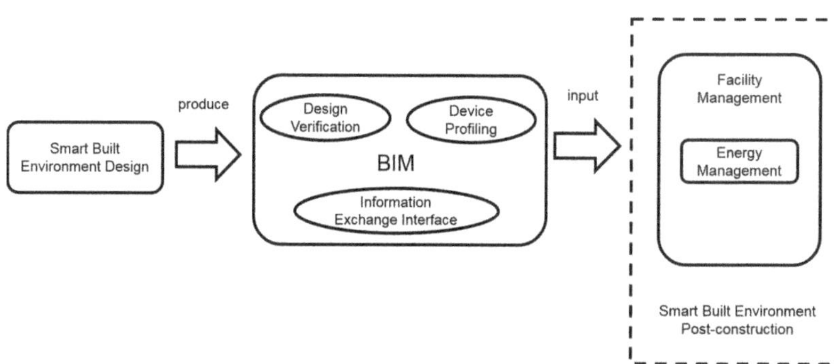

Figure 1. Role of building information modelling (BIM) in smart built environments.

4. Designing Smart Built Environment with BIM

As discussed in the previous section, BIM has the potential to support the entire life-cycle of the SBE. Therefore, we started by investigating how existing BIM platforms can be further developed to better support the design of SBE. To date, the most popular BIM software is the Revit suite from Autodesk and Archicad from Graphisoft. In this work, we have conducted our SBE design in BIM with Revit, but similar work can be conducted as well in Archicad.

4.1. Embedding Smart Objects in BIM

The intention is to profile smart objects with Revit during the SBE design phase. The produced BIM model with smart objects should be correctly exported as an International Foundation Class (IFC) file, so that other BIM tools can parse the information. We adopted three methods for the profiling: IFC shared parameter, family property parameter and the mark tag. The work in [20] demonstrated how sensors can be modelled in Revit, and we took a similar approach, which uses the IFC shared parameter field to indicate the sensor type in IFC, as shown in Figure 2a. For actuators, new properties can be added to the building element that is being controlled by the actuator. Therefore, new BIM artefacts, such as a smart window (e.g., capable of self-actuating to adjust its light transmission properties), can be created using the original window family with added actuator properties.

The IFC shared parameter enables the new smart object family to be a compatible type in BIM, *i.e.*, exportable and parsable in IFC format. However, to fully model an SBE design, we have to provide additional information. For that purpose, we utilize the family property parameters for a type of smart object and mark tag for an individual smart object.

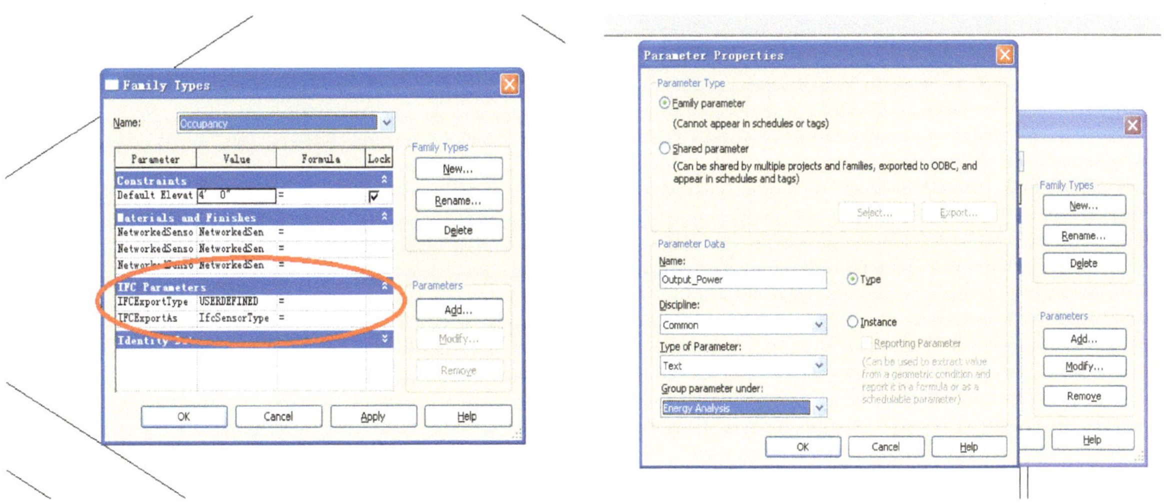

(a) (b)

Figure 2. (**a**) Extending BIM design using the "IFC shared parameter"; (**b**) Adding new parameter properties in Revit.

4.1.1. Spatial Interaction Information

One of the most important factors to consider when designing SBE is the spatial interactions between a smart object and its physical surroundings. For sensors, it has been referred to as the range space or the effective sensing space area [12], which is the foremost element to consider when planning their placement layout during the design of SBE. Different sensors types may have different range space characteristics. For an infrared motion sensor, its range space may have a shape like a sector of a circle. For a temperature sensor, its range space may be spherical-like. For simplicity, we represent the range space of a sensor by a range radius and range angle, and its axis is determined by the surface to which it attaches, as shown in Figure 3.

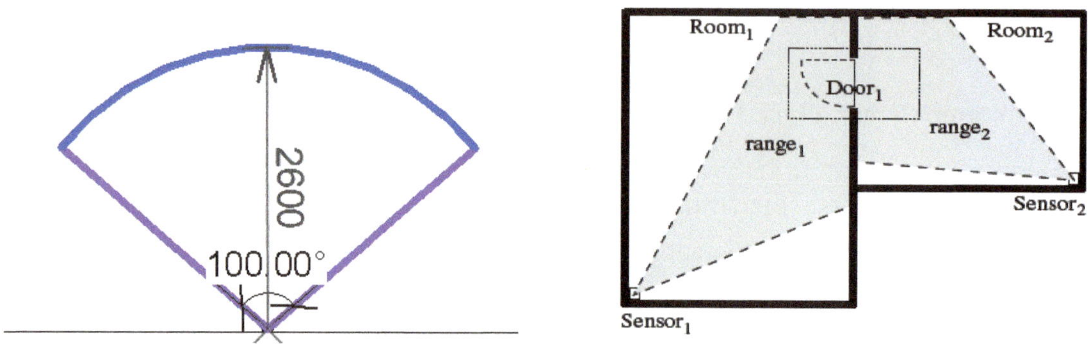

(a) (b)

Figure 3. (**a**) Example of the range space of a sensor with range radius of 2.6 m and a range angle of 100°; (**b**) the door's operational space should be within the range space of the two monitoring sensors.

4.1.2. Informational Exchange Interface

For the information exchange interface, which describes the data input/output behaviour of the smart object, family property parameters are added to the smart object family. Since this work focuses on energy management, we specify parameters in the energy analysis properties.

As illustrated in Figure 2b, in order to enable the BIM file (in IFC format) to provide an information exchange interface for the facility software, family property parameters are designed to be a mapping from IFC text to the device software programming interface, which can be a middleware or another BIM software. Three types of smart object data operations are defined:

- Output: the interface from which external software can read the output (generated power, grid signal, or sensed data) of the device. The format in the IFC file is: "Output_xxxxx".
- Input: the interface from which external software can read the power generation or consumption status of the device. The format in the IFC file is: "Input_xxxxx".
- Control: the interface from which external software can control the operation of the device, the tilting angle of the PV panel/wind turbine, the power consumption of smart appliances compatible with BACnet (Building Automation & Control networks) standard, *etc.* The format in the IFC file is: "Control_xxxxx".

4.1.3. Identification and Service Mapping

Finally, to map a smart object from BIM to an individual real-world device, the mark tag in BIM is used. The BIM software, which parses the produced BIM file, reads the mark tag for the smart object and determines additional runtime information, e.g., the Internet Protocol (IP) address, of the device from a database. This tag is also used to map a smart object to a service descriptor file, which will be detailed in the next section.

Using Revit, we created an SBE design, which is modelled as a smart house with sensors, actuators, smart meters, PV panel and wind turbine. After completing the BIM design, a BIM file in IFC format is generated, as shown in Figure 4.

```
......
#192496= IFCSENSORTYPE('3Ea9KIygfFOf3hy58KLU_a',#52,'Occupancy',$,$,$,(#192495),'211594','Occupancy',.USERDEFINED.);
......
#192534=IFCDISTRIBUTIONCONTROLELEMENT('3Ea9KIygfFOf3hy58KLU_B',#52,'Networked
Sensor:Occupancy:Occupancy:211621',$,'Occupancy',#192533,#192528,'211621',$);
#192535= IFCPROPERTYSINGLEVALUE('Mark',$,IFCLABEL('LivingRoomSensor'),$);
#192537= IFCPROPERTYSINGLEVALUE('Host',$,IFCLABEL('Basic Wall : Interior - Partition'),$);
......
#192540= IFCPROPERTYSINGLEVALUE('RangeRadius',$,IFCLENGTHMEASURE(2600.),$);
#192541= IFCPROPERTYSINGLEVALUE('Output_occupancy',$,IFCINTEGER(0),$);
#192542= IFCPROPERTYSINGLEVALUE('RangeAngle',$,IFCPLANEANGLEMEASURE(99.9999999999999),$);
......
#186580= IFCBUILDINGELEMENTPROXY('31vPjYM8b9$Aof7XNLJOFw',#52,'Wind Power Generator_modified:60" High:60"
High:201370',$,'60" High',#186579,#186574,'201370',.ELEMENT.);
#186581= IFCPROPERTYSINGLEVALUE('Mark',$,IFCLABEL('OutdoorWindTurbine'),$);
#186583= IFCPROPERTYSINGLEVALUE('Output_power',$,IFCINTEGER(0),$);
......
```

Figure 4. Code segment of the BIM IFC file.

4.2. Design Verification Framework for SBE

Recall that SBE serves to enhance the building's efficiency, security and comfort of its human occupants by being aware of the state of the environment and performing autonomous intelligent actions. Such actions performed by the SBE can be viewed as provisioning a type of service to the building owner or occupants. In literature, these services are also referred to as real-world services (RWS) [7]. Designing a set of adaptive RWS with various smart objects could become an integral part of future SBE design.

During the design phase of SBE, uncertainties may arise that could affect the service integrity and performance. For example, there could be doubts about whether the sensors are placed in the best locations to execute their tasks or whether there are conflicting objectives/requirements between different services. If such uncertainties can be verified during the design phase of the SBE, it may reduce the amount of rework necessary after the building is built. Furthermore, it can be useful for as-built verification, e.g., when new smart objects and services are to be introduced to an already constructed SBE. With this motivation, we propose an SBE design verification framework with BIM, as outlined in Figure 5.

In this framework, a three-layer verification is adopted to verify the SBE design. The three layers are: smart object, single service and inter-service performance verification. The input to the framework is the BIM file produced as discussed in Section 4.1 (in IFC format), which contains the design information of the building and associated smart objects. The performance verification criteria are modelled using XML, which are detailed in the following sections.

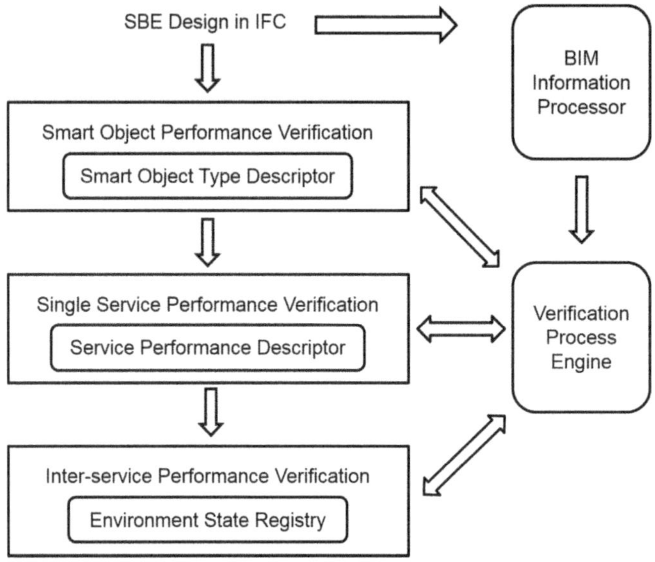

Figure 5. Design verification framework for smart built environments (SBEs) with BIM.

4.2.1. Smart Object Performance

The performance of a smart object can be impacted to different degrees by its surrounding building elements, depending on the type of smart object. For example, a temperature sensor may be mistakenly placed next to a furnace or air conditioner, which can interfere with its proper operation. In [22], it was shown that the performance of an optical sensor can be greatly influenced by the material properties of the target surface that it is sensing. The actual verification criteria used are specific to the type of smart objects. A smart object type descriptor (STD) is created as an external XML file to describe the performance constraining factors of each smart object. An example STD is shown in Figure 6a. The verification engine then parses the description and analyses the spatial interactions between the smart object and its surrounding building elements.

4.2.2. Single Service Performance

An RWS is defined on a set of smart objects and building elements involved in delivering the service to the building occupants/owner/manager. For example, as shown in Figure 3b, a room entry monitoring service is defined on an infrared motion sensor in the room and the door that is within its range space.

Likewise, a service performance descriptor (SPD) is created as an XML file to describe the smart objects and building elements that constitute each service. An example SPD is shown in Figure 6b. Each RWS could be constituted by one or more smart objects. When verifying a service, the verification engine parses the SPD and analyses the smart object's performance constraining factors (from STD) and building element properties, such as operational space and material information (from the IFC file) for their impact on the service performance.

```
<SmartObjectTypeDescriptor>
  <InfraredMotionSensor>
    <AffectingMaterial>
      <Material name = "Glass">
      <Material name = "Plastic Films">
    </AffectingMaterial>
......
  </InfraredMotionSensor>
  <TemperatureSensor>
    <AffectingFurniture>
      <Furniture type = "Furnace" maxDistance = "2m">
      <Furniture type = "AirConditioner" maxDistance = "2m">
    </AffectingFurniture>
......
  </TemperatureSensor>
......
</SmartObjectTypeDescriptor>
```

(a)

```
<ServicePerformanceDescriptor>
  <Service Name = "HouseEntryMotionSensingService">
    <SmartObjects>
      <Sensor type = "InfraredMotionSensor" name =
"LivingRoom_1_InfraredMotionSensor_1">
      <Sensor type = "SoundSensor" name = "LivingRoom_1_SoundSensor_1">
    </SmartObjects>
    <BuildingElements>
      <Furniture type = "Door" name = "HouseEntryDoor">
    </BuildingElements >
......
  </Service>
  <Service Name = "HallwayIlluminationService">
  <SmartObjects>
      <Actuator type = "WindowBlindsController" name =
"HallwayWindowBlindsControler_1">
      <Sensor type = "LightSensor" name = "HallwayLightSensor_1">
    </SmartObjects>
    < BuildingElements >
      <Furniture type = "Window" name = "HallwayWindow_1">
      <Furniture type = "Window" name = "HallwayWindow_2">
    </BuildingElements >
......
  </Service>
......
</ServicePerformanceDescriptor>
```

(b)

Figure 6. (**a**) Example of a smart object type descriptor; (**b**) Example of a service performance descriptor.

4.2.3. Inter-Service Performance

In SBE, each RWS works autonomously in most cases. However, an event may occur that triggers several services, either because the event simultaneously satisfies the trigger conditions of multiple services or the result of one service leads to another, *i.e.*, chain of services [9]. The confluence between services can be complicated and may not be completely anticipated by the designer. Some consequent impacts can be collaborative and beneficial to the service users, while some can be conflicting and degrade the service quality as a whole.

For example, during a hot summer, an occupancy sensor senses a user's presence and instructs the HVAC system to cool the room, while a motion sensor in the room recognizes the same user to be reading a book and instructs the window to roll up its shades to improve illumination. However, the incoming sunshine heats up the room and causes the former cooling service to take more time and consume more energy to achieve its targeted temperature.

Such service conflicts undermine the quality of an SBE design and impact the performance of the conflicted services. In the literature, this conflict of services is sometimes referred to as feature interaction [9], which could be spotted during the design phase by identifying their triggering events and resulting environment state change.

To verify an SBE design against service conflicts, each service registers its triggering/affecting environment factors, which are also profiled in the SPD, with the environment state registry (ESR). The verification engine performs environment state reasoning by analysing the location of smart objects that constitute the different services and the environment factors that they affect. Using the previous example of service conflict, the occupancy sensor and motion sensor may or may not have overlapping range spaces, but they share a common room space. Furthermore, if the room space of two smart objects with common environment factors can be connected through opening a door, a warning should be generated for the design. Once again, BIM provides data on the room space where a smart object resides, and the verification engine queries the ESR and checks against the other services defined on smart objects in the same or adjacent room connected by a door, as shown in Figure 7.

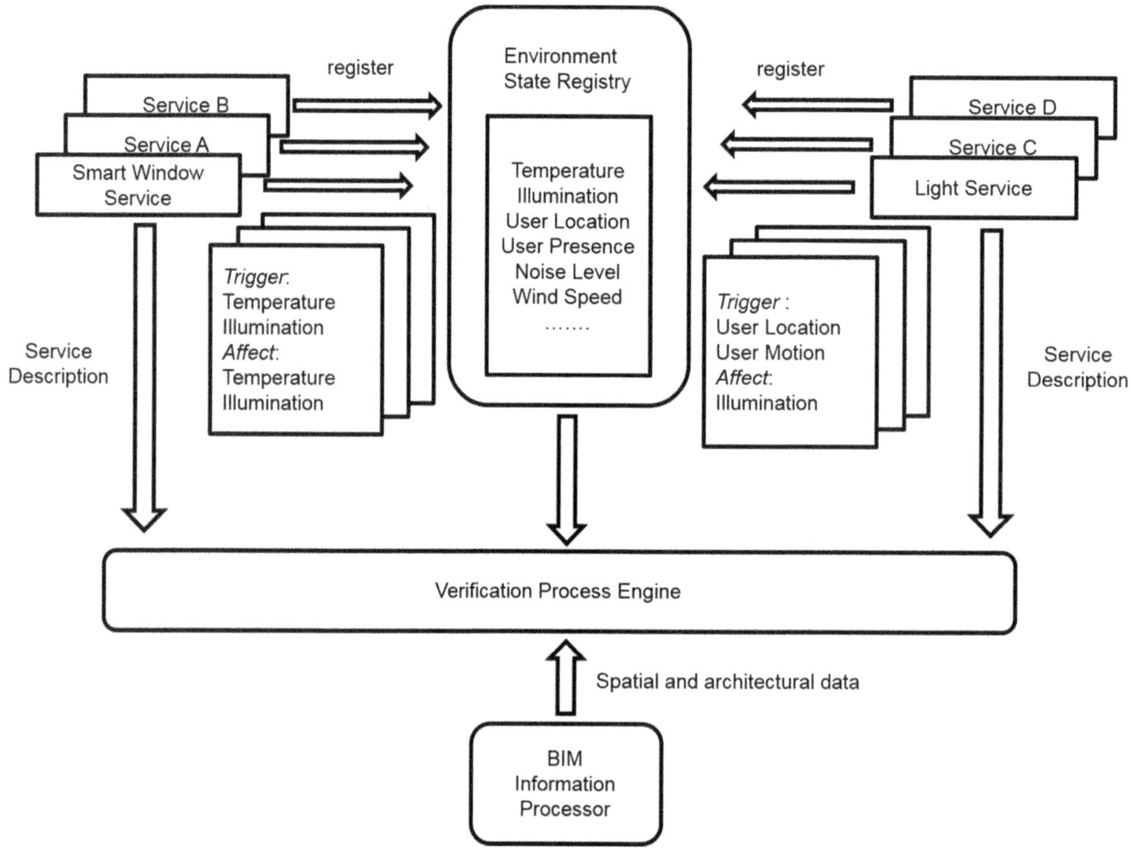

Figure 7. Verification with environment state registry.

5. Integrating BIM in Post-Construction Facility Management

The BIM file generated from Revit and validated using the verification framework in the previous section is an output from the SBE designer or architect during the building design phase. In the post-construction facility management phase, the building manager can apply the information in this BIM file to perform day-to-day building management, including energy management.

To parse the BIM file and read the profiled information, there are two possible options for the design of this research. The first is to extend Revit using a native software development kit (SDK) to perform the energy management task. The second is to develop a standalone BIM tool as a separate energy management engine. After much contemplation, we decided to go with the second option, as we believe that building managers will be more familiar with using a building management system (BMS) for their everyday work than with Revit, which is intended as a computer-aided design (CAD) tool for building designers and architects.

Therefore, we opted for the second option, and a stand-alone BIM tool for energy management has been developed. We developed the tool using the extensible building information modelling [23] toolkit, which provides IFC parsing and 3D presentation utilities.

As different smart objects may have different data input/output interfaces, the properties parameter specified in our BIM design in the previous section provides a convenient way for the BIM tool to handle such low-level operations. An adapter layer is designed in this research to process the requests from BIM software, which parses the IFC file and demands data exchange for the smart object. The mapping/parsing operation is illustrated in Figure 8. The software architecture for our developed BIM tool is shown in Figure 9.

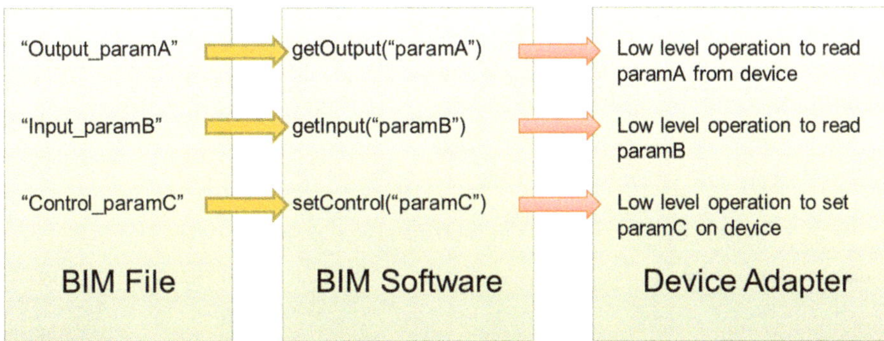

Figure 8. Mapping from the BIM file to the software information interface programming.

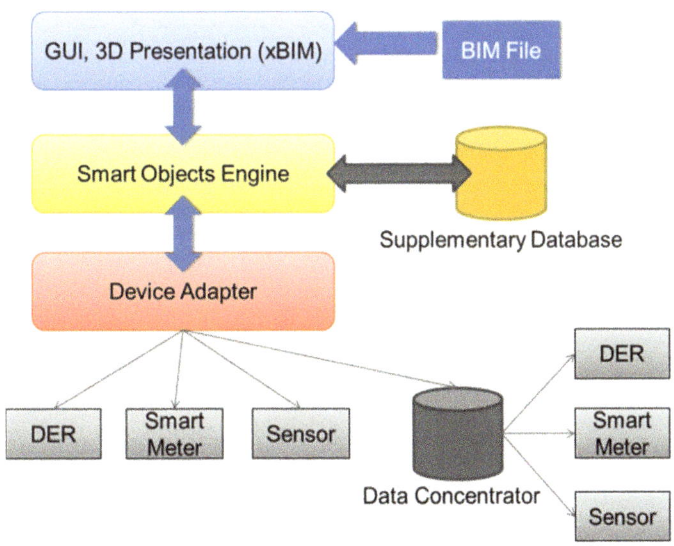

Figure 9. BIM-based energy management platform software architecture. DER, distributed energy resource.

Real-time data from the DERs, smart meters and sensors are collected and stored in a database. Figure 10 shows an instance of the real-time monitoring data from a living room sensor displayed to the user, *i.e.*, home owner or building manager.

With real-time data from the smart objects, energy management and analysis in BIM software are facilitated and achieved. Real-time generation data of on-site DERs show the current energy production capacity and indicate how many loads can be supplied off the power grid. Weather, temperature, building and occupant data from sensors form a view of the present and future energy generation/consumption, as shown in Figure 11. The pricing information from the smart meter allows the building to perform demand response actions in coordination with the power grid.

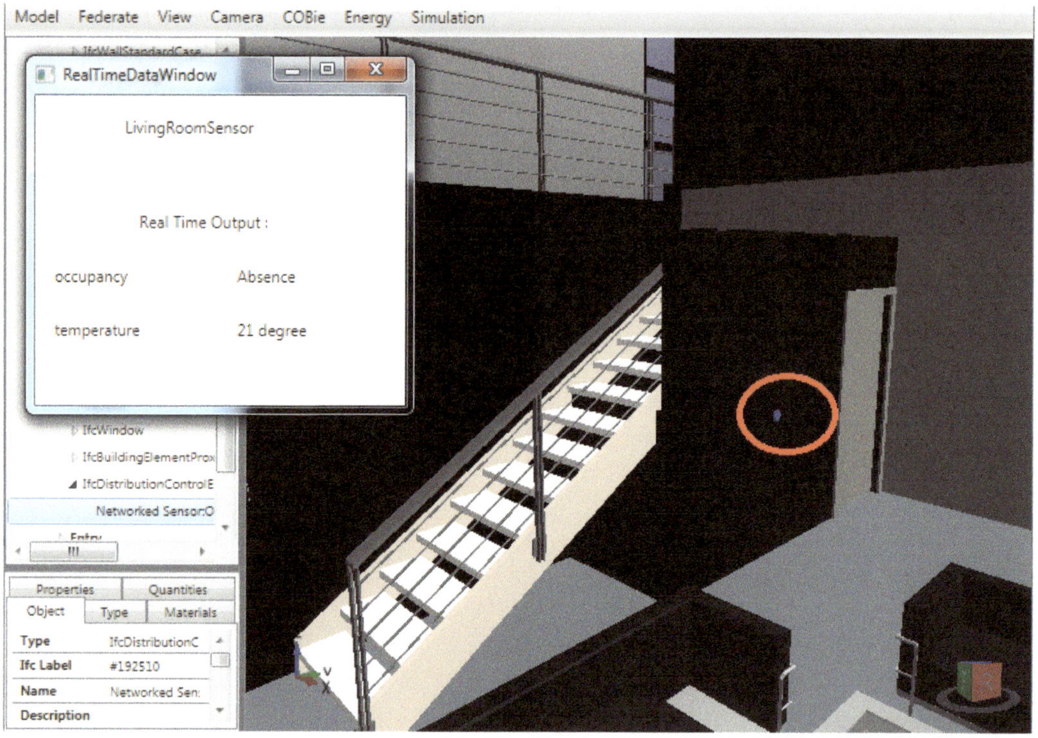

Figure 10. Real-time monitoring data display to the user.

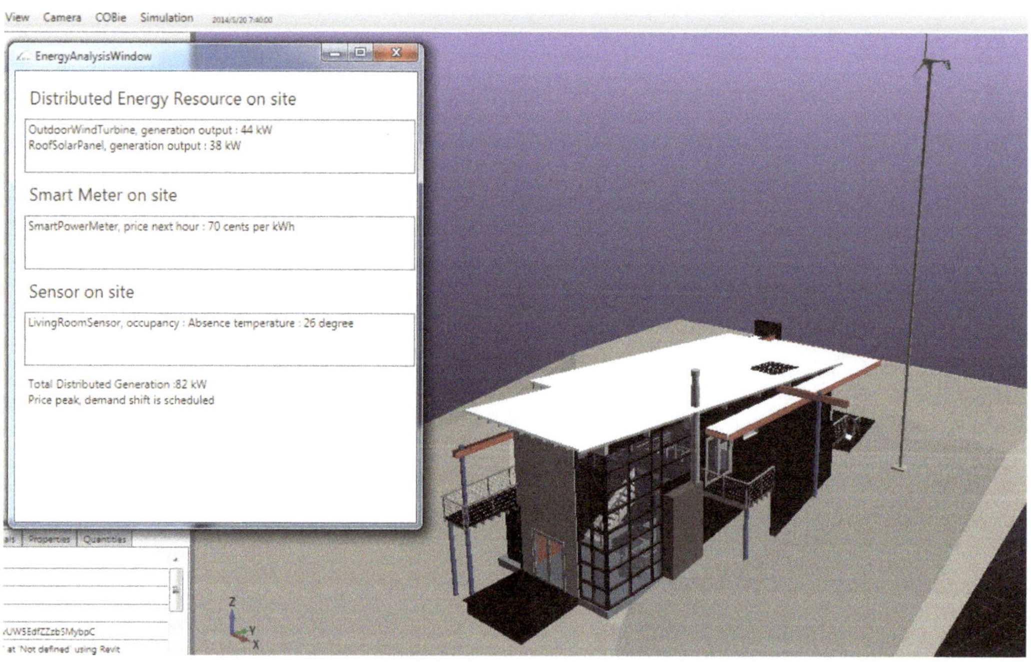

Figure 11. Energy analysis functionality in smart built environment.

In our BIM tool, we demonstrated the energy management functionality through a simulation of demand shifting. When the real-time pricing information from the smart meter reaches a user-defined threshold, the software triggers the demand shift process and transmits control commands to the energy-consuming appliances.

As future work, we are working towards an advanced BIM-based building management platform for smart buildings to host smart grid services and applications, such as virtual power plant and demand response, as shown in Figure 12.

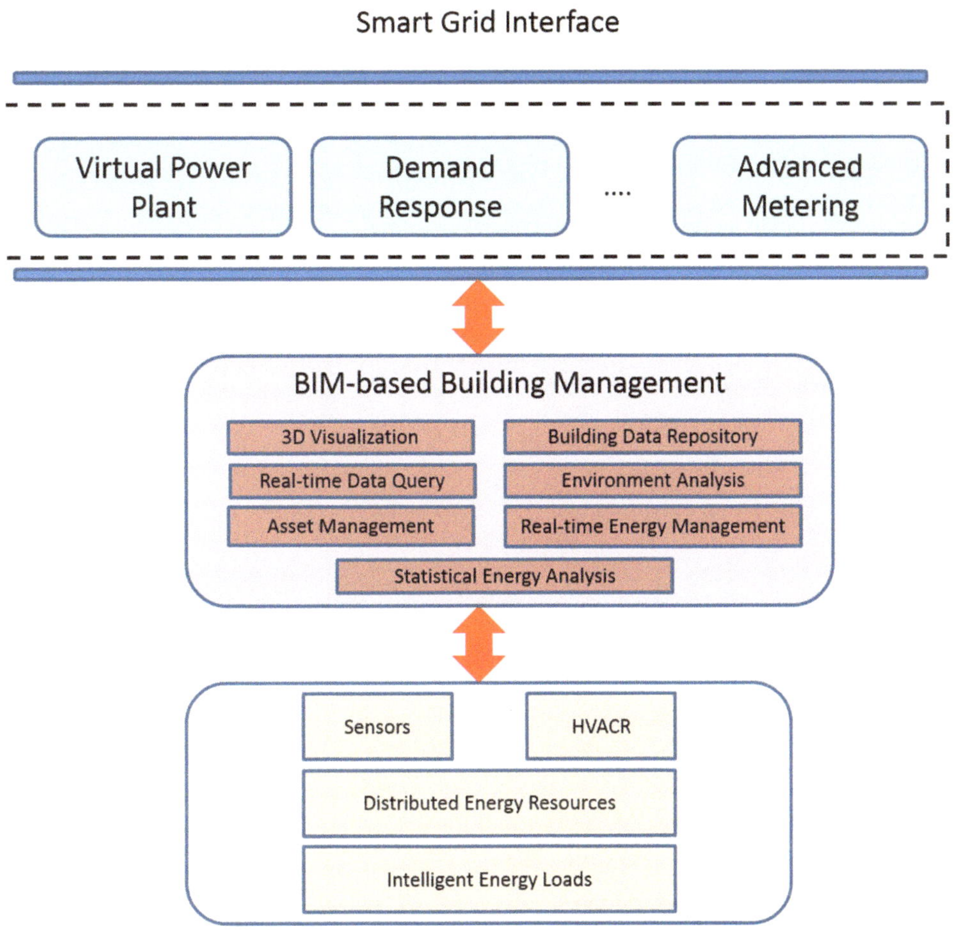

Figure 12. Proposed BIM-based building management for a smart grid.

6. Conclusions

In this paper, the findings and experiences gained from an investigation into how BIM can be developed and utilized for smart built environments are reported. Our work covers the investigation of BIM-based methodologies for pre-construction design and verification and post-construction facility management of smart built environments in the smart grid era. A basic, but functional, prototype of a smart house energy management system using Revit and the xBIM toolkit was also implemented and successfully demonstrated. We are currently extending our work to develop BIM software tools for automated management of BACnet-compliant HVACR (Heating, Ventilating, Air Conditioning and Refrigeration) systems and investigating the information fusion and decision system of smart grid applications for future smart-powered buildings.

Acknowledgments

This work has been supported in part by a research assistantship grant from Auckland University of Technology for Jianchao Zhang. The authors would also like to thank Andrew Douglas for facilitating the access to the software resources used in this research.

Author Contributions

The topic and scope of this research were originally defined in a research proposal by Boon-Chong Seet and Tek Tjing Lie on BIM for smart-powered buildings. Jianchao Zhang conducted the research and implemented the proposed solutions under the supervision of Boon-Chong Seet and Tek Tjing Lie, who contributed their expertise on smart environments and smart grids, respectively. The paper was drafted by Jianchao Zhang and revised by Boon-Chong Seet.

Conflicts of Interest

The authors declare no conflict of interest.

References

1. Sabol, L. Building Information Modelling and Facility Management. In Proceedings of the IFMA World Workplace, Dallas, TX, USA, 15–17 October 2008.
2. Teicholz, P. *BIM for Facility Managers*; John Wiley & Sons: Hoboken, NJ, USA, 2013; p. 332.
3. Talebi, S. Rethinking the Project Development Process through Use of BIM. In Proceedings of the 2nd BIM International Conference on Challenges to Overcome, Lisbon, Portugal, 9–10 October 2014.
4. Nakashima, H.; Aghajan, H.K.; Augusto, J.C. *Handbook of Ambient Intelligence and Smart Environments*; Springer: New York, NY, USA, 2010; p. 1293.
5. International Energy Agency. Sustainable Buildings. Available online: http://www.iea.org/topics/energyefficiency/subtopics/sustainablebuildings/ (accessed on 30 November 2014).
6. Van der Meer, S.; Jennings, B.; Barrett, K.; Carroll, R. Design Principles for Smart Space Management. In Proceedings of the 1st International Workshop on Managing Ubiquitous Communications and Services (MUCS), M-Zones Whitepaper, Waterford, Ireland, 11 December 2003.
7. Umakoshi, K.; Kambayashi, T.; Yoshida, M.; Takemoto, M.; Matsuo, M. S^3: Smart Shadow System for Real World Service and Its Evaluation with Users. In Proceedings of the IEEE/IPSJ 11th International Symposium on Applications and the Internet (SAINT), Munich, Germany, 18–21 July 2011.
8. Dong, J.S.; Feng, Y.; Sun, J. *Sensor Based Designs for Smart Space*; Technical Report; Computer Science Department, National University of Singapore: Singapore, 2004.
9. Inada, T.; Igaki, H.; Ikegami, K. Detecting service chains and feature interactions in sensor-driven home network services. *Sensors* **2012**, *12*, 8447–8464.
10. Guinard, A.; McGibney, A.; Pesch, D. A Wireless Sensor Network Design Tool to Support Building Energy Management. In Proceedings of the First ACM Workshop on Embedded Sensing Systems for Energy-Efficiency in Buildings, Berkeley, CA, USA, 4–6 November 2009.
11. Jeng, T. Toward a ubiquitous smart space design framework. *J. Inf. Sci. Eng.* **2009**, *25*, 675–686.

12. Bhatt, M.; Dylla, F.; Hois, J. Spatio-Terminological Inference for the Design of Ambient Environments. In *Spatial Information Theory*; Springer: Berlin, Germany, 2009; pp. 371–391.
13. Becerik-Gerber, B.; Jazizadeh, F.; Li, N.; Calis, G. Application areas and data requirements for BIM-enabled facilities management. *J. Constr. Eng. Manag.* **2011**, *138*, 431–442.
14. Attar, R.; Hailemariam, E.; Glueck, M.; Tessier, A.; McCrae, J.; Khan, A. BIM-Based Building Performance Monitor. In Proceedings of the SimAUD Conference, Symposium on Simulation for Architecture and Urban Design, Orlando, FL, USA, 12–15 April 2010.
15. Onuma. Onuma System. Available online: http://onuma.com/products/OnumaPlanningSystem.php (accessed on 30 November 2014).
16. Lavelle, M.R.; Onuma, K. Virtual Real Time Information System: Moving to Cloud-Based Building Management. Available online: http://www.lavelleenergy.com/index.php/component/content/article/all-buildings/86-virtual-realtime-information-system (accessed on 30 November 2014).
17. Arnold, G.W. Challenges and opportunities in smart grid: A position article. *Proc. IEEE* **2011**, *99*, 922–927.
18. East, B. Construction-Operations Building Information Exchange (COBie). Available online: http://www.wbdg.org/resources/cobie.php (accessed on 30 November 2014).
19. Park, S.O.; Park, J.H.; Jeong, Y.S. An efficient dynamic integration middleware for cyber-physical systems in mobile environments. *Mob. Netw. Appl.* **2013**, *18*, 110–115.
20. Autodesk. Placing Sensors in the Revit Model. Available online: http://www.digital210king.org/blog.php?p=30 (accessed 30 November 2014).
21. Resendes, S.; Carreira, P.; Santos, A.C. Conflict detection and resolution in home and building automation systems: A literature review. *J. Ambient Intell. Humaniz. Comput.* **2014**, *5*, 699–715.
22. Marza, A.O.; Avram, I.M.; Tamovan, I.G. An Analysis of Target Materials Influence on Optical Sensors Performance. In Proceedings of the IEEE International Conference and Exposition on Electrical and Power Engineering (EPE), 25–27 October 2012; pp. 85–87.
23. The xBIM Toolkit. Available online: http://xbim.codeplex.com/ (accessed on 30 November 2014).

3

Heater Choice, Dampness and Mould Growth in 26 New Zealand Homes: A Study of Propensity for Mould Growth Using Encapsulated Fungal Spores

Mikael Boulic [1,*], Robyn Anne Phipps [2], Malcolm Cunningham [3], Donald John Cleland [1], Pär Fjällström [4], Keiko Abe [5] and Philippa Howden-Chapman [6]

[1] School of Engineering and Advanced Technology, Massey University, Manawatu Campus, Palmerston North 4474, New Zealand; E-Mail: d.cleland@massey.ac.nz

[2] School of Engineering and Advanced Technology, Massey University, Albany Campus, Auckland 0632, New Zealand; E-Mail: r.a.phipps@massey.ac.nz

[3] Building Research Association of New Zealand, Porirua 5381, New Zealand; E-Mail: malcolmcunninghammjc@gmail.com

[4] IVL Swedish Environmental Research Institute, P.O. Box 210 60, SE-100 31 Stockholm, Sweden; E-Mail: par.fjallstrom@ivl.se

[5] Institute of Environmental Biology, JDC Corporation, Kanagawa 243-03, Japan; E-Mail: abekeiko@kamakuranet.ne.jp

[6] Department of Public Health, University of Otago, Wellington 6021, New Zealand; E-Mail: philippa.howden-chapman@otago.ac.nz

* Author to whom correspondence should be addressed; E-Mail: m.boulic@massey.ac.nz

Academic Editor: George Baird

Abstract: The relationship between the use of unflued gas heaters (UGH, $N = 14$) and heat pump heaters (HP, $N = 12$) located in the living rooms, and mould growth on the living room and bedroom walls, of 26 New Zealand (NZ) occupied homes was investigated during winter. Two methods were employed to evaluate the potential of mould growth on walls: (i) measurement of daily hyphal growth rate using a fungal detector (encapsulated fungal spores); and (ii) estimation of fungal contamination based on a four level scale visual inspection. The average wall psychrometric conditions were significantly different between the two heater type groups, in both the living rooms and the bedrooms with the UGH user homes being colder and damper than HP user homes. The UGHs were found to be a significant additional source of moisture in the living rooms which dramatically increased the capacity for fungi to grow on wall surfaces. The average daily hyphal growth rates were 4 and 16 times higher in the living rooms and in the bedrooms of the UGH user homes, respectively. Results from both mould detection methods gave good agreement, showing that the use of a fungal detector was an efficient method to predict the potential of mould growth on the inside of the external walls in NZ homes.

Keywords: New Zealand homes; heater usage; psychrometrics; mould growth; fungal detector; visual mould inspection

---

1. Introduction

1.1. Background

Houses in New Zealand (NZ) are typically damp and cold [1]. The dampness and low indoor temperature is mainly due to two factors: building insulation deficiency (houses constructed before April 1978 were not subject to regulations requiring insulation) and undersized heating systems (only around 5% of NZ dwellings have central heating systems). In winter 2006, when the study was conducted, unflued gas heaters (UGH) were used in 27.7% of the NZ dwellings [2,3]. While UGHs use in NZ has declined since the study to 15.4% of the NZ dwellings [4], they are still in common use and hence the study is still relevant. Further, the results are useful for other countries with similar climates, housing stocks and use of UGHs. In NZ, these UGHs are mainly operated on low or medium setting which gives an estimated average heat output of 2.6 kW [5].

Unflued gas heaters, also called Liquefied Petroleum Gas (LPG) cabinets or portable gas heaters are designed to release combustion gases directly in the indoor environment [6]. In addition, to the chemical pollutants, like nitrogen dioxide [7], released during the combustion process, water vapour is another combustion byproduct which is released at a theoretical rate of 1.6 kg per kg of LPG consumed dependent on the proportion of butane and propane in the bottle. Therefore, the use of an UGH increases the room moisture level [8].

Temperature and moisture levels are critical factors for fungal development. Viable spores are always present in dwellings and germination and subsequent hyphal development can occur if the nutrients, temperature and moisture availability are suitable. The mould and their by-products could be harmful for people, particularly if they are immune deficient or asthmatic [9].

1.2. Methods

The Housing, Heating and Health Study (HHH Study) was a community trial to investigate the relationship between domestic heaters, indoor environment and occupant's health. Four hundred and nine families with a child with doctor diagnosed asthma (index child), and using an UGH or a portable electric heater as their main heater, were enrolled in this interventional study. Baseline health and environmental parameters were measured in all homes during the first winter/spring season. Before the second winter season, the original heater (UGH or portable electric heater) was replaced by a higher capacity and non-indoor polluting heater such as a heat pump heater (HP), for a randomly selected half of the dwellings (*Treatment group*). The *Control group* kept using their original heater during the second winter season and received their replacement heater at the end of the study. Health and environmental measurements were repeated in all homes during the second winter [10].

For a subset (N = 26) of the HHH Study households, who were operating either an UGH or a HP in the living rooms, the propensity for mould to grow and the indoor climate were investigated using a fungal detector (JDC Corporation, Kanagawa, Japan), and temperature/relative humidity (RH) measurements. Visual inspections of mould growth were carried out to validate the fungal detector predictions. This subset study was undertaken during winter in 26 dwellings in the Hutt Valley, a semi-coastal area that is part of Greater Wellington, New Zealand. The dwellings had been insulated (under floor and roof cavity) before the study started, according to the New Zealand Standard 4218:2004 specifications [11]. Twelve households received a HP (*Treatment group*) and 14 households were still using their original UGHs (*Control group*). The level of moisture in homes can be affected by many factors, other than unflued combustion process, like cooking, showering/bathing and drying clothes. However the households were selected with similar other moisture contributors (similar location, similar building construction and insulation level, similar level of occupancy), and it was assumed that the only difference in moisture source will be due to the operation of the heaters. The aim of this study was to investigate the moisture in homes resulting from the use of heater and to determine if a fungal detector could be a useful device to evaluate the suitability of the indoor environment for fungal development during winter time in New Zealand.

For each dwelling, the wall temperature and RH measurements were carried out in the living room and the index child's bedroom, using wall-mounted Hobo® H8 sensors (Onset Computer Corporation, Bourne, MA, USA). The loggers were attached on the interior lining of an external wall at a consistent height of 1.8 m above the floor level (B in Figure 1). The loggers were set to monitor the temperature and the RH continuously every 15 min for up to a 41.4 days' period (maximum memory capacity). The logger thickness placed the sensors about 1 cm away from the wall surface.

In close proximity to this temperature/RH sensor, a fungal detector was located directly on the wall surface (A in Figure 1). The fungal detector consisted of three inclusions of fungal spores; two xerophilic fungi (*Eurotium herbariorum* and *Aspergillus penicilloides*) and one hydrophilic fungus (*Alternaria alternata*) in an encapsulating device (plastic plate, water vapour permeable transparent film and double sided adhesive tape acting as a frame) [12]. Xerophilic fungi grow under relatively dry conditions and hydrophilic fungi need extremely humid conditions to grow. Each fungal inclusion contains 3 μL of spore suspension at 10^6 spores per mL concentration [13]. At the start of exposure, the spores were in a dry state. The fungal detectors were exposed to the living room and bedroom wall climate for an average period of 85 days. At the conclusion of the winter monitoring period, the three fungal inclusions (*Aspergillus penicillioides*, *Eurotium herbariorum* and *Alternaria alternata*) were examined under a microscope and the hyphal lengths were measured.

In addition to the fungal detector assessment method, a visual inspection of mould growth was carried out in the 26 living rooms and bedrooms by the first author. A contamination scale with four graduation levels was used to visually assess the mould level in the living room and in the child's bedroom (M0: "no visible mould", M1: "specks of visible mould", M2: "moderate visible mould patches" and M3: "extensive covered areas").

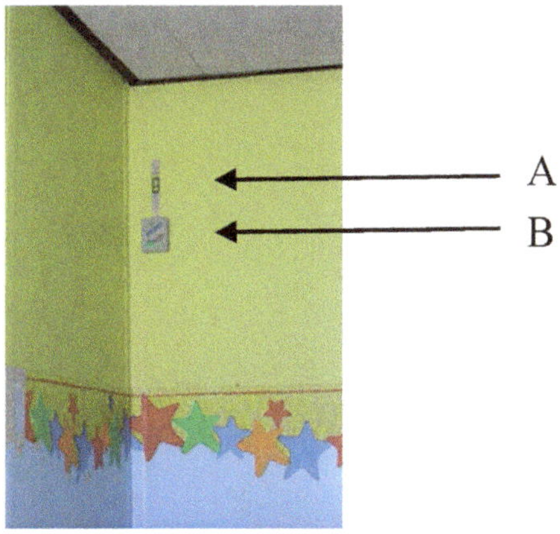

Figure 1. Typical sensor locations on the inside of the external wall in an index child's bedroom (**A**: fungal detector, **B**: temperature and relative humidity logger).

2. Results and Discussion

2.1. The Wall Climate

The temperature/RH loggers operated to full memory capacity (41.4 days). Table 1 shows the average psychrometric conditions in households operating an UGH (N = 14) and in households operating a HP (N = 12).

In the living rooms, the wall climates were on average colder (15.2 °C *vs.* 17.4 °C) and had higher RH (63.6% *vs.* 53.5%) in the UGH group than in the HP group. Similar results were found in the bedrooms (13.9 °C *vs.* 16.0 °C and 68.6% *vs.* 60.9%) for UGH and HP groups respectively. For both groups, the bedrooms were colder and damper than the living rooms. In the living rooms, Table 1 shows higher humidity ratio in the UGH group than in the HP group (6.9 g/kg *vs.* 6.6 g/kg). This result is consistent with UGH being an additional source of moisture for the living room environment. In the bedrooms, for both groups of households, the humidity ratio was similar (6.8 g/kg *vs.* 6.9 g/kg). With a similar level of humidity ratio, the lower average RH for the HP user group appears to be mainly due to a higher average temperature. This result indicates that the HP seemed to be more effective than the UGH for heating more than one room in the house. The moisture released from the UGH operated in the living rooms appears to have little effect on the bedroom humidity ratio.

Households operating an UGH had a higher RH level than households operating a HP, thus the release of water vapour, during the operation of an UGH, was found to be a significant additional source of moisture for the wall surface in the living rooms. Studies found that operating an UGH at a high setting releases around a half litre of water vapour per hour of use [14,15] with an average vapour pressure increase of 0.01 kPa/min [8].

Table 1. Average [95%CI] temperature (T), relative humidity (RH) and humidity ratio in living rooms and bedrooms of households operating either an unflued gas heater or a heat pump heater.

Heater Use in The Living Room	*N*	Living Room			Bedroom		
		T (°C)	RH (%)	Humidity Ratio (g H_2O/kg of Dry Air)	T (°C)	RH (%)	Humidity Ratio (g H_2O/kg of Dry Air)
Unflued Gas Heater (UGH)	14	15.2 14.3–16.2	63.6 60.7–66.5	6.9 6.6–7.1	13.9 12.9–14.8	68.6 64.1–73.1	6.8 6.4–7.2
Heat Pump (HP)	12	17.4 16.0–18.9	53.5 48.9–58.1	6.6 6.3–6.9	16.0 14.8–17.2	60.9 57.6–64.1	6.9 6.6–7.2

Optimal RH is the primary parameter for fungi to grow. When the RH of the air surrounding the building material is in a steady state condition (equilibrium), this RH is called equilibrium RH (ERH) and is equivalent to the water activity (a_W) of the building material when expressed as a fraction. A building material in a steady state condition with a surrounding RH of 70% (ERH = 70%), will have $a_W = 0.70$. This 70% ERH threshold was chosen in Table 2, as it represents the minimal ERH/a_W for xerophilic fungi to start growing [12,16] and two of the three fungal inclusions in the fungal detector contained xerophilic fungi (*Eurotium herbariorum* and *Aspergillus penicilloides*).

Table 2 shows the percentage of time, in the living rooms and in the bedrooms, with a wall RH above 70% for both types of heater being operated in the living rooms during the 41.4 days' period.

Table 2 shows that the bedrooms had greater exposure to RH levels above 70% than the living rooms. The households operating an UGH showed a higher percentage of time in both the living rooms and the bedrooms than the households operating a HP. In fact, the psychrometric conditions were substantially different in the UGH and HP homes which should impact on the capacity for mould to grow on the wall surface.

Table 2. Percentage of time with the wall RH above 70% in the living rooms and in the bedrooms.

Heater Type Used in the Living Room	N	Percentage of Time with RH > 70% (%)	
		Living Room	Bedroom
Unflued Gas Heater (UGH)	14	29	46
Heat Pump (HP)	12	9	18

2.2. Fungal Daily Growth Rate on the Interior Lining of an External Wall Surface

The 52 fungal detectors (one in each living room and one in each bedroom) were exposed for an average period of 85 days with an exposure range between 62 and 109 days. The exception to this was one dwelling where the fungal detectors stayed for 154 days, due to the householders being unavailable at an earlier time for collection of the slides. Following hyphal measurements under microscope, the daily growth rate outliers were removed from the analysis. The lower outliers were identified as values lower than the value of (P75 − ((P75 − P25) × 1.5)) and the upper outliers were identified as values greater than the value of (P75 + ((P75 − P25) × 1.5)). Wilcoxon's rank tests were applied to test if the daily hyphal growth rate (μm/day) was different between households operating an UGH and households operating a HP. Figure 2 shows the average daily hyphal growth rate in the living rooms and in the bedrooms.

In the livings rooms, the daily hyphal growth rates for both xerophilic fungi (*Aspergillus penicilloides* and *Eurotium herbariorum*) were four times higher in the UGH user group than in the HP user group, however these results were not statistically significant due the small sample size. The daily hyphal growth rate for *Alternaria alternata* was slightly higher in households operating a HP than in household operating an UGH (0.90 μm/day (HP) *vs.* 0.77 μm/day (UGH), $p = 0.07$). In the bedrooms, the daily hyphal growth rates for both xerophilic fungi were 7 and 26 times higher in households operating an UGH than in households operating a HP respectively ($p < 0.01$). No significant differences were found for *Alternaria alternata* between both household groups in the daily hyphal growth rates.

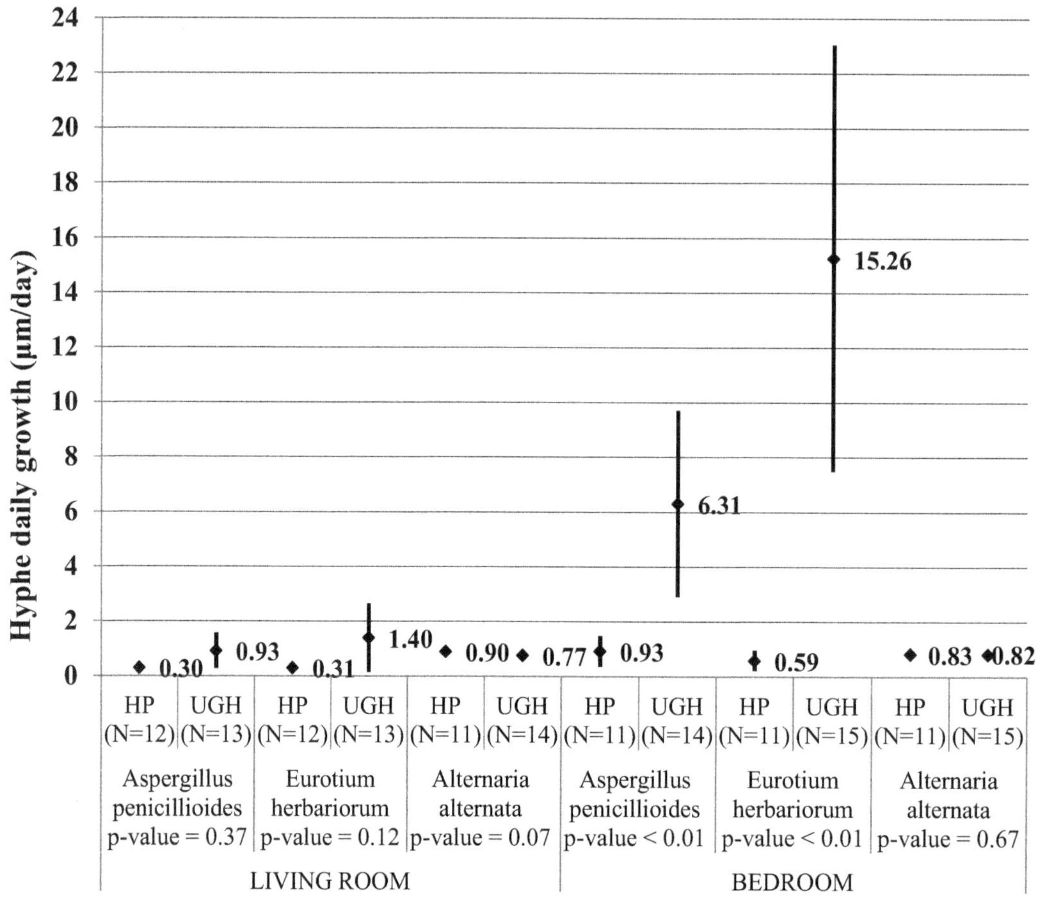

Figure 2. Average daily hyphal growth rate and lower and upper 95% confidence limit in the living rooms and in the bedrooms (outliers removed).

The climate was found to be more suitable for mould development in the bedrooms than in the living rooms. These results are consistent with the percentage of time that the RH was above 70%, being twofold higher in the bedrooms than in the living rooms (Table 2). A factor contributing to this difference in bedroom climates was the effect of the heater operated in the living rooms; the HP seems to be more effective than UGH in heating the bedrooms (forced air circulation due to HP fan). Boulic *et al.* [5] found that UGHs were mainly operated in NZ on low or medium setting which gives an estimated heat output of 2.6 kW. Since most UGH appear to have a nominal heating capacity of 4 kW, that suggests an element of underuse by occupants and as well as a heater capacity issue.

However, the more favourable condition for fungi to grow, in households who operated an UGH, was only true for the two xerophilic fungi *Eurotium herbariorum* and *Aspergillus penicilloides*. Consistent with the fact that these two fungi are the first to react to environment changes, they are considered as first colonizers because they can grow under relatively "dry" conditions ($a_w > 0.70$) whereas hydrophilic fungi like *Alternaria alternata* need very humid conditions ($a_w > 0.90$) to start germination [12,16,17]. Such conditions were infrequent in all studied houses, even those with UGHs. Abe *et al.* [18] found higher hyphal growth rate for *Eurotium herbariorum* in water associated rooms than in other rooms, showing a positive correlation between high moisture source and high hyphal growth rate. However, when comparing *Eurotium herbariorum* and *Alternaria alternata*, the same authors found a higher response for *Alternaria alternata* in water usage area (bathroom) [12].

2.3. Fieldwork Hyphal Development Compared to Laboratory Hyphal Development

Average temperature and RH were plotted on a climograph (isopleths) for *Eurotium herbariorum* (Figure 3), and *Alternaria alternata* (Figure 4). The isopleth for *Aspergillus penicilloides* is similar to the *Eurotium herbariorum* isopleth as both fungi have very similar temperature/RH requirements; therefore the *Aspergillus penicilloides* isopleth is not shown. The temperature/RH values found in the fieldwork were grouped as "UGH living room", "UGH bedroom", "HP living room" and "HP bedroom" and were compared to optimum climate zone for fungal germination which were obtained under constant conditions of temperature/RH in a laboratory. The laboratory data shown in Figure 3 were adapted from published work [13]. The laboratory data shown in Figure 4 were provided by Abe and are unpublished [19].

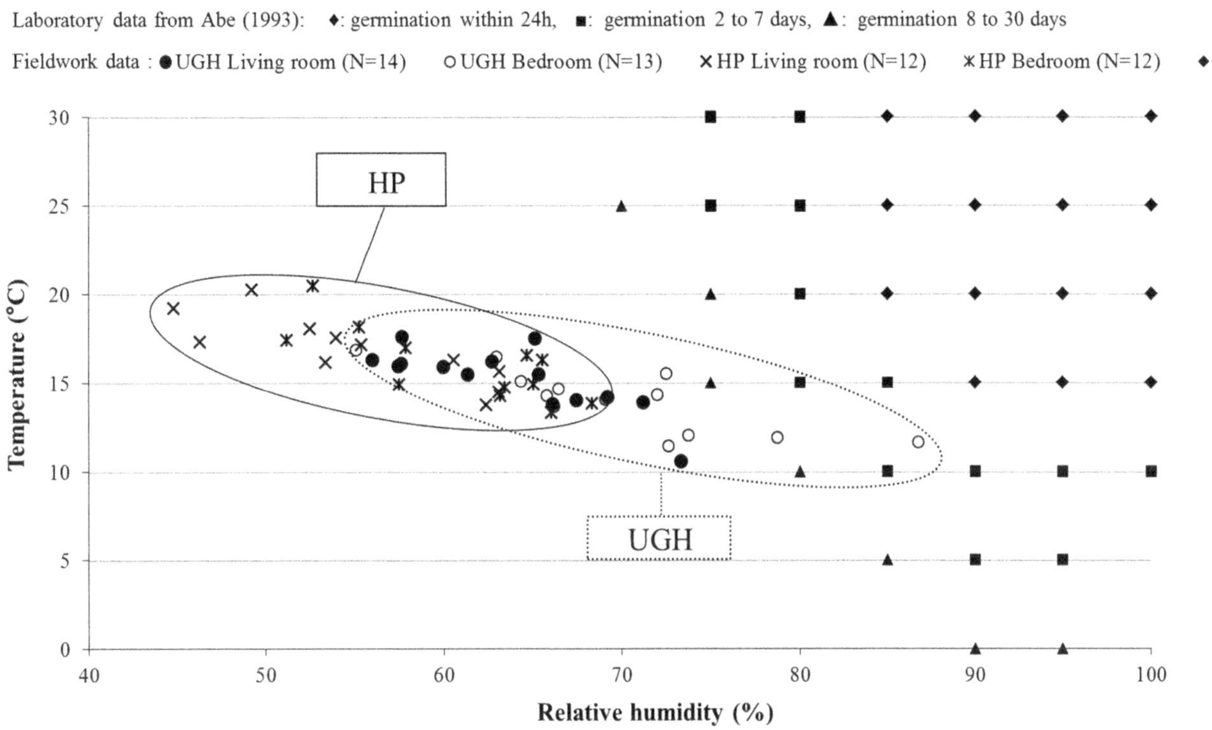

Figure 3. Climograph with germination of *Eurotium herbariorum* spores; fieldwork data compared to laboratory data (adapted from [13]).

In Figure 3, the 8 to 30 days germination zone (black triangle) starts at RH = 70%/T = 25 °C. Figure 3 shows that the average climate in the UGH user group (dashed ellipse) was closer to the germination zone than in the HP user group (plain ellipse). Six bedrooms and two living rooms from the UGH user group had an averaged climate that was either close or within the 8 to 30 days germination zone, whereas no room from the HP user group was within this germination zone. For both groups of households (UGH and HP users), the conditions for mould to grow were more favourable in the bedrooms than in the living rooms.

Figure 4 shows different requirements in terms of temperature/RH for the growth of the hydrophilic fungus *Alternaria alternata*. In Figure 4, the germination zone (black square) starts at RH = 90%. None of the households from either group was within this germination zone. One bedroom from the UGH user group showed an average RH of 86.8% which is close to the germination zone but did not show higher hyphal development than the group's average hyphal length. These results are consistent with the very low daily hyphal growth found in both groups for *Alternaria alternata* (Figure 2).

These results support the previous findings that the daily hyphal growth rates, for both xerophilic fungi (*Eurotium herbariorum* and *Aspergillus penicilloides*), were higher in the bedrooms than in the living rooms and also higher in the households operating an UGH than in households operating a HP. The measured temperature/RH values were well below the laboratory germination zone for the hydrophilic fungus *Alternaria alternata* development (RH = 90%), consistent with a low daily hyphal growth rate found for this fungus.

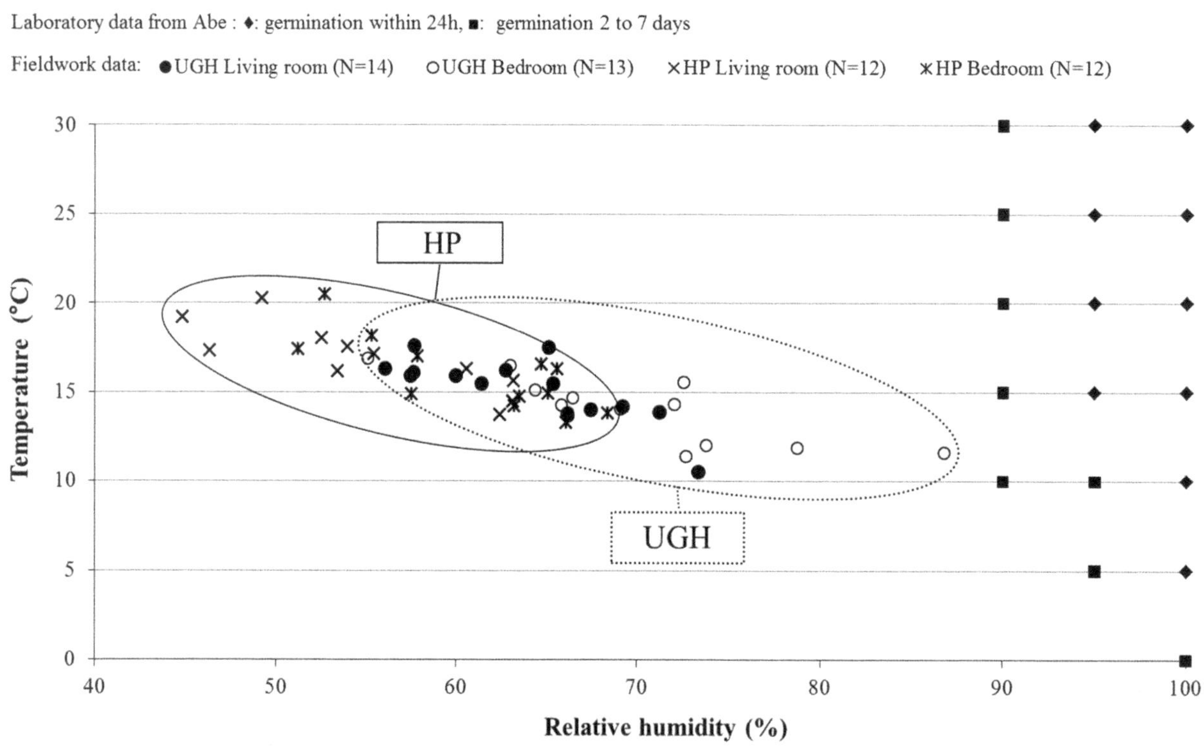

Figure 4. Climograph with germination of *Alternaria alternate* spores; fieldwork data compared to laboratory data (collected by Abe [19]).

2.4. Hyphal Development in Response to Favourable Climate Exposure

The measured daily hyphal growth rate was compared to the time of exposure in favourable psychrometric conditions. This analysis was done using a methodology developed by the Building Research Association of New Zealand [20]. The wall temperature from 0 to 30 °C and the RH from 35% to 100% were divided into 5 °C and 5% RH ranges, respectively. Next, "bins" were created for each 5 °C and 5% RH increment. For example, the temperature and RH combination of 0–5 °C and 35%–40% RH was the first of the 78 bins. Spearman's rank correlation tests were applied to test the correlation between the time of exposure in the defined bin and the measured daily hyphal growth rate for all three fungi.

Living rooms and bedrooms with a high measured hyphal growth rate (above the 75th percentiles) were selected for the analysis as these living rooms and bedrooms gave the best fungal development in response to the psychrometric conditions. A total of 13 rooms were selected consisting of 9 bedrooms and 4 living rooms in order to compare the *Eurotium herbariorum* hyphal development to climate exposure. A total of 14 rooms were selected consisting of 10 bedrooms and 4 living rooms to compare the *Aspergillus penicilloides* hyphal development to climate exposure, and a total of 11 rooms were selected consisting of 6 bedrooms and 5 living rooms to compare the *Alternaria alternata* hyphal development to climate exposure.

For *Eurotium herbariorum* ($N = 13$), the strongest positive correlation value ($R^2 = 0.40$, $p = 0.02$) was detected for the bin 10–15 °C and 80%–85% RH. For *Aspergillus penicilloides* ($N = 14$), the strongest positive correlation value ($R^2 = 0.28$, $p = 0.05$) was detected for the bin 15–20 °C and 85%–90% RH. For *Alternaria alternata*, no significant positive correlation between the measured hyphal growth rate and any climate bin were detected ($p = 0.54$).

Results show that in the rooms where *Eurotium herbariorum* had the largest hyphal development, the RH was above 80% for an average of 4.4 h a day. This result is consistent with a study by Cunningham [20] which found that *Eurotium herbariorum* needed to be exposed for at least 3.6 h per day with a RH above 80% for hyphal development. Similarly, the rooms where *Aspergillus penicilloides* showed the largest hyphal development had a RH above 85% for an average of 1.9 h a day.

Results show that in the rooms where *Alternaria alternata* had the largest hyphal development, the RH was between 35% RH and 70% RH for 77% of the time. This climate was too dry for this hydrophilic fungus to develop. The climate was suitable for *Alternaria alternata* (above 90% RH) for only an average of 0.1 h a day (0.4% of the time). This finding supports the lack of correlation found between the daily growth rate and the climate exposure, and was consistent with the low daily hyphal growth rate found for this fungus (Figure 2).

2.5. Impact of the Heater Choice on the Visible Fungal Level

Figure 5 shows the percentage of living rooms and bedrooms assessed at each of the four mould indexes in the households either operating an UGH or a HP.

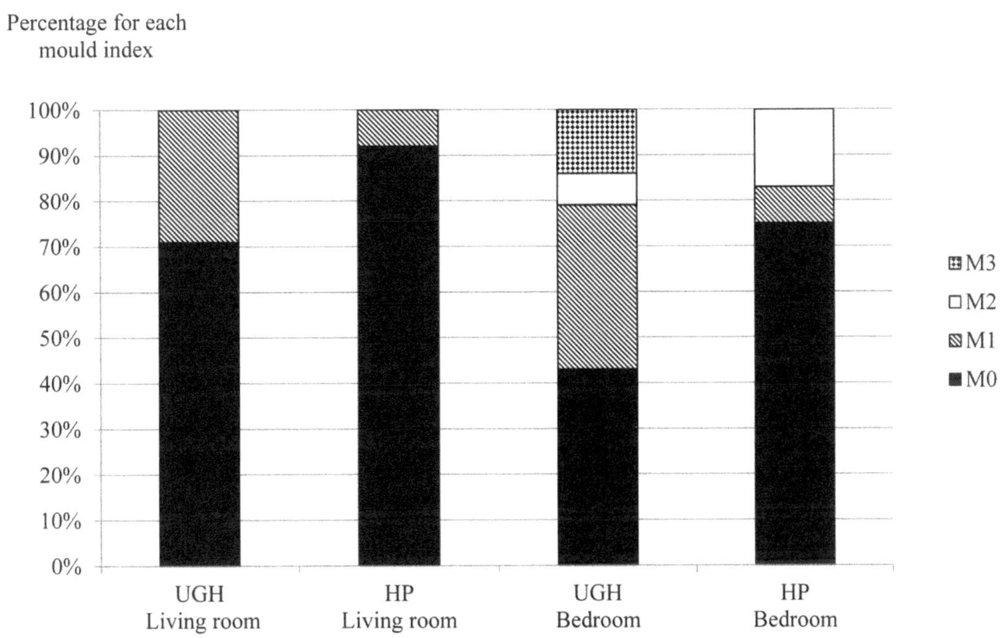

Figure 5. Mould index per heater type (UGH: Unflued Gas Heater or HP: Heat Pump) in the living rooms and in the bedrooms (M0: “no visible mould”, M1: “specks of visible mould”, M2: “moderate visible mould patches”, and M3: “extensive covered areas”).

Figure 5 shows a higher percentage of houses being assessed with the “M0” mould index (no visible mould) where HP have been operated (92% *vs.* 71% in the living rooms and 75% *vs.* 43% in the bedrooms), compared to the houses where UGH have been operated.

Spearman’s rank correlation tests were applied to test the correlation between the visual inspection mould index (M0, M1, M2 or M3) and the measured daily hyphal growth rate for all three fungal inclusions (*Aspergillus penicillioides*, *Eurotium herbariorum* and *Alternaria alternata*).

A positive correlation ($R^2 = 0.16$, $p < 0.01$) was found between the visible mould index in living rooms/bedrooms and the measured daily hyphal growth rate for *Aspergillus penicilloides*. A similar result was found for *Eurotium herbariorum* ($R^2 = 0.16$, $p < 0.01$), however no correlation was found between the visible mould index in living rooms/bedrooms and the measured daily hyphal growth rate for *Alternaria alternata* ($p = 0.39$). In conclusion, for *Aspergillus penicillioides* and *Eurotium herbariorum*, we can assume that the more visible mould, the higher the daily hyphal growth will be. Another study reported a positive correlation between a high visual fungal contamination and a high hyphal growth rate for *Eurotium herbariorum* [18].

A positive correlation was also found between the average wall RH and the visible mould index ($R^2 = 0.16$, $p < 0.01$) and a negative correlation between the average wall temperature and the assessed mould index ($R^2 = (-)\ 0.14$, $p < 0.01$). In the rooms with a contamination level of M0, M1, M2 and M3, the RH was on average 61.5%, 68.2%, 70.4% and 71.5%, respectively. The temperature decrease, correlated to an increase of the visible mould index, is consistent with a RH increase at the same humidity ratio. These results are consistent with a needed threshold of 70% for xerophilic fungi to start growing.

The visible mould level was found to be higher than the level reported from a self-reported NZ telephone survey [2]. The results showed that the visible mould level and the RH level were positively correlated and that the visible mould level and the temperature were negatively correlated. These findings are consistent with a higher percentage of visible mould detected in households where UGHs were operated. Garrett *et al.* [21] reported a correlation between evidence of dampness and visible mould. Another study found fungal levels were positively correlated with basement humidity [22] and negative correlation was found between the temperature of the room and the fungal level [23].

A meta-analysis showed strong associations between home dampness and respiratory/allergy effects, but the mechanisms linking the specific causal dampness and the related agents are still not clarified [24]. Mould and other microbiological organisms are probably the link between dampness and adverse health effects [25].

3. Conclusions

The households operating an UGH experienced a significantly different wall climate than the households operating a HP. In both the living rooms and the bedrooms, the percentage of time with the RH above 70% was much higher in homes where UGHs were operated than in homes where HPs were operated.

The fungal detectors were useful to predict the capacity for three species of fungi to grow on the inside surface of an external wall. The results showed that the bedroom climate was more suitable for xerophilic mould growth than the living room climate with a much higher hyphal development found in houses where UGHs were operated. These results were supported with a positive correlation between the hyphal development and the time of exposure in the germination psychrometric condition zones. However, it was found that the wall RH levels were too low and therefore not suitable for hydrophilic fungus development like *Alternaria alternata*.

Results from the visual inspection of mould growth were found to be very consistent with the predictions from the fungal detector method. A positive correlation was found between visible mould index and the measured daily hyphal growth rate for both xerophilic fungi (*Aspergillus penicillioides* and *Eurotium herbariorum*). However, it should be noted that the visual inspection is a very subjective method of mould assessment which could be influenced by several confounding factors such as the household mould cleaning, the colour of the wall surface and the age of the building material. In conclusion, the fungal detector method is a good complementary method to the visible inspection.

This intervention study had a positive impact on the wall psychrometrics with reduced water availability for mould to grow. The use of UGH was found to be a significant additional source of moisture in the living rooms. It was apparent that UGH user behaviour had an impact on the capacity to heat more than one room in the house. The households were asked to not modify their behaviour in terms of heater usage and indoor activities. The intervention was on the "intention to treat" rather than a "treatment"; consequently the households did not receive any fuel subsidies to encourage the usage of their heater. A follow up study, to investigate the environmental and health effects when a fuel subsidy assists households to pay for a higher consumption of heating could be interesting.

Acknowledgments

The hyphal measurements were undertaken thanks to the microscope facilities in the Institute of Veterinary, Animal, and Biomedical Sciences, Massey University, Palmerston North, New Zealand.

Participating families from Hutt Valley and community workers from Waiwhetu Marae, Kokiri Marae and Mid Valley Primary Health Organisation are thanked for their valuable contribution to the study. This study was funded by Health Research Council of New Zealand and Building Research Ltd.

In addition to the named authors, the Housing and Health Research Programme consisted of: Michael Baker, Chris Bullen, Julian Crane, Chris Cunningham, Sarah Free, Julie Gillespie-Bennett, Ralph Chapman, Bob Lloyd, Sarah Nicholls, Nevil Pierse, David Shields, Helen Viggers, Kristin Wickens and Alistair Woodward.

Author Contributions

The paper is based on Mikael Boulic's PhD research work. Robyn Phipps, Malcolm Cunningham and Don Cleland provided supervision and guidance over this research work. All three supervisors reviewed the paper. Pär Fjällström assisted with the study design, the data collection and the review of the paper. Keiko Abe assisted with the data analysis of the fungal detectors and the review of the paper. This research work was a sub study of the Housing, Heating and Health Study which was led by Philippa Howden-Chapman who assisted with the study design and the review of the paper. In addition to the named authors, researchers from the Housing and Health Research Programme reviewed the paper.

Conflicts of Interest

The authors declare no conflict of interest.

References

1. Isaacs, N.; Camilleri, M.; Pollard, A. Housing, Health and Energy. In *Housing and Health: Research, Policy and Innovation*; Howden Chapman, P., Carroll, P., Eds.; Steele Roberts: Wellington, New Zealand, 2004.
2. Howden-Chapman, P.; Saville-Smith, K.; Crane, J.; Wilson, N. Risk factors for mold in housing: A national survey. *Indoor Air* **2005**, *15*, 469–476.
3. Wilton, E. *Warm Homes Technical Report: Home Heating Methods and Fuels in New Zealand*; Ministry for the Environment: Wellington, New Zealand, 2005.
4. Statistics New Zealand. New Zealand 2013 Census QuickStats about Housing. Available online: http://www.stats.govt.nz/Census/2013-census/profile-and-summary-reports/quickstats-about-housing/heating-fuels.aspx (accessed on 21 November 2014).
5. Boulic, M.; Fjällström, P.; Phipps, R.; Cunningham, M.; Cleland, D.; Howden Chapman, P.; Chapman, R.; Viggers, H. Cold Homes in New Zealand—Heater Capacity or Low Heater Use? In Proceedings of the SB07 NZ Sustainable Building Conference, Auckland, New Zealand, 14–16 November 2007.
6. Francisco, P.W.; Gordon, J.R.; Rose, B. Measured concentrations of combustion gases from the use of unvented gas fireplaces. *Indoor Air* **2010**, *20*, 370–379.
7. Gillespie-Bennett, J.; Pierse, N.; Wickens, K.; Crane, J.; Nicholls, S.; Shields, D.; Boulic, M.; Viggers, H.; Baker, M.; Woodward, A.; *et al.* Sources of nitrogen dioxide (NO_2) in New Zealand homes: Findings from a community randomized controlled trial of heater substitutions. *Indoor Air* **2008**, *18*, 521–528.
8. Francisco, P.W.; Gordon, J.R.; Rose, W.B. Indoor moisture in 30 homes using unvented gas fireplaces. *ASHRAE Trans.* **2009**, *115*, 914–923.
9. Mendell, M.J.; Mirer, A.G.; Cheung, K.; Douwes, J.; Sigsgaard, T.; Bonlokke, J.; Meyer, H.W.; Hirvonen, M.R.; Roponen, M. *Chapter 4: Health Effects Associated with Dampness and Mould. Dampness and Mould: WHO Guidelines for Indoor Air Quality*; World Health Organization—Regional Office for Europe: Copenhagen, Denmark, 2009.
10. Howden-Chapman, P.; Pierse, N.; Nicholls, S.; Gillespie-Bennett, J.; Viggers, H.; Cunningham, M.; Phipps, R.; Boulic, M.; Fjällström, P.; Free, S.; *et al.* Effects of improved home heating on asthma in community dwelling children: Randomised controlled trial. *Br. Med. J.* **2008**, *337*, 852–855.
11. *NZS 4218:2004 Energy Efficiency—Small Building Envelope*; New Zealand Standard: Wellington, New Zealand, 2004.
12. Abe, K. Assessment of home environments with a fungal index using hydrophilic and xerophilic fungi as biologic sensors. *Indoor Air* **2012**, *22*, 173–185.
13. Abe, K. A method for numerical characterization of indoor climates by a biosensor using a xerophilic fungus. *Indoor Air* **1993**, *3*, 344–348.

14. Camilleri, M.; Isaacs, N.; Pollard, A.; Stoecklein, A.; Tries, J.; Jamieson, T.; Pool, F.; Rossouw, P. *Energy Use in New Zealand Households: Report on the Year 4 Analysis for the Household Energy End-Use Project (HEEP)*; Study Report SR 98; BRANZ: Porirua, New Zealand, 2000.
15. Tenwolde, A.; Pilon, C.L. The Effect of Indoor Humidity on Water Vapor Release in Homes. In Proceedings of Thermal Performance of the Exterior Envelopes of Whole Buildings X Conference, Clearwater Beach, FL, USA, 2–7 December 2007.
16. Flannigan, B.; Miller, J.D. Microbial Growth in Indoor Environments. In *Microorganisms in Home and Indoor Work Environments: Diversity, Health Impacts, Investigation and Control*, 2nd ed.; Flannigan, B., Samson, R.A., Miller, J.D., Eds.; Taylor & Francis: London, UK, 2011.
17. Darby, J.A.; Caddick, L.P. *Review of Grain Harvest Bag Technology under Australian Conditions. A Comprehensive Analysis and Field Evaluation of Harvest Bag Technology: Incorporating a Review of Hermetic and Temporary Storage, Control of Insects and Fungi, and Preservation of Grain Quality, under Typical Australian Storage and Handling Conditions*; Technical Report—No. 105; CSIRO Entomology: Clayton South, Australia, 2007.
18. Abe, K.; Nagao, Y.; Nakada, T.; Sakuma, S. Assessment of indoor climate in an apartment by use of a fungal index. *Appl. Environ. Microbiol.* **1996**, *62*, 959–963.
19. Abe, K. Institute of Environmental Biology, JDC Corporation, Kanagawa, Japan. Personal communication, 2006.
20. Cunningham, M.J. *Final Report on an Alternative Solution to NZBC E3 Field Validation and Model Modification*; BRANZ: Porirua, New Zealand, 2001.
21. Garrett, M.H.; Rayment, P.R.; Hooper, M.A.; Abramson, M.J. Indoor airborne fungal spores, house dampness and associations with environmental factors and respiratory health in children. *Clin. Exp. Allergy* **1998**, *28*, 459–467.
22. Dekoster, J.A.; Thorne, P.S. Bioaerosol concentrations in noncomplaint, complaint, and intervention homes in the midwest. *Am. Ind. Hyg. Assoc. J.* **1995**, *56*, 573–580.
23. O'Connor, G.T.; Walter, M.; Mitchell, H.; Kattan, M.; Morgan, W.J.; Gruchalla, R.S.; Pongracic, J.A.; Smartt, E.; Stout, J.W.; Evans, R. Airborne fungi in the homes of children with asthma in low-income urban communities: The inner-city asthma study. *J. Allergy Clin. Immunol.* **2004**, *114*, 599–606.
24. Mendell, M.J.; Mirer, A.G.; Cheung, K.; Tong, M.; Douwes, J. Respiratory and allergic health effects of dampness, mold, and dampness—Related agents: A review of the epidemiologic evidence. *Environ. Health Perspect.* **2011**, *119*, 748–756.
25. Fisk, W.J.; Lei-Gomez, Q.; Mendell, M.J. Meta-analyses of the associations of respiratory health effects with dampness and mold in homes. *Indoor Air* **2007**, *17*, 284–296.

4

Understanding the Importance of Urban Amenities: A Case Study from Auckland

Natalie Allen

School of Architecture and Planning, The University of Auckland, 26 Symonds St, Auckland 1142, New Zealand; E-Mail: natalie.allen@auckland.ac.nz

Academic Editor: George Baird

Abstract: Along with many Pacific Rim cities in Australia and North America, Auckland, New Zealand has enacted an urban growth management strategy premised on two concepts: "liveability" and a "quality compact city". The effective implementation of this strategy will, in part, require higher density housing typologies to be developed within the existing suburban fabric. The urban amenities in a neighbourhood play an important role in providing a sense of liveability for residents. This paper examines these issues by evaluating and reporting on key outcomes from 57 face-to-face qualitative interviews with residents who currently live in medium density housing in four Auckland suburbs; Takapuna, Kingsland, Botany Downs, and Te Atatu Peninsula. Findings consider the trade-offs residents make when choosing to live in medium density housing typologies, how they value the urban amenities in their neighbourhood and the role they think these amenities play in their location satisfaction. Conclusions are drawn around how the resident-derived information may inform the market on the supply side of housing, and comment is made about how these preferences may, or may not, respond to the objectives of the underlying urban management strategies involved.

Keywords: urban intensification; suburban intensification; quality of life; liveability; urban amenities; neighbourhood

---

1. Introduction

This research responds to the urban intensification debate in Auckland by questioning the perceived role of urban amenities in promoting quality of urban life at higher densities in four of the city's suburban neighbourhoods: Takapuna, Kingsland, Albany, and Botany.

Like Auckland, many Pacific-Rim cities in Australia and North America have long imbued compact city principles in to their urban growth management strategies. Cities such as Brisbane, Sydney, Melbourne, Vancouver, and Portland are facing a similar set of development issues related to their growing yet ageing populations and shifting demographics towards smaller households. While a system of common law is in place, governmental and institutional arrangements do differ between the cities [1]. Unlike Auckland, for example, each of the aforementioned cities has both national, state, and local (city-wide) levels of planning governance, whereas Auckland has national and local only. This changes funding streams and methods for implementation. Despite these structural differences, notable similarities exist between the growth management strategies of these cities including their greenbelt ideology that utilises an urban growth boundary to restrict greenfield development at urban peripheries [2] as well as the promotion of networks of higher density mixed-use development clustered around walkable town centres.

There have been various waves of higher density housing trends, across many Pacific-Rim cities in Australasia and North America, including Auckland [3,4]. However, the quality compact goals of these cities largely exist in contradiction to the dominant post-war ethos of suburbia still prevalent despite the considerable regeneration and redevelopment of their downtowns, waterfronts and other former-industrial land. Resident resistance to suburban intensification across these Pacific-Rim cities has been widely noted [5,6]. Writing about traditional aspirational housing ideals both Alves [3] and Randolph [4] in Australia, and Smith and Billig in North America, identify a long-standing preference for suburban single-storey detached dwellings. This trend is referred to as the Quarter-Acre Pavlova Paradise in New Zealand [7]. Alves identifies that research and planning action has been lacking in suburban areas [3]; transforming and intensifying suburbia, while still maintaining the distinct character and perceived liveability of the suburban model, is the next big challenge for urban planners and designers.

In this research it is argued that a key element in the transition to more urbanised environments is related to the extent to which urban amenities have a role in resident perceptions of quality of urban life. Mulligan and Carruthers identify that "*amenities are key to understanding quality of life because they are precisely what make some places attractive for living and working, especially relative to other places that do not have them and/or are burdened with their opposites, disamenities*" [8] (p. 107).

Urban amenities are understood in this research to mean specific urban facilities that contribute to the urban living experience of residents [9]; they are linked to the daily life needs of residents in a neighbourhood. Some examples given by Randall include: "*grocers, convenience stores, access to public transit, schools and professional services [doctor or dentist]*" [10] (p. 47). Gottlieb confirms that "*residential amenities may be defined as place-specific goods or services that enter the utility functions of residents directly*" [11] (p. 1413). Both Mathur and Stein [12] (p. 252) and McNulty *et al.* [13] refer to urban amenities as "quality of life factors" and Howie *et al.*, confirm that "*urban amenities are generally accepted as being important to a household's sense of place*" [14] (p. 235). There are both public sector amenities provided by councils, such as parks, public squares and recreational facilities, as well as private sector amenities such as cafés, restaurants, retail and other goods or service providers.

There are two main reasons given in the literature as to why focusing on the role of urban amenities in the delivery of urban intensification is important. Firstly, in an economic sense, it is argued that a diversity of urban amenities attract economic activity to a city in terms of firms and labour wanting to be located in a place of high amenity value [8,15–17]. In other words, "*the provision of amenities generates urban advantages that perpetuate the concentration of economic activity and population in, and in closer proximity to, them*" [18] (p. 40). Mathur and Stein also confirm that "*the emerging literature on amenities seems to indicate that one of the most effective ways to attract knowledge workers in the regions and promote economic development is the creation of amenities*" [12] (p. 265).

In line with economic reasoning, the second broader reasoning is that the accessibility and convenience of urban amenities contribute to quality of urban life experiences [19–22]. As society changes and evolves so too do people's quality of life requirements and aspirations. There are clear linkages acknowledged in the literature between the provision of varied urban amenities and changing lifestyles and aspirations [1,23,24]; demographic changes for example have a direct impact on the spatial configuration of the city and are closely tied to changing lifestyle preferences. Increasing ethnic diversity through globalisation also contributes to urbanism trends as new city residents bring their own understandings of intensification and the relationship between urban amenities and perceived quality of life. There are also issues of affordability and potential new home owners now being priced out of the market; these buyers may turn to higher densities as a way of entering the property market. Internationally, authors such as Clark [25] have recognised that demographic and ethnographic changes alter the way cities are viewed and experienced and as such, the provision of urban amenities and their integration in to urban areas must also be reconsidered. Randolph [4] in particular, writing about Australia, highlights the need for planners to understand the amenity requirements for higher density neighbourhoods, particularly if more children are going to be living in these urbanised environments, thus increasing the need for schools, child care facilities, and recreational areas. An example given by Schmitz *et al.* is that the increasing number of "work-at-homers" "*often feel isolated in typical suburban communities and would like access to the amenities that are available to downtown office workers*" [26] (p. 6). They consider that options such as the corner coffee shop, lunch bars, a print centre, local gym or recreation area, and retail facilities should be better integrated in to suburban environments. Auckland Council also acknowledges that the city needs "*more housing adjacent to local shops and services, public open spaces and areas with expansive views*" [27] (p. 70). And yet, research into how this might occur and how different amenities are valued by residents is very limited in New Zealand. In this paper, understanding the relationship between urban amenities and perceived quality of life is based on the premise that "*dwellings are important, but so too is the location of the dwelling*" [28] (p. 98).

Whatever the reasons for intensification, the question remains: if suburbia transforms, will these higher density neighbourhoods meet the aspirations and needs of future residents? It is therefore necessary to ask residents about the urban amenities they use and value and the relationships they see between these amenities and their sense of location satisfaction.

2. Method

Guided by a constructivist Grounded Theory approach the researcher conducted 57 hour-long face-to-face structured interviews with local residents from four case study neighbourhoods in Auckland: Takapuna, Kingsland, Botany Downs, and Te Atatu Peninsula (see Figure 1). The structured interviews were recorded and transcribed. The analysis included initial line-by-line manual coding before open and selective coding phases were conducted by the researcher in NVivo. Theoretical coding was the third coding phase in line with Grounded Theory methods.

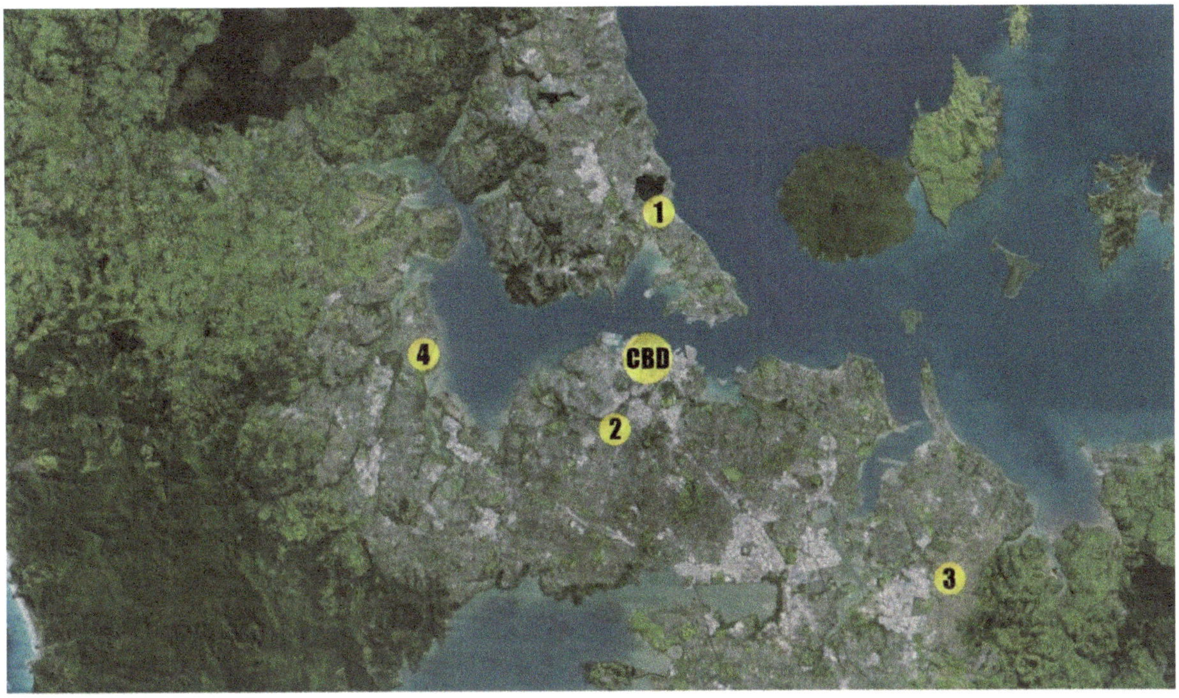

2 kms

2 miles

Key
1. Takapuna
2. Kingsland
3. Botany Downs
4. Te Atatu Peninsula

Figure 1. The four case study neighbourhoods (Source Google Maps).

The chosen neighbourhoods represent both inner and outer fringe belt suburbs [29] that have all experienced strong increases in both rental and owner-occupier medium density developments. They range from being 4.3 to 20 km away from Auckland's Central Business District (CBD) and are located to the north, west, and south-east of Auckland. Within these neighbourhoods, medium density case study developments were chosen as the locations where direct mailbox dropping would occur to attract the interviewees for the study.

The criteria for selecting the developments within the case study neighbourhoods included: that each of the developments was multi-unit, at a Net Residential Density (NRD) greater than 35 dpHA, and in a location that provided local amenities; for example a town centre. This meant a mix of typologies ranging from three- and five-storey apartment complexes to attached townhouses and units. Developments also had to have been established in their communities for a period of more than three years and accessible for interviews. The developments at 130 Anzac St. in Takapuna, 435 New North Road in Kingsland, and

84 Gunner Drive in Te Atatu required approval from building managers to mail box drop as their mail boxes were concealed within the developments.

2.1. Takapuna

Takapuna is a city fringe suburb in the north of Auckland, 9.6 km from the CBD. It has been extensively considered for intensification by North Shore City Council and Auckland Regional Council. It has also been identified as a key growth area in the Auckland Plan [30] and the Unitary Plan [31]. Takapuna was also identified by Fontein [32] as an area that is well placed for market-led intensification due to its access to a range of amenities; including, transit, employment opportunities, mixed use residential and business land, and natural amenities [33] (pp. 33–35). Twenty residents from three developments in Takapuna were interviewed. These developments ranged from townhouses with a Net Residential Density (NRD) of 35.8 dpHA to three-storey apartments that were 60.8 dpHA to even higher density five-storey apartments that were 167.9 dpHA (see Figure 2 and Table 1).

Figure 2. Takapuna case study area: interviewees were from the developments shown in orange, the town centre is shown in yellow. Letters correspond to Table 1. (Source: Google Maps).

Table 1. Information about the chosen developments where residents were approached through direct mailbox drops to be interviewed. (The letters in the column "Map #" correspond to the maps in Figures 2–5).

Case Study Neighbourhood	Site	Map #	Development Lot Size (m^2)	# Units	NRD (Net Residential Density) (dpHA)	Typology	Storeys	Year Built	Total Number of Mail Drops *	Total Number of Interviews
Takapuna	130 Anzac St.	A	7,980	134	167.9	Apartments	5	2007	234	20
	73 Anzac St.	B	6,910	42	60.8	Apartments	3	2005		
	7 Killarney St.	C	4,750	17	35.8	Rowhouses	3	2001		
Kingsland	39 Sandringham Rd.	D	875	25	285.7	Apartments	4	2011	152	13
	435 New North Rd.	E	3,200	90	281.3	Mixed use with Apartments	6	2006		
Botany Downs	Armoy Drive development A	F	7,700	59	76.6	Townhouses and Rowhouses	2	2005	460	14
	Armoy Drive development B	G	29,290	153	52.2	Apartments/Units	2	2005		
	Spalding Rise development	H	29,990	149	49.7	Townhouses and Rowhouses	2	2000–2003		
	Kirikiri Lane development	I	11,200	128	114.3	Townhouses and Rowhouses	3	2005		
Te Atatu Peninsula	84 Gunner Drive	J	1,650	37	224.2	Mixed use with Apartments	6	2006	166	10
	Vinograd Drive development	K	24,130	93	38.5	Townhouses and Rowhouses	2	2001–2004		
	Gunner Drive development	L	4,870	18	37.0	Apartments	2	2000		
–	–	–	–	–	–	–	–	–	1,012	57

Notes: * The number of mail drops differs from the number of units in the development because other developments were chosen as possible sites but returned no residents interested in being interviewed. These development sites were therefore excluded, but the mail drops were not.

2.2. Kingsland

Kingsland is a city fringe suburb, just 4.3 km from the CBD, and one of the top five rental growth suburbs in Auckland [34]. While it has grown in popularity because of its proximity to a range of urban amenities and access to the CBD, it remains predominantly suburban, made up of single-storey detached dwellings, and a noticeably smaller town centre than the other chosen areas. Medium density housing in the area is increasing; although currently these are largely apartments rather than a mix of medium density typologies such as terraced houses and units. Thirteen residents from two developments were interviewed. These developments were both four-storey apartment buildings at densities of 281.3 and 285.7 dpHA (see Figure 3 and Table 1). One was a mixed-use typology with apartments built above the high street shops. In this case it was four storeys on the road side but six storeys on the parallel street where the site contour sloped down.

Figure 3. Kingsland case study area: interviewees were from the developments shown in orange, the town centre is shown in yellow. Letters correspond to Table 1. (Source: Google Maps).

2.3. Botany Downs

Botany Downs is the furthest suburb, at 20 km south-east from Auckland's CBD, considered in this research. It is an area that has seen considerable growth in the last 5–10 years, frequently in the form of two-storey attached dwellings in large scale, partially gated communities. These developments ranged from 49.7 to 114.3 dpHA (see Figure 4 and Table 1). 14 residents were interviewed from four developments in Botany Downs.

Figure 4. Botany Downs case study area: interviewees were from the developments shown in orange, the town centre is shown in yellow. Letters correspond to Table 1. (Source: Google Maps).

2.4. Te Atatu Peninsula

Te Atatu Peninsula is 15.3 km to the west of Auckland's CBD. While it is dominated by low density single-storey detached dwellings its popularity and increasing levels of gentrification have seen attached and semi-attached two-storey townhouses develop to the east of the town centre. 10 residents from three development areas were interviewed. Two of the developments range in density from 37 to 38.5 dpHA. However, the third is a six-storey apartment building, with retail on the ground floor, that has been developed on the main street leading in to the town centre. It has a relatively high density of 224.2 dpHA; considerably greater that the surrounding suburban area that measures between 10 and 20 dpHA (see Figure 5 and Table 1).

Figure 5. Te Atatu Peninsula case study area: interviewees were from the developments shown in orange, the town centre is shown in yellow. Letters correspond to Table 1. (Source: Google Maps).

Table 1 is a summary of information about the chosen developments across the four case study neighbourhoods including the lot size, number of units, Net Residential Density (NRD), their typology, the number of storeys and the year they were built. It also shows the total number of mail drops conducted and the number of interviews conducted in each area. In total 36 females and 21 males responded to the mail-box letter drops. Interviewees were spread between the ages of 23 and 87; the average age of respondents was 44. Of the 57 interviewees, 26 were owner-occupiers and 31 were renters. The average length of dwelling tenure was three years; the shortest time being one month and the longest 13 years. Seven interviewees lived alone, 25 were couples, six interviewees lived with flatmates, and two lived with extended family members. Seventeen had a least one child living at home, of these 10 had children under seven. Only three had two or more children living at home.

Thirty-six of the interviewees were born in New Zealand and nearly half of these had also spent time living overseas where they were exposed to a range of different lifestyles and housing norms. Of the 21 born overseas, only 11 were from countries where English was not their first language. Countries of origin included The United States of America, Australia, Scotland, England, China, Korea, Malaysia, Vietnam, Singapore, India, Portugal, Bulgaria, Slovakia, Serbia, Jordan, Mozambique, and South Africa. Nineteen of the interviewees had had experience living outside of Auckland in areas ranging from Christchurch and Wellington to rural Otago, Napier, Gisborne, Palmerston North, Tauranga, Rotorua, and Tairua. Despite the geographic, demographic, and socio-economic variations between the case study suburbs and the interviewees, clear patterns emerged among the findings.

Interview questions covered household structure, dwelling tenure, current employment and travel habits as well as housing histories and aspirations. Interviewees were asked about the urban amenities they use, how often, their accessibility, and how they valued them. How residents defined quality of life and their perceptions of urban intensification and density were also explored.

3. Results

3.1. Making Medium Density Housing Choices

All the interviewees, excluding two, had had previous experience living in standalone houses; most often it was the housing norm they had experienced the most. Over half, 58% of the interviewees, had also experienced apartment or townhouse living at some stage prior to their current housing experiences; whether it be as students, while living overseas, or while living in a flatting situation away from their regular family home.

When asked about their process for making housing choices 41 interviewees mentioned proximity to urban amenities and the resultant convenience as one of their main reasons for moving to their current housing. Medium density typologies also played a factor, ease of maintenance in particular was stated as affecting housing choices for 26 of the interviewees. Twelve interviewees mentioned trading-off living in a standalone house for an apartment or townhouse because they found them to be warmer and cleaner than the available standalone housing stock they could rent for a similar price point. Safety, security, proximity to work, affordability, and place attachment created due to proximity to friends and family were other reasons frequently mentioned by interviewees as reasons that had influenced their decision making when deciding to live in their current location.

3.2. Future Housing Aspirations

More than half of interviewees saw themselves continuing to live in the same suburb where they currently resided over the coming five years. Where this was not the case the predominant aspirations held were to look at land on the urban fringe for affordability reasons, or to move to a more rural environment as part of long term lifestyle ambitions. Retirees interviewed were split between moving out of Auckland or to fringe areas where they could live quieter lives or staying in apartment typologies, usually in their current neighbourhood, that would be easy for them to maintain or lock up and leave if they planned to travel. Where this was the case proximity to a range of urban amenities and the ability to walk or use public transport were noted as key location drivers to support their active urban lifestyles.

Twenty-two interviewees mentioned their or their partners proximity to work as part of their decision making process around housing and reasoning for wanting to stay in the same area. This largely came down to the notion that interviewees did not want to be "sitting in traffic half the day". Place attachment due to family and friends living in the same area was also frequently mentioned as a factor affecting the interviewees' location aspirations.

When asked about the type of housing they saw themselves living in in five years' time 26 interviewees saw themselves in medium density typologies, including low rise apartments (of three to five storeys) or terraced housing (similar to their current typologies). Twenty-five saw themselves in low density typologies including single detached dwellings on small or large sections and six were unsure what their housing aspirations were at the present time. Of the 25 interviewees who identified an aspiration for low density typologies, five of them simultaneously discussed medium density options as being an equally likely aspiration they would be happy with. This would shift the total number of those who aspired to medium density housing solutions from 26 to 31 respondents. Typologies at the same or higher densities than their current housing scenarios, namely low rise apartments and terraced housing, were also the notably favoured "back-up" plan for interviewees when they were asked where they might be happy to live if their first choice was not available to them for whatever reason.

The concept of "lifestyle" was used by interviewees as a reason to justify both their low and medium density preferences. Those who wanted standalone houses as their first choice usually spoke about wanting space, peace and quiet, the ability to have pets and a garden, and "the ability to cater for acquisitions which you can't do in a smaller apartment". A small number of interviewees who aspired to live in a standalone house did acknowledge that many of the issues they currently had with medium density typologies, particularly related to storage and shared garden space, could be addressed if the apartments or terraced houses were well designed, built from quality materials, and larger than the CBD "shoeboxes" that they currently associated with the idea of "higher density living".

Those who wanted to live in low rise apartments or terraced houses cited proximity to urban amenities as their main reason for choosing this typology over others, stating that they "*like going out and doing things and having everything on our doorstep*". One interviewee commented, "*I'll compromise on smallness... I don't actually mind the smallness, but I like to be close to the city*". When asked why, the interviewee commented about the liveliness of the city contributing to their quality of life because they were always busy and engaged with what was going on around them.

Other reasoning given for preferring medium density typologies included security and their low maintenance nature. One interviewee commented that they wanted to live "*somewhere that is clean, tidy and nice. I really don't want a garden that gets neglected because I don't have the time for it. I'd rather not have it at all*". Another interviewee saw living in an apartment as meaning they could pay off their small mortgage and "make a bit of a nest egg" so that they could travel and enjoy their upcoming retirement.

A number of interviewees who saw themselves living in medium density dwellings over the next 5 years, and in some cases longer term, did temper their aspiration with the concern that many apartments currently available were very small and did not always meet their needs; lack of storage, lack of car parking for multi-bedroom apartments, and small kitchens rather than four burner hobs were the most frequently mentioned concerns. One interviewee spoke of wanting to live in "stacked villas", the villa being a popular traditional standalone housing typology in New Zealand. They wanted a villa on the third or fourth floor; in other words a spatial layout that they were familiar with that had larger shared spaces, more storage, and a larger outside living area than the majority of apartments they had seen previously in Auckland.

Additional reasoning given by those interviewees who stated that they would prefer to live in a standalone house related to their desire to renovate and because they thought that owning a home was part of the "kiwi dream". Fear that the build quality of apartments would be substandard was also cited as a concern by some interviewees. A further concern was that they thought buying an apartment would disadvantage them financially because the capital gains were perceived to be higher for properties with standalone houses. One interviewee who currently owned their apartment commented that they would eventually like to buy a house "out in the suburbs" but that it "*purely would be an investment property and we'd stay living in the apartment*".

3.3. Views about Quality of Life and Urban Amenities

When interviewees were asked directly about how they defined quality of life their responses covered a broad range of issues; the importance of health, income, safety, and broader discussions of freedom, enjoyment and fulfillment were raised. Twenty-four interviewees related their perceptions of quality of life to concepts of neighbourhood and the availability and accessibility of urban amenities. When asked about the perceived role of urban amenities in a neighbourhood, all 57 interviewees linked it to quality of life aspects, in particular to notions of accessibility and the associated convenience or ease this bought to their lives.

When asked to define the term urban amenities seven interviewees thought of them predominantly as being council provided amenities. Examples given included public transport, public pools, community centres, and libraries. Four were unsure of the terms specific meaning. The remaining 43 interviewees thought of urban amenities as all the services and infrastructure they used in their daily lives. One interviewee commenting that "*urban amenities would be all of the things that you value in a community and so that would be the open space, the walkways, the park space, the variety of shops or facilities and our proximity to them*".

Food related amenities, such as supermarkets, cafés, and restaurants, were the urban amenities considered to be the most important by interviewees. Supermarkets were specifically mentioned the most frequently. Recreation amenities and public spaces were also favoured. The areas where residents reported using their local amenities daily and usually accessed them by foot (this was weather dependent) were Takapuna and Botany Downs. Te Atatu Peninsula had similar responses for a variety of food and community amenities but many interviewees commented about their needing to drive for retail amenities. While Kingsland residents were positive about their local café and restaurant culture, walkability for urban amenities such as chemists, clothing stores, dairies and supermarkets was questioned by the interviewees who predominantly drove to neighbouring suburbs for these urban amenities because they were not available to them in Kingsland.

4. Discussion

The initial findings on housing choices in Auckland are in line with the work of Haarhoff and Beattie [35,36], Mead and McGregor [37], and Syme *et al.* [5] who all found that residents living in medium density developments in Auckland rated highly the convenience of living in close proximity to the urban amenities that they valued in their daily lives; including shops, cafés and restaurants, schools, their workplaces, public transport, and public spaces (particularly if they were park and recreation facilities). These studies also all identified that the locations of these urban amenities and their convenience was a significant factor for residents in choosing to live where they did.

A surprising change from previous studies, where standalone houses featured prominently as the favoured typology residents saw themselves living in, was the relatively even divide between the aspirations of residents being low density *versus* medium density typologies. The interviews suggest that the increasing popularity of low rise apartments and terraced housing can be tied to changing lifestyle expectations in the city across a broad spectrum of demographics and lifestages. The shift to favouring similar to higher densities than the interviewees were already living in suggests a disjuncture between the current supply and potential demand for higher density typologies in Auckland's suburbs.

However, the concern that current medium density typologies did not meet the daily life needs of residents, raises questions about how some interviewees might have responded to questions about their housing aspirations if more medium density typology options were available to them within their current suburban areas. The interviewee responses would suggest that a greater acceptance of medium density typologies may occur if concerns about build quality and lack of storage and kitchen space were addressed.

Furthermore, when questioned about how they defined urban amenities the overwhelming majority of interviewees thought of them as all the services and infrastructure they used in their daily lives. This contrasts with much of the planning policy and strategy in New Zealand as well as international literature where amenities are often considered in silos; for example as "natural amenities", "entertainment amenities", or public amenities provided by council. It was therefore found in this research that in order to begin understanding the dynamics of liveability in our current suburban neighbourhoods, "how" residents use and value urban amenities seamlessly across their neighbourhood must be investigated. This means to consider all urban amenities simultaneously as they relate to liveability, whether they be public sector amenities provided by councils, such as parks, public squares and recreational facilities, or private sector amenities such as cafés, restaurants, retail and other goods or service providers. The biggest disjuncture between the findings of this research and current planning policy and strategy in Auckland is just how seamlessly interviewees thought about the spatial relationship between their dwellings and the urban amenities they wanted to live in close proximity to. Current residential zoning, for example, is at odds with this very intimate integration of urban amenities in to the suburban fabric and further research is considerably needed around this issue.

5. Conclusions

A better understanding of how residents are making trade-offs between suburban and urban lifestyle options is critical to understanding the relationship between urban amenities and perceived resident quality of life experiences. Schmitz *et al.* [26] (p. 62) propose that to consider urban amenities is to programme the urban environment to meet residents' needs. By this thinking, it will be the successful integration of urban amenities in to suburban areas that will enable the perceived quality of life experienced by residents to be maintained or even enhanced during the transition to higher density urban environments to accommodate growth rather than the current low density suburban norm. The apparent risk of not considering urban amenities in this way is to misunderstand the nature of contemporary urban life and the effects of changing demographics and household structures on housing choices.

This research has raised questions about the factors that affect how residents make trade-offs between low density and higher density housing choices; and as such it has raised questions about the effect of the current supply on the future demand for housing in Auckland. Most interestingly, it can be concluded that the majority of those interviewed would trade-off standalone living for low-rise apartment or terraced house living in their current neighbourhoods if the medium density options available to them were of a size and spatial layout that were more similar to the traditional New Zealand home and if urban amenities were integrated in to their neighbourhoods in line with the increasing numbers of residents. Higher amenity must go hand-in-hand with higher density if quality of life is to be experienced in tomorrow's urban Auckland.

Acknowledgments

The author would like to thank all the residents who agreed to be interviewed as part of this research.

Conflicts of Interest

The authors declare no conflict of interest.

References

1. Knox, F.; Smith, J. *International Trends and Lessons in Growth Management*; Auckland Regional Council: Auckland, New Zealand, 2007.
2. Jerram, R.; Kvan, T. *Liveable Melbourne*; A Paper Prepared for City of Melbourne Future Melbourne Project by the University of Melbourne; The University of Melbourne: Melbourne, Australia, 2008. Available online: http://www.futuremelbourne.com.au/wiki/pub/FMPlan/S3Resources/Liveable_Melbourne_FINAL.pdf (accessed on 22 January 2015).
3. Alves, T. Medium Density Housing in Melbourne: The Management of Sustainable and Democratic Local Communities under Global Pressure for Increased Urban Efficiencies. In Proceedings of the 19th EAROPH World Planning and Housing Congress and National Housing Conference, Melbourne, Australia, 19–22 September 2004.
4. Randolph, B. *Delivering the Compact City in Australia: Current Trends and Future Implications*; City Futures Research Centre, University of New South Wales: Sydney, Australia, 2006. Available online: http://www.be.unsw.edu.au/sites/default/files/upload/researchpaper6.pdf (accessed on 22 January 2015).

5. Syme, C.; McGregor, V.; Mead, D. *Social Implications of Housing Intensification*; Auckland Regional Council: Auckland, New Zealand, 2005.
6. Searle, G. Sydney's Urban Consolidation Experience: Power, Politics and Community. Available online: http://www.griffith.edu.au/__data/assets/pdf_file/0018/48600/urp-rp12-searle-2007.pdf (accessed on 16 February 2012).
7. Mitchell, A. *The Half-gallon Quarter-acre Pavlova Paradise*; Christchurch Whitcombe and Tombs: Christchurch, New Zealand, 1972.
8. Mulligan, G.; Carruthers, J. Amenities, Quality of Life, and Regional Development. In *Investigating Quality Urban Life, Theory Methods and Empirical Research*; Marans, R., Stimson, R., Eds.; Springer: Dordrecht, The Netherlands, 2011; pp. 107–134.
9. Kelly, A. Securing urban amenity, does it coincide with biodiversity conservation at the local government level? *Australas. J. Environ. Manag.* **2006**, *13*, 243–253.
10. Randall, T. Preferences of suburban residents in Thunder Bay, Ontario towards neighbourhood intensification and rediversification. *Can. J. Urban Res.* **2008**, *17*, 28–56.
11. Gottlieb, P. Residential amenities, firm location and economic development. *Urban Stud.* **1995**, *32*, 1413–1436.
12. Mathur, V.; Stein, S. Do amenities matter in attracting knowledge workers for regional economic development? *Pap. Reg. Sci.* **2005**, *84*, 251–269.
13. McNulty, R.J.D.; Penne, L. *The Economics of Amenity: Community Futures and Quality of Life*; Partners for Livable Places: Washington, DC, USA, 1985.
14. Howie, P.; Murphy, S.; Wicks, J. An application of a stated preference method to value urban amenities. *Urban Stud.* **2010**, *47*, 235–256.
15. Auckland Council. *A New Kind of Compact: Design-Led Part 3*; Auckland Council: Auckland, New Zealand, 2011.
16. Hansen, H.; Winther, L. Amenities and Urban and Regional Development: Critique of a New Growth Paradigm. In Proceedings of the Regional Studies Association Annual International Conference, Pécs, Hungary, 24–26 May 2010.
17. Ministry for the Environment. *People, Places, Spaces: A Design Guide for Urban New Zealand*; Ministry for the Environment: Wellington, New Zealand, 2002.
18. Partridge, M.M.O.; Alasia, A. Canadian cities as regional engines of growth: Agglomeration and amenities. *Can. J. Econ.* **2007**, *40*, 39–68.
19. Johnson Gardner Consulting. *An Assessment of the Marginal Impact of Urban Amenities on Residential Pricing*; Portland Metro: Portland, 2007. Available online: http://www.reconnectingamerica.org/assets/Uploads/JohnsonGardner-Urban-Living-Infra-Research-Report.pdf (accessed on 22 January 2015).
20. Ministry for the Environment. *Building Competitive Cities: Reform of the Urban and Infrastructure Planning System*; Ministry for the Environment: Wellington, New Zealand, 2010.
21. Rappaport, J. Consumption amenities and city population density. *Reg. Sci. Urban Econ.* **2008**, *38*, 533–552.
22. Wheeler, S. *Planning for Sustalnability*; Routledge: New York, NY, USA, 2004.
23. North Shore City Council. *Good Solutions Guide for Mixed Use Development in Town Centres*; North Shore City Council: Auckland, New Zealand, 2005.

24. Parliamentary Commissioner for the Environment. *The Cities and Their People: New Zealand's Urban Environment*; Parliamentary Commissioner for the Environment: Wellington, New Zealand, 1998.
25. Clark, T.; Lloyd, R.; Wong, K.K.; Jain, P. Amenities drive urban growth. *J. Urban Aff.* **2002**, *24*, 493–515.
26. Schmitz, A. *New Shape of Suburbia: Trends in Residential Development*; Urban Land Institute: Washington, DC, USA, 2003.
27. Auckland Council. *Technical Report: Centres and Corridors Workstream*; Auckland Council: Auckland, New Zealand, 2011.
28. Beacon Pathway Ltd. *The Determinants of Tenure and Location Choices of 20–40 Year Old Households in the Auckland Region*; The Centre for Housing Research New Zealand: Wellington, New Zealand, 2010.
29. Gu, K. Exploring the fringe belt concept in Auckland, an urban morphological idea and planning practice. *N. Z. Geogr.* **2010**, *66*, 44–60.
30. Auckland Council. *Auckland Plan*; Auckland Council: Auckland, New Zealand, 2012.
31. Auckland Council. *Draft Unitary Plan*; Auckland Council: Auckland, New Zealand, 2013.
32. Fontein, P.; Ray, A.; Robinson, T. *Auckland Plan, Total Auckland Development Potential, Final Report*; Auckland Council: Auckland, New Zealand, 2011.
33. Auckland Council. *A New Kind of Compact: Design-Led Part 1*; Auckland Council: Auckland, New Zealand, 2011.
34. Realestate Investar. 50 Fastest Growth Suburbs in New Zealand. Available online: http:\\www.realestateinvestar.co.nz (accessed on 4 April 2012).
35. Haarhoff, E.; Beattie, L. Trading Suburbia for an Urban Lifestyle: Obstacles and Opportunities in Auckland, New Zealand. In Proceedings of the 43rd Conference of the Urban Affairs Association, San Francisco, CA, USA, 2–5 April 2013.
36. Haarhoff, E.; Beattie, L.; Dixon, J.; Dupuis, A.; Lysnar, P.; Murphy, L. *Future Intensive: Insights for Auckland's Housing*; The University of Auckland: Auckland, New Zealand, 2012.
37. Mead, D.; McGregor, A. *Intensive Housing Demand and Supply Issues*; Auckland Regional Council: Auckland, New Zealand, 2007.

5

Integrating Simplified and Full Life Cycle Approaches in Decision Making for Building Energy Refurbishment: Benefits and Barriers

Xabat Oregi [1,*], Patxi Hernandez [1], Cristina Gazulla [2] and Marina Isasa [2]

[1] Energy and Environment Division, Tecnalia Research and Innovation, Tecnalia, Azpeitia 20730, Spain; E-Mail: patxi.hernandez@tecnalia.com

[2] UNESCO Chair in Life Cycle and Climate Change (ESCI—Pompeu Fabra University), Barcelona 08003, Spain; E-Mails: cristinagaz@gmail.com (C.G.); marina.isasa.sarralde@gmail.com (M.I.)

* Author to whom correspondence should be addressed; E-Mail: xabatoregi@gmail.com

Academic Editor: Elma Durmisevic

Abstract: The life cycle assessment (LCA) method is a powerful tool that can serve to aid decision making regarding the environmental benefits of refurbishment projects. However, due to the relative complexity of LCA studies, simplified LCA methodologies are frequently used, focusing on just some of the building life cycle phases or a reduced number of indicators. The most common and widespread simplification is to only evaluate the differences a refurbishment project makes on the operational energy use of the building. This paper compares the results of applying full LCA, simplified LCA and operational energy use assessment in a refurbishment case study. Results show that simplified LCA methodologies including building use phase and product manufacturing phase can generally be sufficiently accurate to aid decision making for building energy refurbishment, as other building life cycle phases related to transport of products, on site construction, deconstruction or end of life represent a generally negligible part of the total life cycle impacts, both in terms of resource use or environmental impacts. Barriers and benefits of applying simplified LCA approaches to building energy refurbishment projects are subsequently discussed.

Keywords: refurbishment; life cycle assessment (LCA); simplified life cycle assessment; building refurbishment; embodied energy; life cycle energy performance; sensitivity analysis

1. Introduction

Buildings are one of the world's largest energy-consuming sectors, accounting for nearly 30% of final global energy consumption, reaching 40% in the European Union (EU) [1]. With new constructions adding at most 1% a year to the EU existing stock [2], there is large potential for improving the energy performance of the other 99% of the building stock, making "energy refurbishment" a top priority in current EU and national policies. In the last decade, a remarkable number of studies have focused on quantifying the environmental improvement potential of buildings. Some of these studies [3] focus on the quantification of energy consumption during the use phase of the building, without taking into account the environmental impacts, resources needed and the waste and emissions generated from the building processes. In the cases in which a life cycle approach has been applied on an assessment, there is a general consensus that the use phase contributes more than 80%–85% share in the total life cycle energy use of buildings [4–7]. The study carried out by Karimpour *et al.* [8] shows that when considering the time value of carbon, in relation to emission targets, the embodied energy (total energy required for the extraction, processing and manufacture of building materials) can represent up to 35% of the future emissions target of a building in a mild climate. In buildings with a very low energy use in the operational phase, logically other phases of the life cycle like raw materials supply, product manufacturing, transport and installation, and end of life would have a much larger relative impact, becoming the only energy related impact in the so called "zero energy buildings". In that sense, Cellura *et al.* [9] have emphasized the embodied energy of the building as a key issue to not be neglected in the exhaustive evaluation of the primary energy demand of low energy buildings, whereas according to Lützkendorf *et al.* [10] the embodied energy of a passive house with PV installation could be 44% higher than its operation energy for a 60 years lifespan.

From these observations it is clear and generally accepted that when undertaking an environmental assessment of a new building, it is necessary to evaluate all its phases with a life cycle approach, and recent standardization efforts from ISO TC 59/SC17 [11] or CEN TC 350 [12] have developed a methodology and standards to apply this approach in a structured way. However, the effort needed for such analysis is relatively large, and the discussion about whether the evaluation of all building phases is also necessary in building energy refurbishment projects arises. As buildings are extremely complex systems entailing an enormous amount of products, systems, stages and processes, a fully-fledged application of the life cycle assessment (LCA) methodology is still not widely applied, and analysts usually apply some simplifications to reduce the amount of time and facilitate the interpretation of the results. Cabeza *et al.* [13] and *Chau et al.* [14], provided a review on comparing life-cycle-based methodologies in the building sector.

In this context and based on a building case study, the main objective of this paper is to verify whether different simplified methodologies currently applied for analyzing building refurbishment strategies are sufficiently reflective of a comprehensive application of the LCA, in terms of reducing resource use and environmental impacts of existing residential buildings. Several sensitivity analyses trying to capture a wide range of potential scenarios have been developed in order to assess the influence of key parameters in the comparison between the different methodologies.

2. Methodology and Case Study

With the aim of identifying to what extent simplifications may influence the results of a decision-making process, three different methodologies, namely LCA, simplified life cycle assessment (LCA-Si) and operational stage assessment (OSA), are applied to the same case study, an existing block of apartments in San Sebastian (Spain). Section 2.1 explains LCA, LCA-Si and OSA methodologies and how are they used by the authors to assess each refurbishment strategy. The case study context, including climate conditions and existing building construction characteristics, are evaluated in Section 2.2, also presenting a diagnosis of the baseline energy performance. Section 2.3 analyzes the selected refurbishment strategies, which have been designed with the objective of decreasing the final heating consumption of the building. Finally, Section 2.4 presents the calculation details for each life cycle phase of the refurbishment strategy.

2.1. Difference between LCA, LCA-Si and OSA Methodologies

According to the European Commission Communication on Resource Efficiency Opportunities in the Building Sector [15], the LCA methodology is currently the best framework available to assess the potential environmental impacts of any activity, product or service without geographical, functional or time limits, since it quantifies the environmental impact of the inputs and outputs along its whole life cycle, including the extraction of raw materials, production process, use and end of life stages.

LCA is standardized by ISO 14040 [16] and 14044 [17] standards, and consists of four phases. The first phase is to define the goal and scope of the assessment, which serves as a description of the type of study. The scope of the study determines which processes should be included in the inventory phase of the assessment. In the second phase, the life cycle inventory (LCI) includes information on all of the environmental inputs and outputs associated with a product or service, *i.e.*, material and energy requirements, as well as emissions and waste. The third phase is the impact assessment, where the potential contribution of each substance to predefined environmental impact categories is calculated. Once the impact has been calculated, the fourth and final step of the assessment is the interpretation, where the results of the calculations are summarized and discussed. LCA can choose from several methods to quantify the environmental performance of a product, system or process. For example, the CML method (Centrum voor Milieukunde Leiden) [18] uses indicators at midpoint level showing direct potential impact on the environment situated halfway along the chain of causes and effects. On the other hand, the Eco-indicator method [19] includes endpoint indicators, which represent the ultimate consequences of the environmental impact for humans and ecosystems.

Specifically for the construction sector, new standards, such as EN 15978:2011 [20], already define the different phases of a building life cycle and a number of indicators and methods used to declare the results of the analysis, which are midpoint level (e.g., Global Warming Potential in kg CO_2 equivalent, or Non Renewable Primary Energy Use, in MJ). A description of the building stages defined by EN 15978 is shown in Figure 1.

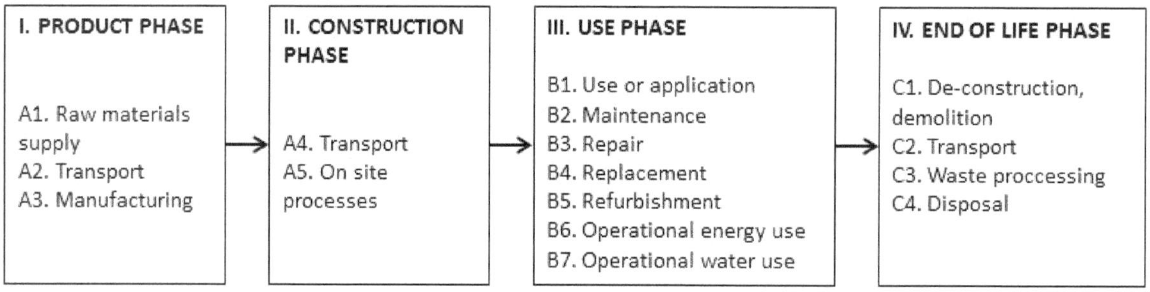

Figure 1. Building stages defined by EN 15978 [20].

Despite the standardization efforts, it is difficult to find published work including all the described life cycle stages, and instead most studies have focused on just some of the stages, *i.e.*, product phase (A1-3) and operational energy use stage (B6). Table 1 shows that very few authors considered all the LCA phases in their studies, being a general trend the omission of some life cycle modules such as on site processes (A5) or maintenance (B2). These omissions are due mainly to the lack of information, the difficulty of predicting future scenarios and the relatively low impact in comparison to the whole life cycle, which according to previous studies [21,22] is less than 1% for the life cycle energy use. Only some studies assess the replacement phase (B4), which is directly related to the estimated service life (ESL) of each product used which may have a remarkable influence in achieving Life Cycle Zero Energy Buildings (LC-ZEB) [23]. Regarding the calculation method used, it is worth mentioning that a process LCI approach is generally used, except for few studies such us Stephan *et al.* [24,25] or Kofoworola *et al.* [26], which use the hybrid LCI analysis.

In order to evaluate the relation between different simplifications and to analyze the influence of these simplifications in the evaluation of a building energy refurbishment, the authors will compare three calculation methodologies: the first methodology is the full LCA methodology (see Table 2), which obviously considers all life cycle stages. The second methodology is a simplified LCA (LCA-Si), focusing only on the evaluation on the product, replacement and operational energy use phases, as applied by Hernandez *et al.* [23] prior to the definition of the concept of Zero Energy Building with a life cycle approach. Finally, the third methodology is an operational stage assessment (OSA), where the evaluation focuses only on the reduction of the impact during the operational stage of the building. This third methodology is much simpler and generally more accessible to a wider range of technicians, as can be linked with building energy performance evaluation and with building energy rating schemes.

Table 1. Building Life Cycle Stages in relation to existing studies.

Building Life Cycle Stages	Erlandsoona *et al.*, 2004 [27]	Junnilla, 2004 [28]	Citherlet *et al.*, 2007 [29]	Zabalza *et al.*, 2009 [30]	Utama *et al.*, 2009 [31]	Kofoworola *et al.*, 2009 [26]	Blom *et al.*, 2010 [32]	Blengini *et al.*, 2010 [33]	Gustavsoon *et al.*, 2010 [34]	Hernandez *et al.*, 2010 [23]
Product phase (A1-3)	X	X	X	X	X	X	X	X	X	X
Transport (A4)	-	X	X	-	X	X	X	X	X	-
On site processes (A5)	-	X	-	-	X	X	-	X	-	-
Maintenance (B2)	-	X	X	-	-	X	X	X	-	-
Replacement (B4)	-	-	X	-	X	-	X	X	-	X
Operational energy use (B6)	X	X	X	X	X	X	X	X	X	X
End of life phase (C1-4)	-	X	X	-	-	X	X	X	-	-

Building Life Cycle Stages	Oritz *et al.*, 2010 [35]	Dodoo *et al.*, 2010 [36]	Malmqvist *et al.*, 2011 [37]	Rossi *et al.*, 2012 [38]	Stephan *et al.*, 2012 [24]	Ramesh *et al.*, 2012 [39]	Stephan *et al.*, 2013 [25]	Mosteiro *et al.*, 2014 [40]	Dodoo *et al.*, 2014 [41]	-
Product phase (A1-3)	X	X	X	X	X	X	X	X	X	-
Transport (A4)	X	X	-	X	X	X	X	X		-
On site processes (A5)	X	X	-	-	X	X	X	X	X	-
Maintenance (B2)	X	-	-	-	-	-	-	-	-	-
Replacement (B4)	-	-	-	-	X	X	X	X	-	-
Operational energy use (B6)	X	X	X	X	X	X	X	X	X	-
End of life phase (C1-4)	X	X	-	X	-	-	-	X	X	-

Table 2. Building Life Cycle Stages in relation to the three methods considered: Life Cycle Assessment (LCA), simplified LCA (LCA-Si) and Operational Stage Assessment (OSA).

Building Life Cycle Stages	Building Stages	LCA	LCA-Si	OSA
A1–3	Product phase	X	X	-
A4	Transport	X	-	-
A5	On site processes	X	-	-
B2	Maintenance	X	-	-
B4	Replacement	X	X	-
B6	Operational energy use	X	X	X
C1–4	End of life phase	X	-	-

2.2. *Case Study*

In order to assess to what extent simplifications in the life cycle calculations may help or mislead decision-making processes, the three methodologies are applied for selecting the best energy refurbishment option for the same building: A residential building built in 1962 in San Sebastian (Spain).

The analysis takes into account the different life cycle phases of the building, starting from the existing scenario and considering a range of refurbishment strategies. A Reference Service Life (RSL_b) of 50 years from the date of refurbishment is considered, a value often used by default, since it is generally difficult to foresee the real life span of a building [37].

The functional equivalent used to compare the different options is the building itself and the results are expressed per year and per unit of useful heated floor area, meeting the conditions of design requirements (thermal comfort, *etc.*).

From the range of environmental and resource indicators that are used on common LCA, this study will focus for simplicity on one indicator, which is "Use of non-renewable primary energy resources" (in MJ-Eq/functional unit) according to the CML method [18]. A sensitivity analysis for using a range of other environmental indicators on the evaluation is presented in Section 4 of this paper.

2.2.1. Building Context

With a total net floor area of 9484 m^2 and a heated surface of 8574 m^2, the building consists of a commercial ground floor and 9 residential floors (with 12 apartments on each floor), which are heated by a centralized natural gas heating system (see Figure 2). All apartments are naturally ventilated and no cooling or renewable energy systems are installed. *U*-values (W/(m^2·K)) of the building envelope before its refurbishment include cavity wall façade 1.12 W/(m^2·K), reinforced concrete deck with ceramic finish 2.34 W/(m^2·K), reinforced concrete first floor slab 1.79 W/(m^2·K), monolithic glazing 5.77 W/(m^2·K) and aluminum frame 4.2 W/(m^2·K). These values do not meet the minimum requirements [42] specified by the current national building regulations.

The building is located in San Sebastian, where the annual average temperature is 14 °C. In summer, the daily average temperature is below 20 °C, so cooling systems are generally unnecessary, particularly if measures such as solar shading or night cooling are implemented. In winter, the daily average temperature is about 10 °C, justifying the need for heating systems.

Figure 2. Picture of the case study building.

2.2.2. Baseline Operational Energy Use

During their use phase, buildings require operational energy for meeting the demand for heating, cooling, hot water, ventilation, lighting and the use of appliances. However, refurbishment projects such as the one discussed in this paper are focused on the building envelope and aimed at reducing the operational energy use of buildings for heating and cooling, and do not directly affect other building end-uses.

Using the Design Builder [43] software and the International Weather Files for Energy Calculation [44] for the city of San Sebastian, the final energy demand for heating the building before refurbishment was estimated. Being an interface for Energy Plus, Design Builder is a dynamic energy simulation tool that generates detailed data about the energy performance of a building during 1 year by using real weather data as well as temporal aspects such as solar radiation, thermal mass or user occupancy. The building model developed replicates the real geometry, including overhangs, setbacks and the surrounding buildings. Parameters like occupancy rate, schedules, and internal gains have been estimated following current Spanish regulations [42]. For example, a general occupancy value of 0.03 people/m^2 was considered and the common areas (stairs and portals) and ground floor were treated as unheated without occupation or internal gains.

The heating system is composed of a centralized natural gas installation with a nominal performance of 0.92. Thermostats are set at 21 °C from 4 p.m. until 11 p.m. for the heating period (from 30 September to 31 May). The baseline operational heating demand calculated according to these parameters is 264 MJ/(m^2·a). This demand is multiplied by the system performance and by 1.13, which is the conversion factor from natural gas to Non-Renewable Primary Energy use (NRPE) in Spain, taken from the Ecoinvent 3.0 database [45]. The resulting baseline operational Non Renewable Primary Energy (NRPE) use for the building is 324 MJ/(m^2·a). For the variables in Equation (1), refer to Table 3.

$$B_{B6} = \sum_{m=1}^{m} \frac{ED_b}{\rho_m} \times CF_m / HS \quad (1)$$

Table 3. Variables applied during the calculation of the Non Renewable Primary Energy (NRPE) of each building life cycle phase.

Acronym	Description	Unit
B_{B6}	Operational energy use of the baseline building	NRPE (MJ)
CE_{A5}	Energy use associated to construction waste treatment of each refurbishment strategy	NRPE (MJ)
CF_m	Conversion factor of the energy source m (MJ non renewable primary energy/MJ final energy)	NRPE (MJ)
D_m	Transport distance of product m	km
DW_m	Transport distance of waste m from site to waste management facility	km
ED_b	Baseline operational annual energy demand	MJ
ED_m	Operational annual energy demand for the refurbished building	MJ
EE_m	Embodied energy of material or system m applied during the refurbishment, expressed per unit of material or system (in kg, m^2 or m^3)	NRPE (MJ)
ESL_m	Estimated service life of the material m	years
HS	Heated surface	m^2
IA_t	Energy used in the transportation per t and km	NRPE (MJ)
IEE_{A1-3}	Initial embodied energy of each refurbishment strategy	NRPE (MJ)
IE_{A4}	Initial energy used in the transportation of each refurbishment strategy	NRPE (MJ)
IE_{C1-4}	Energy use associated to the end of life stage of each refurbishment strategy	NRPE (MJ)
RB_{B6}	Operational energy use of the refurbished building	NRPE (MJ)
$REE_{B4(A1-3)}$	Recurrent embodied energy of each refurbishment strategy	NRPE (MJ)
$RE_{B4(A4)}$	Recurrent energy use associated to the transport of each refurbishment strategy	NRPE (MJ)
RSL_b	Reference service life of the building	years
Q_m	Quantity of materials and systems in each refurbishment strategy	kg-m^2-m^3
Q_{mt}	Quantity of material and system in each refurbishment strategy	t
WP_m	percentage of waste generated in each refurbishment strategy	%
WT_m	Energy use of the waste treatment process per kg	NRPE (MJ)
ρ_m	Performance of the energy generation system m	%

2.3. Energy Refurbishment Strategies

In order to reduce the environmental impact related to the building's heating, different energy refurbishment strategies have been evaluated during this study, taking into account that the area where the case study is located shows no historic, urban or architectonic restrictions, allowing direct refurbishment actions on all the envelope elements.

Refurbishment strategies will be applied to the case study in two efficiency levels. The basic efficiency level is based on restoration strategies that enforce the minimum thermal requirements determined by the existing regulations and standards. The advanced efficiency level strategies improve the thermal properties adding insulation to very high values such as those used in standards like the Passive House [46].

Another variable considered for defining refurbishment strategies has been the type of material used, which is important for the life cycle performance particularly in relation to their embodied energy. The strategies have been subsequently divided also by the employment of low and high embodied energy materials. Table 4 details the strategies chosen.

Table 4. Set of energy refurbishment strategies applied in the case study.

Strategy	Efficiency Level	Embodied Energy of Construction Products	Strategy ID
Window replacement (1)	Basic (b)		1b
	Advanced (a)		1a
Ventilated façade (2)	Basic	Low (l)	2bl
		High (h)	2bh
	Advanced	Low	2al
		High	2ah
External Insulation System (3)	Basic	Low	3bl
		High	3bh
	Advanced	Low	3al
		High	3ah
Internal (4)	Basic	Low	4bl
		High	4bh
	Advanced	Low	4al
		High	4ah

The first strategy focuses on the replacement of all existing windows with a new frame and glazing. The windows for the basic energy efficiency level (1b) consist of a double glazing (2.7 W/(m^2·K)) and aluminum frame (2.9 W/(m^2·K)), meeting the minimum thermal requirements for refurbishments in Spain. The windows for advanced level (1a) consist of a low-emissivity coated glazing (1.4 W/(m^2·K)) and wooden frames (1.2 W/(m^2·K)). The second solution is a ventilated facade system, which is composed of an aluminum substructure, a layer of insulation and a ceramic outlayer. The third strategy is an external insulation system composed of an insulation layer and mortar outlayer. Finally, the fourth strategy is an indoor thermal improvement solution consisting of a layer of insulation and plasterboard.

According to the efficiency level parameter, different insulation thicknesses are proposed for basic and advanced levels. The projected insulation thicknesses for the basic efficiency energy level are 5 cm for the façade, 8 cm for the deck and 6 cm for the first floor slab. The thicknesses proposed for the advanced energy efficiency level are 25, 30 and 15 cm, respectively. Regarding the aluminum profile of the ventilated façade, 10.2 cm^2 of aluminum per m^2 is projected for the basic level and 49.6 cm^2 for the advanced level. Strategies with low embodied energy insulation apply wooden fiber and high embodied energy strategies apply Extruded Polystyrene (XPS) insulation. Regarding the other products that make up the three systems (outlayer, mortar and plasterboard), their properties and quantities are maintained in all cases. It should also be highlighted that secondary products that form part of these strategies, such as screws, sealants, glues are not considered within the scope of the study.

2.4. Calculation of Non Renewable Primary Energy (NRPE) Use for Each Life Cycle Phase

2.4.1. Input Data

Product Phase (A1–3), Initial Embodied Energy. The embodied energy or initial energy input associated with the production phase of each product and system has been calculated applying Equation (2) (for variables, refer to Table 3) using process data from Ecoinvent [45] and GaBi [47] databases as well as Environmental Product Declarations (EPDs) issued by manufacturers (see Table 5).

$$IEE_{A1-3} = \sum_{m=1}^{m} EE_m \times Q_m \Big/ (RSL_b \times HS) \tag{2}$$

Transportation from Production Unit to Point of Use (A4). The transportation of building materials to construction sites involves a variety of transportation modes. Equation (3) describes the calculation of the NRPE use for this stage (for variables, refer to Table 3). Due to the lack of detailed transportation data for each of the materials used to the building site, three different distances are considered (see Table 5): 50 km (distribution within the province), 120 km (distribution within the region) and 300 km (distribution within the same country). The 300 km value is based on a standard parameter defined in EeB Guide [48], which sets up an average transportation distance in Europe.

$$IE_{A4} = \sum_{m=1}^{m} D_m \times Q_{mt} \times IA_t \Big/ (RSL_b \times HS) \tag{3}$$

Construction Process (A5). Energy use associated with the construction stage of a new envelope is not considered since it typically represents less than 1% of the life cycle energy demand of the building [21]. However the transportation and end of life treatment processes of the inert and non-hazardous wastes that will be managed in a landfill are considered. For this case study, it is considered that 3% of the total final product will be wasted during their placement and transported 50 km to the landfill.

$$CE_{A5} = \sum_{m=1}^{m} [(Q_{mt} \times WP_m) \times DW_m \times IA_t] + [(Q_{mt} \times WP_m) \times WT_m] \Big/ (RSL_b \times HS) \tag{4}$$

Replacement (B4). Recurrent Embodied Energy (A1–3). As described in Equation, the recurrent embodied energy represents the sum of energy inputs associated with the energy required to manufacture and replace refurbishment materials across the building's service life. Service life and durability of materials are among the most important factors affecting the recurrent embodied energy. In accordance with ISO 15686-8:2008 [49], construction materials and systems usually do not possess the same Estimated Service Life (ESL_m) as the building Reference Service Life (RSL_b) and may require one or multiple replacements over the building's service life. The lower the service life of a material, the greater the quantity of material required for ongoing maintenance and repair and therefore the greater the embodied energy associated with manufacturing and installing replacement materials throughout a building's life. Table 5 shows the values of ESL_m applied in this case study, selected according to the information obtained by different EPDs or databases such as National Association of Home Builders (NABH) [50].

$$REE_{B4\,(A1-3)} = \sum_{m=1}^{m} EE_m \times Q_m \times \big((RSL_b/ESL_m) - 1\big) \Big/ (RSL_b \times HS) \tag{5}$$

$RSL_b/ESL_m = 1$, No replacement (same or longer material SL than the RSL_b).

$RSL_b/ESL_m > 1$, Replacement is necessary.

Replacement (B4). Recurrent Transportation from Production Site to Point of Use (A4). Due to the replacement of some products during the RSL_b of the case study, the environmental impact related with the transportation process of the additional products is estimated using Equation (6). The same transport distances of the initial transport phase have been used.

$$RE_{B4\,(A4)} = \sum_{m=1}^{m} D_m \times Q_{mt} \times IA_t \times \left((RSL_b/ESL_m) - 1\right) \Big/ (RSL_b \times HS) \tag{6}$$

Operational Energy Use (B6). As explained before, in this case study focused on refurbishment of the building envelope, the operational energy use only includes the heating of the building during its useful life (see Equation (7)). Other end-uses (hot water production, use of appliances, illumination, *etc.*), despite being very relevant in an overall building energy use evaluation will not be affected by the proposed refurbishment. The new heating demands of the refurbished building scenarios have been calculated with the same methodology as the baseline operational energy use, *i.e.*, through building energy simulations with Design Builder software.

$$RB_{B6} = \sum_{m=1}^{m} \frac{ED_m}{\rho_m} \times CF_m/HS \tag{7}$$

End of Life (C1–4). Energy consumption of the end-of-life stage is usually not considered since it typically represents less than 1% of the life cycle energy of buildings [51]. However, in order to assess all building phases, during this study the transportation of the products (initial and recurrent) to the waste treatment facility and their management is evaluated (see Equation (8)). For this case study, road transportation during 50 km to the landfill for inert and non-hazardous waste is considered. Being beyond the scope of the study, potential recycling or reuse of the aluminum used in the frame of the windows (strategy 1) and in the ventilated facade sub-structure (strategy 2) is not included.

$$IE_{C1-4} = \sum_{m=1}^{m} (Q_{mt} \times DW_m \times IA_t) + (Q_{mt} \times WT_m) \Big/ (RSL_b \times HS) \tag{8}$$

Table 5 shows information about the parameter applied to evaluate the NRPE use for each product and process that participates in different refurbishment strategies, allowing the calculation of the impact of each Life Cycle Phase.

Table 5. Non Renewable Primary Energy (NRPE), Transport characteristics, and Estimated Service Life of the materials (ESL_m) for products and processes included in the analysis.

Product/Process	Non Renewable Primary Energy Use		Transport	ESL_m	
	Value	Data Source	(km)	Years	Data Source
Double glazed	463 (MJ/m^2)	INIES—FDES SGG Climaplus [52]	Truck-120	30	Environmental Product Declaration (EPD)
Triple glazed	707 (MJ/m^2)	INIES—FDES. SGG Climatop [53]	Truck-300	30	EPD
Aluminum frame	1852 (MJ/m^2)	GaBi. Aluminum wing profile, powder coated PE (2010)	Truck-50	20	National Association of Home Builders (NAHB)

Table 5. *Cont.*

Product/Process	Non Renewable Primary Energy Use		Transport	ESL$_m$	
	Value	Data Source	(km)	Years	Value
Wood frame	206 (MJ/m^2)	ENVIRONDEC-Cormo [54]	Truck-300	30	EPD
Aluminum sub-structure	118 (MJ/kg)	GaBi. Aluminum extrusion profile PE (2010)	Truck-120	50	NAHB
Insulation (high)	92.4 (MJ/kg)	Ecoinvent. Polystyrene, extruded (XPS), at plant	Truck-50	50	NAHB
Insulation (low)	5 (MJ/m^3)	GaBi. Lightweight wood fibers panel PE (2010)	Truck-300	50	NAHB
Outlayer	256 (MJ/m^2)	GaBi. Ceramic façade panels—NBK Ceramic PE (2008)	Truck-120	50	NAHB
Mortar	1.3 (MJ/kg)	Ecoinvent. Cement mortar, at plant	Truck-50	35	NAHB
Plasterboard	54 (MJ/m^2)	Ecoinvent. Gypsum plaster board, at plant	Truck-120	30	NAHB
Transport-truck	0.8 (MJ/(t·km))	GaBi. Articulated lorry (40t) incl. fuel ELCD (2005)	-	-	-
Landfill	0.2 (MJ/kg)	GaBi. Landfill for inert matter (construction waste) (2010)	-	-	-

2.4.2. Sample Calculation of Non Renewable Primary Energy (NRPE) Use for one of the Energy Refurbishment Strategies

As an example, in this section, the environmental calculation of a proposed rehabilitation strategy (1a + 2ah) is explained. This strategy focuses on replacing existing windows with 204 m^2 of new wooden frames (11,526 kg) and 1018 m^2 of triple glazing (34,917 kg). The façade will be refurbished with a ventilated façade system, which consists of 5581 m^2 of 25 cm XPS insulation (60,275 kg), 12,830 kg of Aluminum sub-structure with a section of 49.6 cm^2 and 4409 m^2 of ceramic façade panels (60,274 kg). Through this information, and the values previously defined in Table 5 (B_{B6}, RSL_b, ESL_m, EE_m, D_m, Q_m, Q_{mt}, IA_t, DW_m, CF_m and ρ_m), the authors have performed the calculation of the NRPE use of each of the life cycle phases of this strategy.

Figure 3 shows that the NRPE use of the building after applying the refurbishment strategy 1a + 2ah is reduced from 324 MJ/(m^2·a) to 92 MJ/(m^2·a), reaching a reduction of 72%, which is mostly due to the operational energy use phase (69.6 MJ/(m^2·a)) and the Initial Embodied Energy phase (20.93 MJ/(m^2·a)).

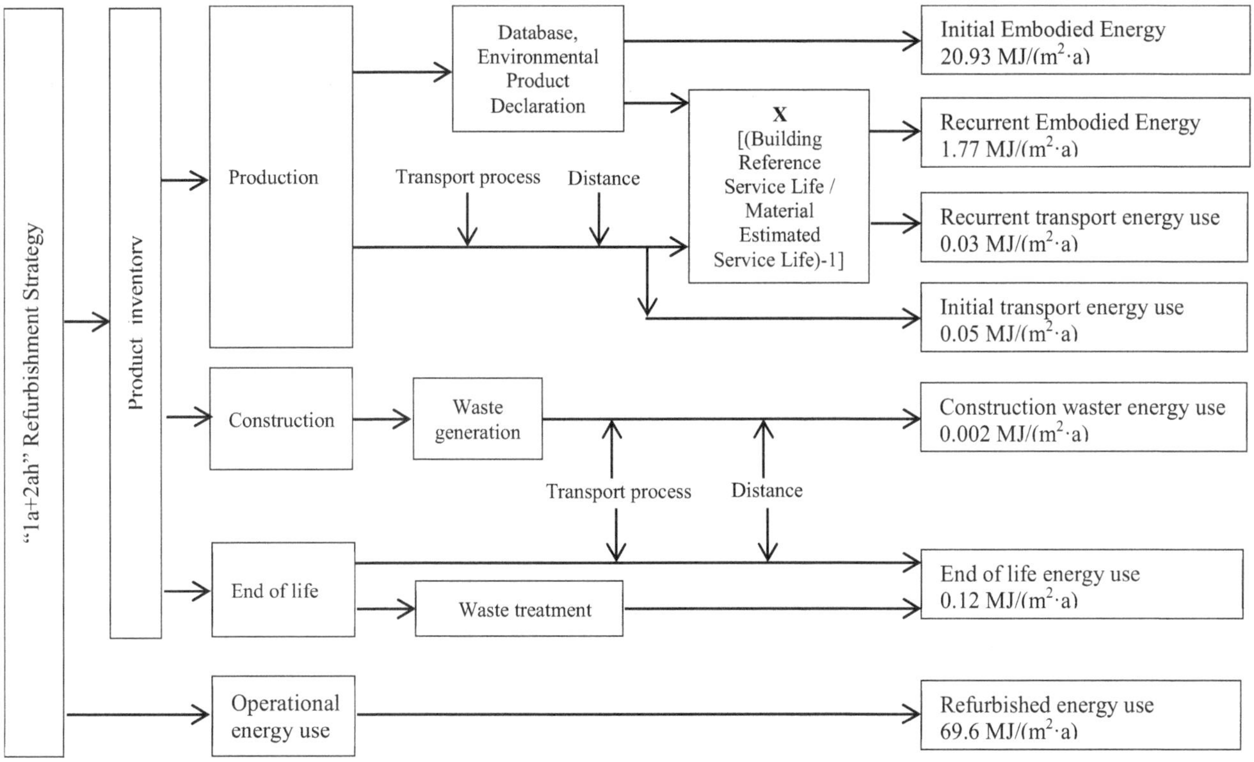

Figure 3. Scheme of inputs and Non Renewable Primary Energy (NRPE) use of each life cycle phase of the 1a + 2ah energy refurbishment strategy.

3. Results

3.1. Life Cycle Energy Performance Evaluation—NRPE Results

Applying the same calculation procedure for all energy rehabilitation strategies, the NRPE use of each phase was obtained. Figure 4 shows, in comparison with the baseline scenario, the decrease on the NRPE during the operational stage and the increase of NRPE derived from the production, transportation, construction, replacement and end of life phases of the products and systems applied in each refurbishment strategy.

For the particular building studied, Table 6 shows that the refurbishment option with the highest NRPE use reduction (*i.e.*, 77%) calculated with the LCA methodology is strategy "1a + 3al", which increases the building envelope insulation level and replaces the current glazing with double-glazed low-emissivity coated window. Results show that the influence of the constructive characteristics and embodied energy parameters is very low, being the advanced level refurbishment strategies the most effective and efficient to reduce the baseline NRPE use. As for the refurbishment option with the highest reduction on energy use according the OSA methodology (only considering operational stage of the building), "1a + 2al", "1a + 2ah", "1a + 3al", and "1a + 3ah" strategies present the same result, as differences amongst them occur in other life cycle stages.

As expected the reduction of life cycle NRPE is much larger than the energy consumed by the different refurbishment strategies. The ratio between the increase and decrease on NRPE reach up to 15%, for strategies using products with higher embodied energy (*i.e.*, 2ah strategy). Consequently, the ratio has lower values (up to 0.12%) when the products applied have lower embodied energy (*i.e.*, 3bl strategy).

NRPE attributed to the replacement phase represents less than 5% of the decrease on operational NRPE, whereas NRPE attributed to the transportation (initial and recurrent), construction and end of life phases (neither of which is covered by the LCA-Si method), do not reach 0.19% of the NRPE reduction. Detailed values for the NRPE indicator for each of the refurbishment strategies and each of the calculation methodologies are presented in Table 6.

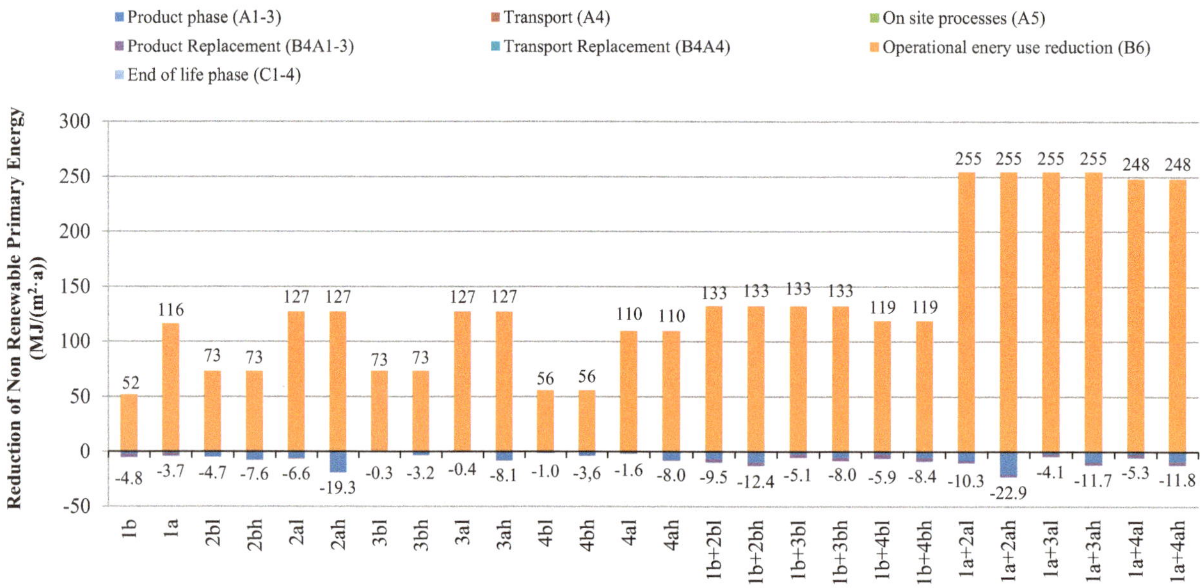

Figure 4. Reduction of Non Renewable Primary Energy use (MJ/(m^2·a)) from each refurbishment strategy in relation to the baseline. Positive values indicate a reduction of energy use on the operational phase. Negative values represent the energy use on the rest of the life cycle phases.

Table 6. Reduction of Non Renewable Primary Energy use of each refurbishment strategy according to Life Cycle Assessment (LCA), simplified LCA (LCA-Si) and Operational Stage Assessment (OSA) (MJ/(m^2·a)).

Strategy	LCA	LCA-Si	OSA	Strategy	LCA	LCA-Si	OSA
1b	47.1	47.1	51.9	**1b + 2bl**	123.1	123.1	132.6
1a	112.6	112.7	116.2	**1b + 2bh**	120.2	120.2	132.6
2bl	68.6	68.6	73.3	**1b + 3bl**	127.4	127.4	132.6
2bh	65.7	65.7	73.3	**1b + 3bh**	124.6	124.6	132.6
2al	120.8	121.3	127.4	**1b + 4bl**	113.3	113.3	119.2
2ah	108.2	108.3	127.4	**1b + 4bh**	110.7	110.7	119.2
3bl	73.0	73.0	73.3	**1a + 2al**	244.4	245.0	254.7
3bh	70.1	70.1	73.3	**1a + 2ah**	231.8	232.0	254.7
3al	127.0	127.1	127.4	**1a + 3al**	250.6	250.8	254.7
3ah	119.4	119.4	127.4	**1a + 3ah**	243.0	243.2	254.7
4bl	54.7	54.7	55.7	**1a + 4al**	243.3	243.9	248.4
4bh	52.1	52.1	55.7	**1a + 4ah**	236.7	237.0	248.4
4al	108.3	108.7	109.7	-	-	-	-
4ah	101.7	101.8	109.7	-	-	-	-

3.2. Comparison of LCA, LCA-Si and OSA Methodologies

In order to evaluate the influence of the methodological simplifications in a building energy refurbishment decision making process, the difference between the OSA and LCA-Si from a full LCA is expressed as a relative difference (see Equations (9) and (10)) in Figure 5.

$$LCA - Si_{relative\ difference\ (\%)} = ([LCA - LCAsi]/LCA) \times 100 \tag{9}$$

$$OSA_{relative\ difference\ (\%)} = ([LCA - OSA]/LCA) \times 100 \tag{10}$$

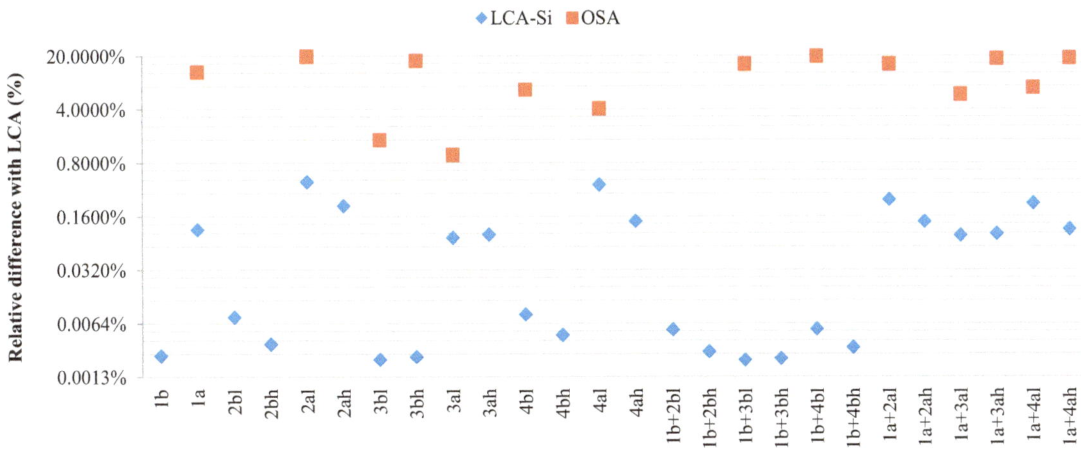

Figure 5. Relative difference of LCA-Si and OSA with LCA.

The values of Figure 5 shows that for this case study and for the selected NRPE indicator, the difference between the results obtained with LCA and LCA-Si methodologies is below 0.39%. However, the results obtained with the OSA methodology in comparison with LCA present differences varying from 0.34% (for 3al strategy) to 15% (for 2ah strategy). Higher differences occur in the case of strategies using products with higher embodied energy associated.

4. Sensitivity and Uncertainty Evaluation

With the aim of drawing a general conclusion that offers an overview of residential building refurbishment actions, a sensitivity analysis tackling different relevant aspects is conducted in this section. In order to simplify and facilitate the reading of the results, the study only reflects the results for the strategies 2ah and 3al, which presented the largest and lowest percentage differences on the results when analyzed respect the LCA methodology.

4.1. Reference Service Life of the Building (RSL_b)

The annualized NRPE use has been assessed for different RSL_b values: 25, 50 (current scenario), 75 and 100 years. Table 7 shows that the difference between the results obtained by LCA and OSA methodologies increases when RSL_b is lower, up to a difference of 30.24% in strategy 2ah. This difference falls to 0.36% for strategies applying products with lower embodied energy and when the same RSL_b (25 years) is applied. Regarding the difference between the LCA and LCA-Si methodologies, the lower RSL_b value, and the greater the difference. However differences between these two methodologies are in all cases below 1%.

Table 7. Sensitivity analysis results for refurbishment strategies 2ah (advanced efficiency level ventilated façade system with high embodied energy products and 3al (advanced efficiency level external insulation system with low embodied energy products).

Non Renewable Primary Energy use Reduction (MJ/(m²·a))							
Variable	**2ah Strategy**			**Variable**	**3al Strategy**		
	LCA	**LCA-Si**	**OSA**		**LCA**	**LCA-Si**	**OSA**
Reference Service Life Building (*RSLb*)							
RSL_{25}	89.1	89.1	127.4	RSL_{25}	127.1	127.1	127.4
RSL_{50}	108.2	108.3	127.4	RSL_{50}	127.0	127.1	127.4
RSL_{75}	114.5	114.7	127.4	RSL_{75}	127.0	127.1	127.4
RSL_{100}	117.8	117.9	127.4	RSL_{100}	127.1	127.2	127.4
Estimated Service Life Materials (*ESLm*)							
ESL_{half}	88.9	89.1	127.4	ESL_{half}	126.8	126.9	127.4
$ESL_{current}$	108.2	108.3	127.4	$ESL_{current}$	127.0	127.1	127.4
$ESL_{50\ years}$	108.2	108.3	127.4	$ESL_{50\ years}$	127.2	127.3	127.4
Product Transport Distance (*Dm*)							
Dm_{50}	108.2	108.3	127.4	Dm_{50}	127.1	127.1	127.4
Dm_{120}	108.2	108.3	127.4	Dm_{120}	127.0	127.1	127.4
Dm_{300}	108.1	108.3	127.4	Dm_{300}	127.0	127.1	127.4
Dm_{500}	108.1	108.3	127.4	Dm_{500}	126.9	127.1	127.4
Dm_{1000}	107.9	108.3	127.4	Dm_{1000}	126.8	127.1	127.4
Dm_{2000}	107.7	108.3	127.4	Dm_{2000}	126.6	127.1	127.4
Dm_{5000}	106.8	108.3	127.4	Dm_{5000}	125.8	127.1	127.4
Climate Zone							
Current	108.2	108.3	127.4	**Current**	127.0	127.1	127.4
Cold	233.8	233.9	253.1	**Cold**	252.7	252.8	253.1
Warm	59.3	59.4	78.6	**Warm**	78.1	78.2	78.6
Life Cycle Inventory							
LCI-Process	108.2	108.3	127.4	**LCI-Process**	127.0	127.1	127.4
LCI-Hybrid	49.7	50.2	127.4	**LCI-Hybrid**	125.6	126.2	127.4
Operational Energy Use							
+20%	131.7	131.8	151.0	**+20%**	150.6	150.7	151.0
−20%	72.8	72.9	92.1	**−20%**	91.7	91.8	92,1

4.2. Estimated Service Life of the Products (ESL_m)

ESL_m values for each product and system where defined for the baseline scenario (see Table 5). However, depending on the real use and maintenance practices, these values may be reduced or increased and, therefore, two additional ESL_m are assessed: Reducing the lifetime of all products by half and increasing their life by up to 50 years. Table 7 shows that the reduction of the ESL_m increases the difference between the LCA and OSA methodologies, obtaining a difference of up to 30.2% in 2ah strategy. On the other hand, for the same strategy, ESL_m 50 year values reduce the difference to 15.11%. For those refurbishment strategies using products with a lower embodied energy the difference decreases but continues below 0.5%.

4.3. Transportation Distance

Transportation distance values from the production gate to the building site have been changed to 50, 120, 300, 500, 1000, 2000 and 5000 km. Table 7 clearly shows that except in very long distances (5000 km or more), the transportation distance of the products does not alter the initial results obtained from the different methodologies. Furthermore, although all products and systems are transported from 5000 km, the overall NRPE use of the building will only be 0.01% higher than in a scenario where all products and systems are transported within a distance of 50 km. Although these results reflect that transport distance is practically irrelevant for reducing the NRPE use, it is worth noting that transport may have remarkable contributions to other environmental impact categories related to air quality or noise, as well as in economic terms where transport costs are also relevant.

4.4. Climate Zone

In order to check the influence of applying the same refurbishment strategies in other climate zones, two additional climate zones are evaluated: Warm climate (south of Italy, Palermo) and cold climate (Norway, Oslo). Due to the relevance of cooling consumption in warm climates, in this case the operational energy includes energy consumption for both heating and cooling indoor spaces (see Equation (7)). The cooling system consists of individual air-air electric installations with a nominal performance of 3 (ρ_m value). Thermostats are set at 26 °C from 12 a.m. until 8 p.m. for the cooling period (from 1 June to 31 August). To calculate the cooling related Non-Renewable Primary Energy use (NRPE), the demand is multiplied by the system performance and by a conversion factor value of 2.34 corresponding to an average value for the electricity mix in Europe [55].

Table 7 shows that in countries with a cold climate, refurbishment strategies lead to high reductions of NRPE during the use phase, and therefore the influence of other life cycle stages is reduced. The maximum difference between LCA and OSA results to 8.5% (strategy 2ah) for cold climates. In warm climates with a lower heating demand, the difference between LCA and OSA results to 32% (strategy 2ah), because the NPRE increase due to the use of products is relatively higher.

4.5. Uncertainty on Data on Embodied Energy of Products

Most current studies, including the present one, use environmental data from LCI databases and/or environmental product declarations, which are based on the LCA methodology for calculating the environmental inputs and outputs of each process involved in the corresponding supply chain (bottom-up technique). This process based LCA approach may conduct to a large associated uncertainty [56] due to the truncation error derived from data scarcity at a certain point of the upstream supply chain. Crawford [57] has shown that this error can be up to 87% of the embodied energy of building products, and has proposed the use of input-output LCA analysis, a top-down technique that establishes a link between economic transactions and the energy intensity of economic sectors, or hybrid LCA analysis, combining process and input-output analysis, to capture all the upstream processes. Crawford [58], Crawford and Stephan [59] and Stephan and Stephan [60] have shown that input-output-based hybrid analysis can produce embodied energy figures around four times higher than process analysis, for the same building. In order to assess the effect of a potential underestimation of the embodied energy values, a new scenario has been calculated multiplying the energy values by an average coefficient of 4.03 [60].

When data derived from hybrid LCA studies is used instead of process-based data, the difference between LCA and OSA increases, reaching a maximum difference of 60.5% for strategies with highest embodied energy values. This large difference on the results respond to the much high values of NRPE for building products that are obtained through LCI-hybrid methods. With these values, the overall reduction of NRPE for the LCI-Hybrid method is less than half the reduction obtained with the LCI-Process method for the LCA and LCA-Si.

4.6. Uncertainty in Relation to Occupancy Schedules and User Behavior

Occupancy schedules may significantly vary the results of operational energy use. This parameter adds a remarkable uncertainty to the results, as it has a critical role when estimating energy loads in residential buildings, as shown for example by Topouzi [61]. Indeed, Pettersen [22] and Juodis *et al.* [62] have shown that user behavior could vary the primary operational energy associated with space heating by ±15%–20%. Therefore, two new scenarios have been added to take these issues into account, assuming that the heating demand of the baseline scenario can be 20% higher and 20% lower than initially calculated due to occupancy and behavioral issues. Table 7 shows that when the operational energy consumption increases, the difference between LCA and OSA decreases, and *vice versa.*

4.7. Sensitivity Analysis for a Combination of Different Factors

In order to explore additional potential situations, 32 new refurbishment scenarios were assessed by combining extreme values for the different parameters evaluated separately in the previous sections (see Table 8). Taking into account the low influence of the material transportation distance (less than 0.01%), this parameter was not included in the combination of different factors.

For simplicity, Figure 6 shows the relative difference with LCA of the results obtained with LCA-Si and OSA for those refurbishment strategies with the highest and the lowest difference in their parameter values: 2ah and 3al, respectively.

As presented in Figure 6, for almost all the additional scenarios differences obtained using LCA and LCA-Si methodologies are lower than 3%. However, new scenarios highlight the differences between LCA/LCA-Si with OSA, which can reach up to 319% in buildings located in warm climates and by the application of hybrid embodied energy data. The second lecture of the Figure 6 is related to the embodied energy of applied products. The results show that the relative difference between LCA and OSA in refurbishment strategies with low embodied energy products is lower that 4%, except in scenarios such as C1 and C9 (with a relative difference of 4.1%, 5.4% respectively). However, in refurbishment strategies with high embodied energy, the difference between LCA and OSA in considerable, increasing the importance of obtaining objective and actual information about each product, and therefore showing the relevance of the application of hybrid data, which can have much higher embodied energy values.

Regarding service life, in energy refurbishment projects with high RSL_b and low ESL_m values, the difference between these LCA and OSA methodologies is greater than 35% (C1, C3, C5, C17 and

C51). However, in climate zones with high operational heating energy demand, this difference is decreased to 5.6% (C23).

Table 8. Definition of new energy refurbishment scenarios by the combination of different Reference Service Life of buildings (RSL_b), Estimated Service Life of material (ESL_m), Climate zone (*Cl*), type of Embodied Energy data (*EE*) and Operational Energy use uncertainties (*OE*).

Scenario ID	RSL_b (years)		ESL_m (years)		*Cl*		*EE*		*OE*	
	100	25	Half	50	Warm	Cold	Hybrid	Process	−20%	+20%
C1	X	-	X	-	X	-	X	-	X	-
C2	X	-	-	X	X	-	X	-	X	-
C3	X	-	X	-	-	X	X	-	X	-
C4	X	-	-	X	-	X	X	-	X	-
C5	X	-	X	-	X	-	-	X	X	-
C6	X	-	-	X	X	-	-	X	X	-
C7	X	-	X	-	-	X	-	X	X	-
C8	X	-	-	X	-	X	-	X	X	-
C9	-	X	X	-	X	-	X	-	X	-
C10	-	X	-	X	X	-	X	-	X	-
C11	-	X	X	-	-	X	X	-	X	-
C12	-	X	-	X	-	X	X	-	X	-
C13	-	X	X	-	X	-	-	X	X	-
C14	-	X	-	X	X	-	-	X	X	-
C15	-	X	X	-	-	X	-	X	X	-
C16	-	X	-	X	-	X	-	X	X	-
C17	X	-	X	-	X	-	X	-	-	X
C18	X	-	-	X	X	-	X	-	-	X
C19	X	-	X	-	-	X	X	-	-	X
C20	X	-	-	X	-	X	X	-	-	X
C21	X	-	X	-	X	-	-	X	-	X
C22	X	-	-	X	X	-	-	X	-	X
C23	X	-	X	-	-	X	-	X	-	X
C24	X	-	-	X	-	X	-	X	-	X
C25	-	X	X	-	X	-	X	-	-	X
C26	-	X	-	X	X	-	X	-	-	X
C27	-	X	X	-	-	X	X	-	-	X
C28	-	X	-	X	-	X	X	-	-	X
C29	-	X	X	-	X	-	-	X	-	X
C30	-	X	-	X	X	-	-	X	-	X
C31	-	X	X	-	-	X	-	X	-	X
C32	-	X	-	X	-	X	-	X	-	X

Finally, climate zone and operational energy use demand is analyzed together because are directly linked. Generally, when the baseline operational energy demand is lower (warmer climates), the possibility to improve the energy performance by the refurbishment strategies is lower, increasing the importance of other building life cycle phases, and therefore increasing the relative difference between LCA and OSA methodologies.

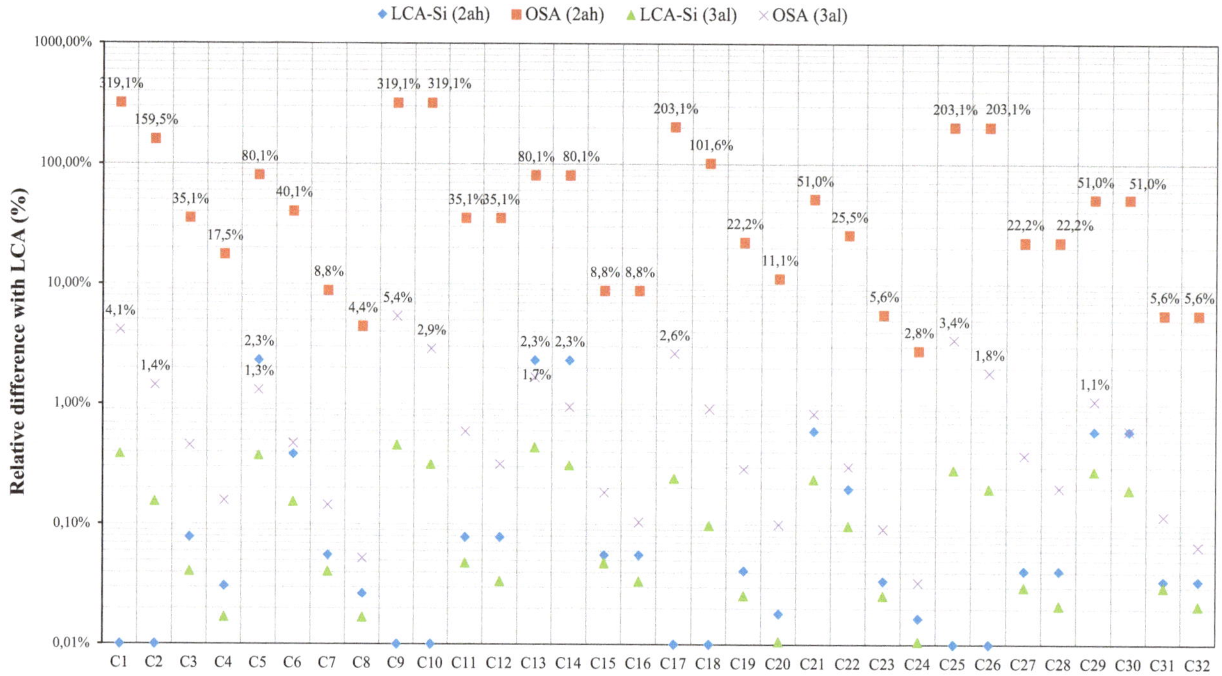

Figure 6. Relative difference with LCA for 2ah (advanced efficiency level ventilated façade system with high embodied energy products) and 3al (advanced efficiency level external insulation system with low embodied energy products) refurbishment strategies.

4.8. Other Indicators and Environmental Impact Categories

For simplicity, previous sections have presented only the NRPE results for the different refurbishment scenarios and using different calculation methodologies. However, as previously mentioned, a number of indicators need to be used in LCA to avoid potential shifting of impacts between different categories. Various previous studies as shown in Table 9 have studied different environmental impact categories.

Table 10 presents the results for additional environmental impact categories calculated for the case study. Differences between LCA and simplified methodologies are similar to the previously calculated for NRPE for indicators ADP and GWP, as it could be expected taking into account the strong correlations between these categories. For other frequently used environmental indicators such as ODP, POCP or AP, which are part of the indicators set in the CEN TC 350 standards [12], there is also a large similarity on the differences between methodologies, as the life cycle impacts on these categories are still strongly linked to the use of fossil fuels. For other impact categories not considered in this study or in the recent standardization efforts, for example those related to toxicity, further research is needed on the impact of all the different life cycle phases as there is currently not enough data available and results could differ in this case.

Table 9. Environmental impact categories used in previous studies.

Environmental impact categories	Erlandsoona *et al.*, 2004 [27]	Junnilla, 2004 [28]	Citherlet *et al.*, 2007 [29]	Zabalza *et al.*, 2009 [30]	Utama *et al.*, 2009 [31]	Kofoworola *et al.*, 2009 [26]	Blom *et al.*, 2010 [32]	Blengini *et al.*, 2010 [33]	Gustavsoon *et al.*, 2010 [34]	Hernandez *et al.*, 2010 [23]
Resource use, primary energy, PE	-	-	-	X	X	X	-	X	X	X
Depletion of Abiotic resources, elements, ADP-elements.	-	-	-	-	-	-	X	-	-	-
Global warming, GWP	X	X	X	X	-	-	X	X	X	-
Ozone Depletion, ODP	-	-	-	-	-	-	X	X	-	-
Photochemical Ozone Creation, POCP	X	-	X	-	-	-	X	X	-	-
Acidification for soil and water, AP	X	X	X	-	-	-	X	X	-	-
Eutrophication, EP	X	X	-	-	-	-	X	X	-	-
Environmental impact categories	**Oritz *et al.*, 2010 [35]**	**Dodoo *et al.*, 2010 [36]**	**Malmqvist *et al.*, 2011 [37]**	**Rossi *et al.*, 2012 [38]**	**Stephan *et al.*, 2012 [24]**	**Ramesh *et al.*, 2012 [39]**	**Stephan *et al.*, 2013 [25]**	**Mosteiro *et al.*, 2014 [40]**	**Dodoo *et al.*, 2014 [41]**	-
Resource use, primary energy, PE	-	X	-	-	X	X	X	X	X	-
Depletion of Abiotic resources, elements, ADP-elements.	X	-	-	-	-	-	-	-	-	-
Global warming, GWP	X	-	X	X	-	-	-	X	-	-
Ozone Depletion, ODP	X	-	-	-	-	-	-	X	-	-
Photochemical Ozone Creation, POCP	-	-	-	-	-	-	-	-	-	-
Acidification for soil and water, AP	X	-	-	-	-	-	-	X	-	-
Eutrophication, EP	-	-	-	-	-	-	-	X	-	-

Table 10. Results of the analysis for other impact indicators expressed for m^2 and year. For abbreviations, refer to Table 9.

Environmental impact reduction							
Environmental Impact Indicators (EII)	**2ah Strategy**			**EII**	**3al Strategy**		
	LCA	**LCA-Si**	**OSA**		**LCA**	**LCA-Si**	**OSA**
ADP fossil fuels (MJ/(m^2·a))	95	96	110	**ADP**	110	110	110
GWP (kg CO_2/(m^2·a))	5.5	5.6	8.2	**GWP**	8.1	8.1	8.2
ODP (kg CFC 11/(m^2·a))	2.5×10^{-7}	2.5×10^{-7}	2.1×10^{-6}	**OPD**	2.1×10^{-6}	2.1×10^{-6}	2.1×10^{-6}
POCP (kg ethene/(m^2·a))	1.3×10^{-3}	1.3×10^{-3}	1.8×10^{-3}	**POCP**	1.8×10^{-3}	1.8×10^{-3}	1.8×10^{-3}
AP (kg SO_2-eq/(m^2·a))	2.4×10^{-2}	2.4×10^{-2}	2.9×10^{-2}	**AP**	2.9×10^{-2}	2.9×10^{-2}	2.9×10^{-2}

5. Discussion

Simplifications of the LCA methodology can have important implications in decision-making process for selecting the most appropriate energy refurbishment solutions of existing buildings. The exercise previously developed, and especially the results obtained from the sensitivity analysis, provide useful information about how simplifications can bring insignificant or very remarkable uncertainties in comparison to a fully-fledged LCA methodology.

Differences between LCA and Simplified LCA are negligible in the case of assessing the use of Non-Renewable Primary Energy (NRPE) when comparing different refurbishment solutions oriented to reduce the energy consumption of existing buildings. The use of OSA does, however, have important limitations. For warm climates and ambitious goals for reducing the energy consumption of buildings by using materials and systems with a high energy embodied values, OSA is an oversimplified methodology and can mislead the decision making for choosing solutions with the best life cycle environmental performance. If products with low embodied energy are used, OSA, which studies only the reduction on energy use during the operational stage of the building, might be of sufficient accuracy. Of course, it should then be decided and benchmarked which materials are considered low embodied energy. In the case of colder climates or situations with drastic reductions on operational energy use due to a refurbishment strategy, values obtained with the OSA methodology will be very similar to those yielded by an LCA, as relative importance of products and other building life cycle phases is lower. In addition, OSA is not recommended if construction products and systems with short Estimated Service Life are used.

Regarding the quantity and type of environmental indicators used on the analysis, it has to be considered that when applying LCA to compare different options, commonly there is not a specific alternative that scores the best in all the impact categories. Therefore, the decision maker needs to select the best option according to his/her environmental values and objectives and apply some kind of weighting among the different impact categories. This study is focused on the NRPE Indicator as it generally offers a good correlation with other related impact categories (such as Global Warming and Abiotic Depletion of fossil fuels) and, on the other hand, captures the consequences of applying energy efficiency strategies. Other environmental impact categories not so correlated to the use of fossil use (such as toxicity) are not well represented by this indicator.

The source of data used in the LCA has a remarkable influence increasing the uncertainty of the results, as significant differences exist in the calculations using data derived from process inventories (e.g., from Ecoinvent and GaBi databases, for instance, or from EPDs) or from hybrid Input-Output analysis This observation is in line with previous studies from Crawford [57] and Stephan *et al.* [25] which have demonstrated that the truncation error associated to process-based life cycle inventories may significantly affect the final results.

Regarding data associated to the construction stage, the main barrier is again related to the lack and uncertainty of environmental information, which largely depends on project specific parameters and are difficult to assess for a set of different strategies. As for data associated for transportation, technicians usually have information on the exact location of the distributor of the product or system applied in the refurbishment project, but in many cases the manufacturing location is unknown, being the impacts associated to transport from manufacturer to distributor difficult to calculate. Due to the lack of information about future waste management processes, the uncertainty of the end of life stage is also very large.

Finally, it is important to mention that despite heating being one of the major contributors to the operational energy consumption of buildings, cooling of indoor spaces (especially in warm climate zones) should not be left out of the scope of the assessment of energy refurbishment in any case (LCA, LCA-Si or OSA) as, for some scenarios, it can have a significant influence on the decision-making process.

6. Conclusions

The refurbishment of the existing building stock with the aim of reducing its operational energy consumption is being fostered for different reasons and by different actors in the EU. The intense standardization efforts on the assessment of the sustainability of buildings carried out by the European Committee for Standardization Technical Committee 350 [12] suggests that the application of the life cycle approach will be more common in the near future. In this context, simplified LCA (LCA-Si) could be seen as a first step towards a comprehensive and extensive application of LCA within the construction sector. On the other hand, depending on the objectives pursued and the resources available, the practical application of the LCA methodology could be neither necessary nor possible, and in some circumstances an analysis of the energy reduction derived from the refurbishment during the operational stage (OSA) could be sufficient. As proved through a complete case study, in the case of assessing different refurbishment strategies oriented to reduce the energy consumption of buildings, a simplified LCA study may underestimate the absolute energy values, but it generally allows identifying the most (life-cycle) efficient solution. However, considering only the operational energy use (OSA) may lead to the selection of less efficient solutions.

The application of the fully-fledged LCA methodology still faces important challenges, and the construction industry needs to play a key role in the calculation and publication of data to facilitate accurate assessments of building energy refurbishment projects.

Acknowledgments

This paper is part of the PhD research of Xabat Oregi Isasi who thanks the Basque Government for the financial support given for his PhD research. All the authors acknowledge the contribution of reviewers to improve the text.

Author Contributions

The paper is based on Xabat Oregi's PhD research work. Oregi was responsible for the research design, data collection and analyses. Patxi Hernandez and Cristina Gazulla provided guidance and supervision respectively. Marina Isasa provided additional insights.

Conflicts of Interest

The authors declare no conflict of interest.

References

1. International Energy Agency (IEA). *CO_2 emissions from fuel combustion Highlights*; IEA: Paris, France, 2013; p. 600.
2. Power, A. Does demolition or refurbishment of old and inefficient homes help to increase our environmental, social and economic viability? *Energy Policy* **2008**, *36*, 4487–4501.
3. IEE Project TABULA (2009–2012). Available online: http://episcope.eu/iee-project/tabula/ (accessed on 5 February 2015).
4. Sharma, A.; Saxena, A.; Sethi, M.; Shree, V.; Varun. Life cycle assessment of buildings: A review. *Renew. Sustain. Energy Rev.* **2011**, *15*, 871–875.
5. Ramesh, T.; Prakash, R.; Shukla, K.K. Life cycle energy analysis of buildings: An overview. *Energy Build.* **2010**, *10*, 1592–1600.
6. Chang, Y.; He, C.Q.; Tang, B.J.; Wei, Y.M. China's energy consumption in the building sector: A life cycle approach. *Energy Build.* **2015**, *94*, 240–251.
7. Richman, R.; Pasqualini, P.; Kirsh, A. Life-cycle analysis of roofing insulation levels for cold storage buildings. *J. Archit. Eng.* **2009**, *15*, 55–61.
8. Karimpour, M.; Belusko, M.; Xing, K.; Bruno, F. Minimising the life cycle energy of buildings: Review and analysis. *Build. Environ.* **2014**, *73*, 106–114.
9. Cellura, M.; Guarino, F.; Longo, S.; Mistretta, M. Energy life-cycle approach in net zero energy buildings balance: Operation and embodied energy of an Italian case study. *Energy Build.* **2014**, *72*, 371–381.
10. Lützkendorf, T.; Foliente, G.; Balouktsi, M.; Wiberg, A.H. Net-zero buildings: Incorporating embodied impacts. *Build. Res. Inf.* **2014**, *43*, 62–81.
11. ISO/TC 59/SC 17. *Sustainability in Buildings and Civil Engineering Works*; British Standards Institution: London, UK, 2014.
12. CEN TC 350. European Committee for Standardization, Technical Committees. Sustainability of Construction Works. Available online: http://portailgroupe.afnor.fr/public_espacenormalisation/centc350/index.html (accessed on 30 April 2015).
13. Cabeza, L.F.; Rincón, L.; Vilariño, V.; Pérez, G.; Castell, A. Life cycle assessment (LCA) and life cycle energy analysis (LCEA) of buildings and the building sector: A review. *Renew. Sustain. Energy Rev.* **2014**, *29*, 394–416.
14. Chau, C.K.; Leung, T.M.; Ng, W.Y. A review on life cycle assessment, life cycle energy assessment and life cycle carbon emissions assessment on buildings. *Appl. Energy* **2015**, *143*, 395–413.

15. European Commission. Resource Efficiency Opportunities in the Building Sector. COM (2014) 445 Final. Brussels, 1.7.2014. Available online: http://ec.europa.eu/environment/eussd/pdf/Sustainable BuildingsCommunication.pdf (accessed on 5 February 2015).
16. UNE-EN ISO 14040:2006. *Environmental Management—Life Cycle Assessment—Principles and Framework*; The International Organization for Standardization (ISO): Geneva, Switzerland, 2010.
17. UNE-EN ISO 14044:2006. *Environmental Management—Life Cycle Assessment—Requirements and Guidelines*; The International Organization for Standardization (ISO): Geneva, Switzerland, 2010.
18. Guinée, J.B.; Gorrée, M.; Heijungs, R.; Huppes, G.; Kleijn, R.; de Koning, A.; van Oers, L.; Sleeswijk, A.W.; Suh, S.; Helias, A.; *et al. Life Cycle Assessment—An Operational Guide to the ISO Standards*; Centre for Environmental Studies (CML), Leiden University: Leiden, The Netherland, 2011.
19. Goedkoop, M.; Spriensma, R. *The Eco-indicator 99. A Damage Oriented Method for Life Cycle Assessment*; Pre-Consultants: Amersfoot, The Netherlands, 2001.
20. EN 15978:2011. Sustainability of Construction Works—Assessment of Environmental Performance of Buildings—Calculation Method. Available online: http://standards.cen.eu/dyn/www/f?p=204:110:0::::FSP_PROJECT:31325&cs=16BA443169318FC086C4652D797E50C47 (accessed on 5 February 2015).
21. Wadel, G.; López, F.; Sagrera, A.; Prieto, J. Refurbishment considering environmental bimpact reduction targets: A test case for a multiple-family dwelling in the area of Playa de Palma, Mallorca. *Inf. Constr.* **2011**, *63*, 89–102
22. Crowther, P. Design for Disassembly to Recover Embodied Energy. In Proceedings of the 16th International Conference on Passive and Low-Energy Architecture, Melbourne, Australia, 22–24 September 1999.
23. Hernandez, P.; Kenny, P. From net energy to zero energy buildings: Defining life cycle zero energy buildings (LC-ZEB). *Energy Build.* **2010**, *42*, 815–821.
24. Stephan, A.; Crawford, R.H.; Myttenaere, K. Towards a comprehensive life cycle energy analysis framework for residential buildings. *Energy Build.* **2012**, *55*, 592–600.
25. Stephan, A.; Crawford, R.H.; Myttenaere, K. A comprehensive assessment of the life cycle energy demand of passive houses. *Appl. Energy* **2013**, *112*, 23–34.
26. Kofoworola, O.; Gheewala, S. Life cycle energy assessment of a typical office building in Thailand. *Energy Build.* **2009**, *4*, 1076–1083.
27. Erlandssona, M.; Levin, P. Environmental assessment of rebuilding and possible performance improvements effect on a national scale. *Build. Environ.* **2004**, *39*, 1453–1465.
28. Junnila, S. The Environmental Impact of an Office Building throughout its Life Cycle. Ph.D. Thesis, Helsinki University of Technology, Espoo, Finland, 2004.
29. Citherlet, S.; Defaux, T. Energy and environmental comparison of three variants of a family house during its whole life span. *Build. Environ.* **2007**, *42*, 591–598.
30. Zabalza, I.; Aranda, A.; Scarpellini, S. Life cycle assessment in buildings: State-of-the-art and simplified LCA methodology as a complement for building certification. *Build. Environ.* **2009**, *44*, 2510–2520.
31. Utama, A.; Gheewala, S.H. Indonesian residential high rise buildings: A life cycle energy assessment. *Energy Build.* **2009**, *41*, 1263–1268.

32. Blom, I.; Itard, L.; Meijer, A. Environmental impact of dwellings in use: Maintenance of façade components. *Build. Environ.* **2010**, *45*, 2526–2538.
33. Blengini, G.A.; Di Carlo, T. The changing role of life cycle phases, subsystems and materials in the LCA of low energy buildings. *Energy Build.* **2010**, *42*, 869–880.
34. Gustavsson, L.; Joelsson, A. Life cycle primary energy analysis of residential buildings. *Energy Build.* **2010**, *42*, 210–220.
35. Ortiz-Rodriguez, O.; Castells, F.; Sonnemann, G. Life cycle assessment of two dwellings: One in Spain, a developed country, and one in Colombia, a country under development. *Sci. Total Environ.* **2010**, *408*, 2435–2443.
36. Dodoo, A.; Gustavsson, L.; Sathre, R. Life cycle primary energy implication of retrofitting a wood-framed apartment building to passive house standard. *Resour. Conserv. Recycl.* **2010**, *54*, 1152–1160.
37. Malmqvist, T.; Galumann, M.; Scarpellini, S.; Zabalza, I.; Aranada, A.; Llera, E.; Diaz, S. Life cycle assessment in buildings: The ENSLIC simplified method and guidelines. *Energy* **2011**, *36*, 1900–1907.
38. Rossi, B.; Marique, A.F.; Glaumann, M.; Reiter, S. Life-cycle assessment of residential buildings in three different European locations, basic tool. *Build. Environ.* **2012**, *51*, 395–401.
39. Ramesh, T.; Prakash, R.; Shukla, K.K. Life cycle energy analysis of a residential building with different envelopes and climates in Indian context. *Appl. Energy* **2012**, *89*, 193–202.
40. Mosteiro-Romero, M.; Krogmann, U.; Wallbaum, H.; Ostermeyer, Y.; Senick, J.S.; Andrews, C.J. Relative importance of electricity sources and construction practices in residential buildings: A Swiss-US comparison of energy related life-cycle impacts. *Energy Build.* **2014**, *68*, 620–631.
41. Dodoo, A.; Gustavsson, L.; Sathre, R. Lifecycle primary energy analysis of low-energy timber building systems for multi-storey residential buildings. *Energy Build.* **2014**, *81*, 84–97.
42. Spanish Technical Building Code, CTE, 2013. Available online: http://www.codigotecnico.org/ (accessed on 5 February 2015).
43. Design Builder Simulation Tool. Available online: http://www.designbuilder.es (accessed on 5 February 2015).
44. ASHRAE International Weather Files for Energy Calculations 2.0 (IWEC2). Available online: https://www.ashrae.org/resources--publications/bookstore/iwec2 (accessed on 5 February 2015).
45. Ecoinvent Centre. Swiss Centre for Life Cycle Inventories. Düberdorf, Suiza. Available online: http://www.ecoinvent.org/database/ecoinvent-version-3/ (accessed on 5 February 2015).
46. International Passive House Association. Available online: http://www.passivehouse-international.org/index.php?page_id=150 (accessed on 5 February 2015).
47. GaBi. Product Sustainability Performance Solutions by PE International. Available online: http://www.gabi-software.com/databases/gabi-databases/ (accessed on February 2015).
48. Wittstock, B.; Gantner, J.; Lenz, K.; Saunders, T.; Anderson, J.; Carter, C.; Gyetvai, Z.; Kreißig, J.; Braune, A.; Lasvaux, S.; *et al.* EeBGuide Guidance Document. Part B: Buildings. Operational Guidance for Life Cycle Assessment Studies of the Energy-Efficient Buildings Initiative 2011. Available online: http://www.eebguide.eu/?page_id=704 (accessed on 5 February 2015).

49. ISO 15686–8:2008. Buildings and Constructed Assets—Service-life Planning—Part 8: Reference Service Life and Service-Life Estimation. Available online: http://www.iso.org/iso/iso_catalogue/catalogue_tc/catalogue_detail.htm?csnumber=39070 (accessed on February 2015).
50. National association of Home Builders (NAHB) & and Bank of America. Study of Life Expectancy of Home Materials, 2007. Available online: https://www.nahb.org/ (accessed on 5 February 2015).
51. Winistorfer, P.; Chen, Z.; Lippke, B.; Stevens, N. Energy consumption and greenhouse gas emissions related to the use, maintenance, and disposal of a residential structure. *Wood Fiber Sci.* **2007**, *37*, 128–139.
52. Saint-Gobain Glass. Declaration Environnementale et Sanitarie. SGG CLIMAPLUS. AFNOR Registrierungsnummer N° 08–257:2011. Available online: http://www.inies.fr/Upload/2261_sgg_climaplus_4s_fr_v2.2_05_2012.pdf (accessed on February 2015).
53. Saint-Gobain Glass. Declaration Environnementale et Sanitarie SGG CLIMATOP AFNOR: 08–259:2011. Available online: http://epd.saint-gobain-glass.com/ (accessed on February 2015).
54. CORMO. S3 Window in Solid Swedish Pinewood Double Leaves. Available online: http://gryphon.environdec.com/data/files/6/9218/epd142_20130408.pdf (accessed on February 2015).
55. *A Comparison of Emissions Factors for Electricity Generation*; Department of Energy & Climate Change: London, UK, 2013.
56. Treloar, G.J. Extracting embodied energy paths from input-output tables: Towards an input-output-based hybrid energy analysis method. *Econ. Syst. Res.* **1997**, *9*, 375–391.
57. Crawford, R.H. Validation of a hybrid life-cycle inventory analysis method. *J. Environ. Manag.* **2008**, *88*, 496–506.
58. Crawford, R.H. *Life Cycle Assessment in the Built Environment*; Spon Press: London, UK, 2011.
59. Crawford, R.; Stephan, A. The Significance of Embodied Energy in Certified Passive Houses. In Proceedings of the International Conference on Construction and Building Materials, Copenhagen, Denmark, 13–14 June 2013; pp. 473–479.
60. Stephan, A.; Stephan, L. Reducing the total life cycle energy demand of recent residential buildings in Lebanon. *Energy* **2014**, *74*, 618–637.
61. Topouzi, M. The Effect of "Default User" Inputs in Modelling Tools and Methods in Energy Use. In Proceedings of the PLEA 2011—27th International Conference on Passive and Low-Energy Architecture, Louvain-la-Neuve, Belgium, 13–15 July 2011; pp. 35–40.
62. Juodis, E.; Jaraminiene, E.; Dudkiewicz, E. Inherent variability of heat consumption in residential buildings. *Energy Build.* **2009**, *11*, 1188–1194.

Field Observation of Cooling Energy Savings Due to High-Reflectance Paints

Hideki Takebayashi [†,*] and Chihiro Yamada [†]

Department of Architecture, Graduate School of Engineering, Kobe University, 1-1 Rokkodai, Nada, Kobe 657-8501, Japan; E-Mail: 120t066t@stu.kobe-u.ac.jp

† These authors contributed equally to this work.

* Author to whom correspondence should be addressed; E-Mail: thideki@kobe-u.ac.jp

Academic Editor: Fitsum Tariku

Abstract: Cooling energy savings in a building with the roof coated with high-reflectance paint are examined. It is difficult to recognize such savings using the data observed hourly. Among the factors assumed to affect cooling energy load are: (1) internal heat generation, (2) set temperature, and (3) weather conditions. By analyzing the relationship between indoor-outdoor air temperature difference (averaged) and electric power consumption (integrated) of the air conditioner in the building over a day, the reduction in electric power consumption due to the use of a high-reflectance paint coating is estimated at approximately 72 $Wh \cdot m^{-2} \cdot day^{-1}$.

Keywords: high-reflectance paint; cooling energy saving; roof

---

1. Introduction

Nowadays, high-reflectance paint is gaining popularity in Japan for mitigating the heat island phenomenon. The method for evaluating high-reflectance paint is established in the Japanese Industrial Standards (JIS) [1,2]. Although the effect of such paint on the mitigation of the heat island phenomenon has been evaluated in many studies [3], few evaluate the cooling energy savings in a building actually coated with such paint, because such savings in a typical office or residential building with high thermal insulation performance are generally insignificant [4]. However, it cannot be ignored in a building with low thermal insulation performance, such as a factory, warehouse, or gymnasium. In this study, we analyzed the cooling energy savings due to the use of high-reflectance paint in a real building under use.

The cooling energy savings in a building brought about by the use of heat island mitigation technologies such as green wall, green roof, and high-reflectance paints have been studied by several researchers [5,6]. In these studies, the researchers conducted controlled experiments using the same two prefabricated houses without internal heat generation, to compare the difference in the energy consumption with or without heat island mitigation technologies under same weather conditions. However, building owners or managers require data on cooling energy savings caused by the high-reflectance paints in actual building sites.

Akbari *et al.* [7] analyzed cooling energy savings in real buildings under use in California, namely, a shop, school, and refrigerated warehouse. On the basis of the measurement results for summer, they concluded that the savings produced were 70, 42–48, and 57–81 $Wh \cdot m^{-2} \cdot day^{-1}$, respectively. We have conducted similar experiments in Japan.

The introduction of high-reflectance paint to buildings without high thermal insulation performance, such as factories, warehouses, and gymnasia, is one of the energy saving measures. During implementation, it is necessary to analyze the effect of introducing the high-reflectance paint. In this study, we analyzed the cooling energy savings produced in a real building based on energy measurements.

2. Outline of Building and Measurement Case Study

General office buildings have relatively higher ratio of window area and smaller ratio of roof area to the total envelope surface area. Moreover, they have significant internal heat generated from the appliances used in them, and so on. Therefore, the ratio of the cooling load owing to heat conduction from the roof to the total cooling load is low. In this study, the test building is a small (60 m^2) two-story research building (with its first floor raised on pilotis) at Kobe University (see Figure 1 and Table 1). The test building is expected to offer significant potential for achieving cooling energy savings by deploying high-reflectance paint, owing to its small window area (15%) and large roof area (25%–30%) ratios to the total envelope surface area.

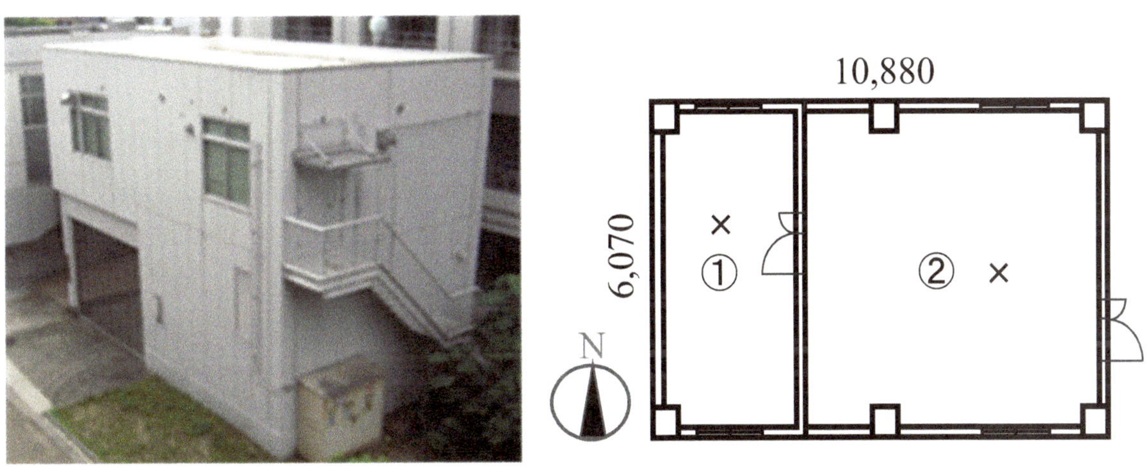

(a) (b)

Figure 1. (**a**) Facade and (**b**) Plan (second floor) of test building (x: Measurement point in Figure 2).

Table 1. Characteristics of objective rooms.

Room	Roof-to-Envelope Ratio	Wall-to-Envelope Ratio	Window-to-Envelope Ratio	Set Temperature
Room 1	20.7%	38.2%	5.3%	26 °C before 15:00, 8 August 2011
Room 2	24.9%	36.8%	4.1%	25 °C after 15:00, 8 August 2011 25 °C through 2012

The experiment was conducted in rooms 1 and 2. The air conditioners in both rooms were continuously running for 24 h. The roof consists of a concrete slab (with 100 mm thickness) above the suspended ceiling (with 700 mm depth and 2730 mm height from the floor) without any heat insulating material. We measured the below and above surface temperatures of the slab and suspended ceiling, room temperature (at 1200 mm height), and power consumption of the air conditioners. The cross section of the point of measurement of temperature is shown in Figure 2. Waterproof membrane was installed on the slab. To avoid the influence of insolation, the roof surface temperature was measured using an infrared thermometer and those of the other elements were measured using a thermistor thermometer. The certificate cooling capacities of the air conditioners in rooms 1 and 2 are 7.1 and 14.5 kW, respectively. The power consumption was measured using a clamp-type power meter (HIOKI3168-98) at the power distribution panel. We obtained the temperature and power data at one-minute intervals, and averaged the temperature values and integrated the power consumption values over 10 min.

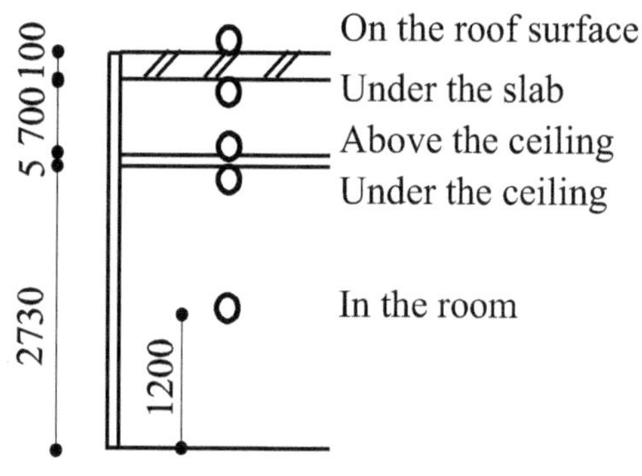

Figure 2. Cross section of temperature measurement points.

We conducted measurements for the periods, 12 July–26 September 2011 and 12 July–26 September 2012. High-reflectance paint was introduced on 2 August 2011. The cooling energy savings were evaluated by comparing the measurements before and after painting. The amount of horizontal solar radiation and average air temperature on typical summer days in Kobe are about 24–28 $MJ \cdot m^{-2} \cdot day^{-1}$ and about 28 to 32 °C, respectively. The albedo of the objective paint is 86.9% (300–2500 nm wavelength according to JIS K 5602 [1]). The albedo measured by the pyranometer (which comprises white and black boards and measures solar radiation) was 16.9% before, 86.9% immediately after, and 76.1% after one year, of painting [8]. These are averaged values of the measurement results at three points on the roof surface (east, middle, and west).

The results of temperature measurement under the slab, and above and under the suspended ceiling in room 1; the outdoor air temperature; and the solar radiation are shown in Figure 3. The temperature under the slab ranged from 40 to 27 °C and the maximum temperature was observed at around 18:00 owing to the thermal storage of the slab. The maximum temperature reduction on the roof surface was approximately 30 K after painting. The cooling energy consumption and solar radiation measurement results are shown in Figure 4. It is difficult to recognize the cooling energy savings by comparing the two values in a typical day even if the solar radiation values are similar to each other.

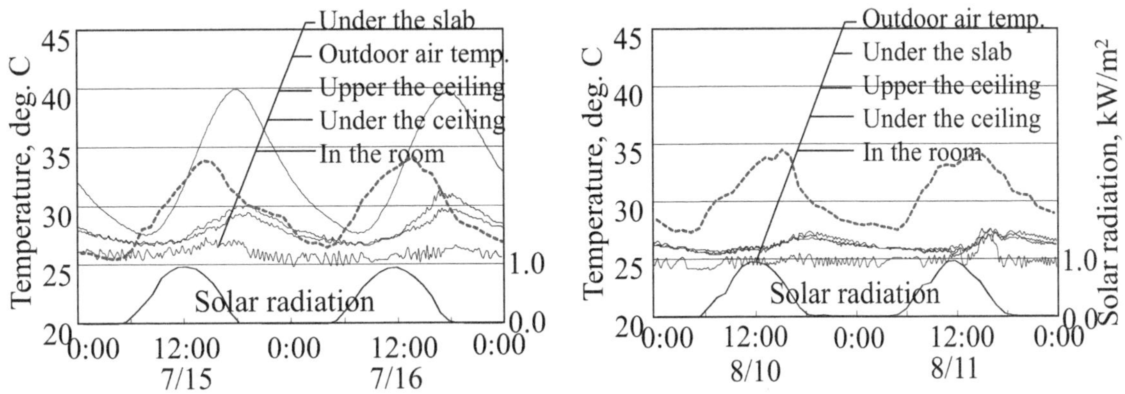

Figure 3. Measurement results for temperature under slab, and above and under suspended ceiling in room 1; outdoor air temperature; and solar radiation. **left**: Before painting; **right**: After painting.

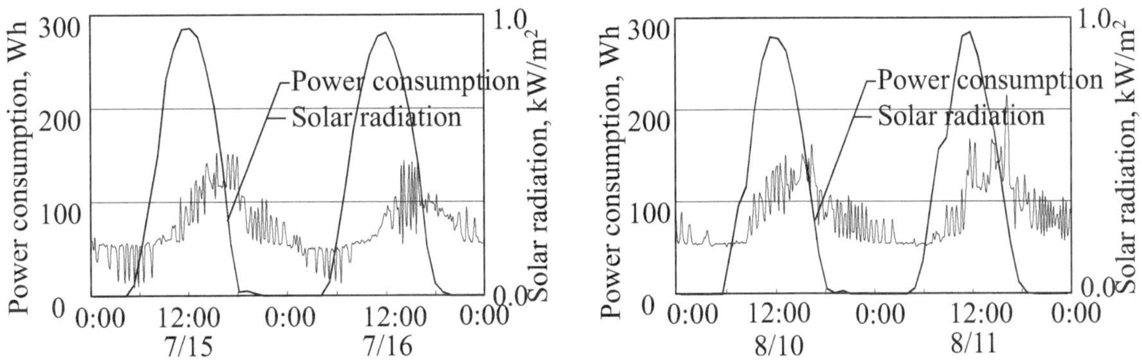

Figure 4. Measurement results for cooling energy consumption and solar radiation. **left**: Before painting; **right**: After painting.

3. Method of Evaluating Cooling Energy Savings due to High-Reflectance Paint

We analyzed the cooling energy savings by observing the difference in the cooling power consumption before and after painting. The hourly relationships between the outdoor-to-room air temperature difference and the cooling power consumption are shown in Figure 5. The black and white points represent the data before and after painting, respectively. Sol-air temperature is calculated by $T_a + a \times I/\alpha$, where T_a is air temperature (°C); a is solar absorptance (-); I is solar radiation (W/m^2); α is convective heat transfer coefficient (W/m^2/K). After painting, the sol–air temperature changes significantly (influenced by solar radiation) but the power consumption does not change as much. The air temperature difference is averaged and the power consumption is integrated over 1 h. Since the temperature under the slab reaches its maximum around the evening, it is difficult to recognize the relationship between the temperature difference of outdoor considering solar radiation and room conditions and power consumption using hourly data in the right of Figure 5.

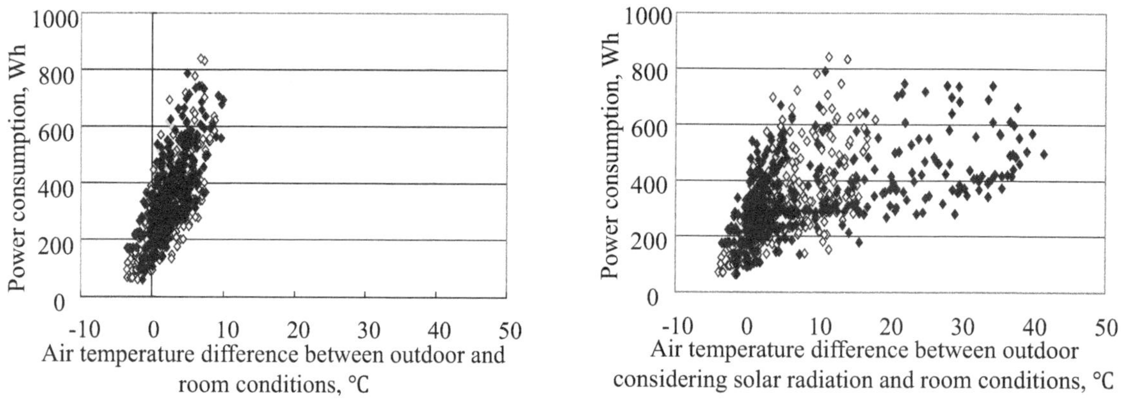

Figure 5. Hourly relationship between outdoor-to-room air temperature difference and cooling power consumption. black: Before painting; white: After painting; **left**: Air temperature difference without considering solar radiation effect; **right**: Air temperature difference considering solar radiation effect.

We analyzed the relationship between the daily averaged outdoor-to-room air temperature difference and the daily integrated cooling power consumption by referring to the method of Akbari *et al.* [7]. The evaluation of cooling energy savings due to high-reflectance paint is mainly influenced by the following factors:

- internal heat generation;
- set temperature of the air conditioner;
- weather condition (air temperature, solar radiation).

As the object of this study is the evaluation of cooling energy savings in a real building under use, we did not instruct the room occupants on its use. Although the measurement results are affected by above three factors, the differences in the internal heat generation and the set temperature of air conditioner are not so large compared to those outlined in the reference method. In the test building, experimental devices such as constant temperature drying machine (1.4 kW, maximum use time: 5 h/day), mass spectrometer (2.8 kW, maximum use time: 10 h/day) are used irregularly by the day. The ratio of internal heat generation to total cooling load was estimated to be about 15% to 20% on average.

The cooling power consumption influenced by the internal heat generation, set temperature of the air conditioner, and weather condition is derived as follows:

$$E = A \times I + B \times \Delta T + C \tag{1}$$

where E is the daily integrated cooling power consumption (Wh day^{-1}); I is the daily integrated solar radiation (Wh day^{-1}); ΔT is the difference between the daily averaged outdoor and room temperatures (in K); A is the coefficient related to absorptivity; B is the coefficient related to thermal conductance; and C is the internal heat generation. The cooling power consumption is directly proportional to these coefficients. As per the method, we should use the difference of enthalpy instead of ΔT since the power consumption includes the latent heat process, and consider the coefficient of performance of the air conditioner. However, in this study, we measured E, I, and ΔT and assumed that B and C do not change after painting. The influence of the set temperature of the air conditioner is reflected in ΔT, and that of the weather condition is reflected in ΔT and I. In the following chapter, we analyze the measurement results derived using Equation (1).

4. Examination of Cooling Energy Savings due to High-Reflectance Paint

The daily relationship between the outdoor-to-room air temperature difference and the cooling power consumption is shown in the left of Figure 6. Since the slopes of the regression equations in both before- and after-painting cases are similar, it can be concluded that cooling power consumption is influenced by the outdoor air temperature. Additionally, it is influenced by the set temperature of the air conditioner, which was changed during the measurement period. The daily relationship between the solar radiation and the intercept of the regression equation (in the left of Figure 6) is shown in the right of Figure 6. Influences by ΔT and I are discussed respectively based on the left and right of Figure 6. The power consumption does not change after painting even when the daily integrated solar radiation value is high. When this factor is relatively higher, the cooling energy savings are about 1.6 kWh day^{-1} (72 Wh m^{-2} day^{-1}).

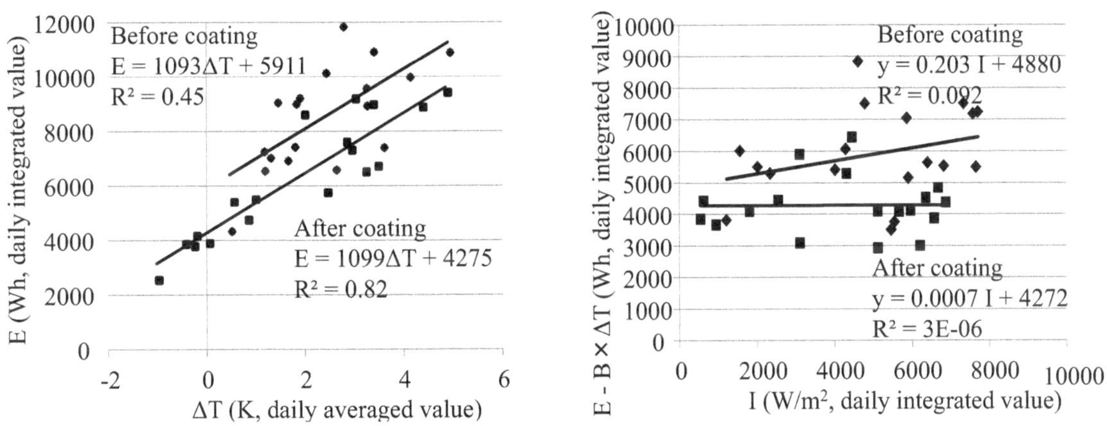

Figure 6. Relationship between (1) daily averaged outdoor-to-room air temperature difference and daily integrated cooling power consumption (**left**) and (2) daily integrated solar radiation and cooling consumption (**right**).

The value of B before painting is mostly the same as that after painting (in the left of Figure 6), and hence the influence of ΔT on power consumption is assumed to be similar before and after painting. The value of A after painting is nearly zero (in the right of Figure 6), and hence the influence of solar radiation on the cooling power consumption is assumed to be negligible after painting. Since internal heat generation was not controlled during measurement, the value of C is different before and after painting (in the right of Figure 6). Table 1 lists the characteristics of the objective rooms. Although the test building is actually in use, the analysis results have captured the approximate characteristics of its cooling energy consumption.

The regression equations of the cooling power consumption by air temperature difference between outdoors and rooms 1 and 2 are listed in Table 2 for three scenarios of painting: (1) before,

(2) immediately after, and (3) one year after. The total cooling power consumption savings are divided by the respective floor areas (22.5 m^2 for room 1 and 43.5 m^2 for room 2) and expressed in the table. Since the slopes of the regression Equation (B) in the three scenarios are relatively similar in each room, the influence of ΔT on the cooling power consumption is similar in each room. For room 1, the values of B are almost similar for both immediately and one year after painting. However, the difference of energy savings between the immediately and one year after painting is not significant due to a little lower determination coefficient of each regression equation. The cooling energy savings in room 2 are lower than that in room 1. The ratio of internal heat generation to the total cooling load in room 2 is higher than that in room 1 because the former's floor area is approximately twice the latter's. Since the values of B for room 2 are lower than those for room 1, room 2 is not heavily influenced by the external weather condition.

Table 2. Regression equations for cooling power consumption due to air temperature difference between outdoor and room condition for the scenarios (1) before, (2) immediately after, and (3) one year after painting, in rooms 1 and 2.

Room	Condition	Regression Equation	Energy Savings
Room 1	Before painting *	$E = 48.6\ \Delta T + 262.7\ (R^2 = 0.45)$	$72\ Wh \cdot m^{-2} \cdot day^{-1}$
	After painting **	$E = 50.3\ \Delta T + 190.8\ (R^2 = 0.82)$	$91\ Wh \cdot m^{-2} \cdot day^{-1}$
	One year after painting ***	$E = 48.0\ \Delta T + 171.6\ (R^2 = 0.82)$	
Room 2	Before painting *	$E = 26.7\ \Delta T + 151.3\ (R^2 = 0.54)$	$35\ Wh \cdot m^{-2} \cdot day^{-1}$
	After painting **	$E = 27.6\ \Delta T + 116.6\ (R^2 = 0.60)$	$9\ Wh \cdot m^{-2} \cdot day^{-1}$
	One year after painting ***	$E = 26.9\ \Delta T + 142.8\ (R^2 = 0.77)$	

* before painting: From 12 July to 31 July 2011; ** after painting: From 3 August to 26 September 2011; *** one year after painting: From 12 July to 26 September 2012.

5. Conclusions

In this study, we analyzed the cooling energy savings brought about by high-reflectance paint in a building actually under use, based on measured results. Considering the factors assumed to influence cooling energy load (namely, internal heat generation, set temperature of the air conditioner, and weather condition), we analyzed the relationship between (1) the daily integrated solar radiation and cooling power consumption values, and (2) the daily averaged outdoor-to-room temperature difference and cooling power consumption.

From (2), the cooling power consumption savings brought about by the high-reflectance paint were estimated at approximately 1.6 kWh day^{-1} (72 Wh m^{-2} day^{-1}). Similar results were confirmed one year after painting. However, the difference of energy savings between the immediately and one year after painting is not significant due to a little lower determination coefficient of each regression equation. Although the results measured in a building actually under use were considered, the approximate energy savings could be satisfactorily evaluated. However, when ΔT is relatively higher, the cooling power consumption tends to be significantly higher. Therefore, we may have to consider an improvement in the coefficient of performance of the air conditioner in such a case. In this study, the cooling energy saving effect by high-reflectance paint is mainly discussed. For example, reduction of mean radiation temperature by lowering the ceiling surface temperature should be evaluated in future study.

Acknowledgments

This study was supported by the Center for Environmental Management, Kobe University.

Author Contributions

The paper is based on Chihiro Yamada's master's degree program. Chihiro Yamada was responsible for data collection and analyses. Hideki Takebayashi managed the analysis.

Conflicts of Interest

The authors declare no conflict of interest.

References

1. *Determination of Reflectance of Solar Radiation by Paint Film*; Japanese Industrial Standards, JIS K 5602; Japanese Standards Association: Tokyo, Japan, 2008
2. *High Solar Reflectant Paint for Roof*; Japanese Industrial Standards, JIS K 5675; Japanese Standards Association: Tokyo, Japan, 2011
3. Takebayashi, H.; Kondo, Y. Development of evaluation tool for public benefit—Simple evaluation system of cool roof for proper promotion (Part 2). *AIJ J. Technol. Des.* **2010**, *33*, 589–594.
4. Kondo, Y.; Nagasawa, Y.; Irimajiri, M. Reduction of solar heat gain of building, urban area and vending machines by high reflective paint. *Trans. Soc. Heat. Air Cond. Sanit. Eng. Jpn.* **2000**, *78*, 15–24.
5. Kawashima, H.; Kato, C.; Yarita, Y.; Kurooka, H.; Ishima, T. Field experimental results on the rooftop green with the textile mats—The power saving effect of air-conditioner in summer. *J. Heat Isl. Inst. Int.* **2011**, *6*, 1–7.
6. Yamasaki, M.; Mizutani, A.; Ohsawa, T. Cooling load reduction effect of green roof and green wall in the case of building with thermal thin wall. *AIJ J. Technol. Des.* **2009**, *29*, 155–158.
7. Akbari, H.; Levinson, R.; Rainer, L. Monitoring the energy-use effects of cool roofs on California commercial buildings. *Energy Build.* **2005**, *37*, 1007–1016.
8. Murata, Y.; Sakai, K.; Kanamori, H.; Takebayashi, H.; Matsuo, Y.; Moriyama, M.; Yoshida, A.; Nishioka, M.; Yano, N.; Shimizu, R.; *et al.* A study on insolation reflectivity measurement of the cool painting surface: A proposal of measuring method by the 2 point correction and accuracy testing on the horizontal surface. *J. Environ. Eng.* **2008**, *632*, 1209–1215.

7

Modern Housing Retrofit: Assessment of Upgrade Packages to EnerPHit Standard for 1940–1960 State Houses in Auckland

Paola Leardini [†,*] and Manfredo Manfredini [†]

School of Architecture and Planning, The University of Auckland, 26 Symonds Street, 1010 Auckland, New Zealand; E-Mail: m.manfredini@auckland.ac.nz

[†] These authors contributed equally to this work.

* Author to whom correspondence should be addressed; E-Mail: p.leardini@auckland.ac.nz

Academic Editor: George Baird

Abstract: New Zealand state housing includes a significant portion of problematic buildings constructed after the public housing scheme launched in 1936. Most of these houses are still uninsulated, thus, cold, draughty, mouldy, and progressively decaying; however, as they are fundamental elements of the country's culture, society, and environment, and are built with good quality materials and sound construction, they are suitable candidates for effective energy upgrades. This paper presents findings of a study on problems and opportunities of retrofitting the state houses built between 1940 and 1960 in the Auckland region. It advocates strategic national policies and initiatives for retrofitting, based on more challenging performance thresholds. The research defines and virtually implements an incremental intervention strategy including different retrofit packages for a typical 1950s stand-alone house. Indoor and outdoor environmental parameters were monitored over a year, and data used to establish a base case for thermal simulation. The upgrade packages were then modelled to assess their impact on the house's thermal performance, comparing heating requirements and comfort of various insulation and ventilation options. The paper reports on effective ways of preserving the integrity of such a house, while improving its thermal performance to the EnerPHit standard, and discusses the benefits of introducing this holistic approach into New Zealand retrofit practice.

Keywords: New Zealand state house; energy retrofit; EnerPHit; Passive House Planning Package; stand-alone house

1. Introduction

According to the International Energy Agency, buildings are responsible for 32% of the total final energy consumption and around 40% of the primary energy consumption in most of its member countries [1]. The growing awareness of the impact of the built environment on energy security that started with the 1973 oil crisis has made building energy efficiency a common target of many national energy conservation policies. It also has produced a wide variety of analytical, certification and rating tools aimed at assessing and enhancing the energy performance of new buildings. The effect of these measures, though, has been limited, since these instruments only partially address the activities of renovation, repair and maintenance of existing buildings, which constitute a large part of construction activity in developed countries. An increased effort towards consistent inclusion of the existing stock in policies and programmes for the improvement of energy efficiency is crucial to achieve tangible savings in building-related energy consumption and CO_2 emission reduction. This is particularly relevant, as in many urban transformations retrofitting is often more effective than demolition and reconstruction in both economic and environmental terms as demonstrated by studies using Life Cycle Analysis [2]. Further benefits need to be accounted when the retrofit involves buildings with historic or cultural value: heritage retrofit not only contributes to the development of local culture—enhancing the quality of the urban environment while retaining its character and heritage—but also supports local economies, since this kind of retrofit requires a wide array of skilled labour on site. Overall, the conservation of existing buildings positively influences local communities, promoting resilient and sustainable urban living patterns.

In New Zealand, retrofit activity has proved to be non-cyclical (while new building construction is characterized by sharp highs and lows) [3] and has exceeded new construction, especially since the global financial crisis. According to Building Research Association New Zealand (BRANZ) already in 2008 alterations and additions (including approximately 33,000 residential consents [4]) represented about one-third the total value of new dwellings. This activity is expected to increase, due to the significant ageing of the New Zealand housing stock and the need for earthquake strengthening and other urgent maintenance interventions. Current residential buildings will constitute the largest portion of the country's housing stock in coming decades [5,6]: BRANZ estimates 85% will be still in use by 2025 [7]. However, most houses do not have sufficient insulation to provide comfortable living: about 65% were built before 1978, when mandatory insulation was first introduced with the standard NZS 4218P: 1977 *Minimum Thermal Insulation Requirements for Residential Buildings*. Research found that even when upgraded according to current local practice, average winter indoor air temperatures are extremely low [8], and frequently fall below the limit of 18 °C recommended by the World Health Organisation [9]. As housing is internationally recognised as a key determinant of health and the country has one of the highest incidences of asthma worldwide (with low socio-economic groups disproportionately affected [10]), the New Zealand government has committed to improve the nation's health introducing relevant policies, as well as subsidies to enhance the performance of houses, particularly those of low-income households. The state housing stock includes a significant portion of problematic houses, built in the 1940s and 1950s under the government-funded social housing scheme introduced in 1936. Most of these are managed by Housing New Zealand Corporation (HNZC), the largest public landlord in the country, which recently focused on upgrading its housing stock—especially in Auckland where 43% of state housing is concentrated [11]. Historically, due to the mild climate of this region, the extent of this problem and the urgency of actions required have been often underestimated. However, the socio-economic relevance of these interventions was confirmed by studies on the relationship between the density of state housing and area deprivation status [12] that highlight the key role of the house in improving living conditions of disadvantaged social groups [13].

This paper presents the most recent results of a study undertaken at the University of Auckland to assess the performance of pre-1978 low density state housing in the Auckland region with regard to occupant health and comfort, as well as energy efficiency, and to identify better retrofit strategies and practices. It aims to delineate a path to "nearly zero" energy housing—borrowing goals and definition of the recent European Directive 2010/31/EU (EPBD recast). The research focuses on the 1940 to 1960 housing cohort, one of the most suitable for effective renovation due to its pre-eminence and homogeneous composition [14]. This cohort is mainly composed of suburban detached or semi-detached houses occupied by low-income households.

This paper briefly discusses the social and cultural value of this house type. Then, it critically reviews past and current retrofit practice, highlighting the limits of its market-driven approach, and advocates future strategic national policies and initiatives for retrofitting, based on more challenging performance thresholds. The paper reports on effective ways of preserving the integrity of a typical stand-alone state house in Auckland, while improving its thermal performance to the EnerPHit Standard [15], and discusses the benefits of introducing this holistic approach into New Zealand retrofit practice.

2. Intangible Cultural Heritage Value of Mid-Century State-Supported Housing

Mid-twentieth century state-supported housing is an important part of New Zealand's cultural patrimony, reflecting the fundamental values, knowledge and techniques that characterise the transition of the country to the post-colonial era. These houses and settlements, with associated physical and social infrastructures, not only produced a significant portion of the country's urban fabric, but constituted the benchmark for the overall production of houses and urban extensions, both in terms of spatial patterns and technological solutions (Figure 1).

Today, they distinctively emerge in the contemporary city, since their material consistency and experiential dimensions express an approach to the construction of the living environment that substantially differs from those of antecedent and following periods. Beyond the peculiar morphological features that make these houses immediately recognisable (such as the 32° pitch roof and the standard windows placed just under the soffit), what marks their difference is their consistent sets of conception, representations and practices. Their paradigm still actively engages contemporary life, to some even representing the main reference and framework for both their activity and identity [16]. The numerous projects completed with state support through direct intervention (by the Housing Division of the

Ministry of Works) or financial means (through the State Advances Corporation), constitute a diffuse presence in contemporary New Zealand cities [17]. They oppose the progressive fragmentation of the current city with the integration model framed in the social welfare politic introduced in 1935 by the first Labour government. This model becomes more and more important in the current urban reorganisation towards sustainable environments for resilient communities, showing its efficacy through the vitality of its environments. Indeed, it successfully elaborates subtle articulations of fundamental integrative elements concerning the reciprocal relations between private and public spheres, urban and natural environments, senses of independence and belonging, and community and multitude. These forms of integration are testimony to the modern re-interpretation of the extraordinary effort of colonial New Zealand society to concretise the shared "happy dream of the imagination" envisioned by Edward Gibbon Wakefield in its theory of colonisation [18]. Its mid-century revision underpinned the development of a prosaic formulation of an "overall culture of family and home" [17] (p. 204); a concrete utopia deployed through a narrative structured around four key elements: family life, bucolic existence, cohesive community and diversity. On the one hand, the idea of providing the family with a retreat in the domestic idyll associated with the reconnection of urban and natural life further elaborated the garden suburb model derived from the British garden city that was introduced in New Zealand in 1919. It is worth noting that this model has become the desired norm of New Zealanders, orienting housing choices and preventing the development of denser urban forms, notwithstanding urban sprawl issues that have been recognised since the 1950s [19]. On the other hand, the support for individual paths of "emancipation through social identification" and the recognition of "originality in self-expression" led to the establishment of habitats of multiplicity where integration was to achieve through the manifestation of difference.

The construction of diversified domestic landscapes through the adoption of the traditional type of the detached home, confirmed as a fundamental hinge of people's life, was further developed by combining individuality with the latest functionalist precepts of rational layout and access to sun, air and nature. Coherent responses to specific environments, conceptions and practices were implemented with an idiosyncratic range of spatialisation forms that were modulated through an extremely high number of patterns, types and variations, and resulted in the creation of never repeated places and landscapes. The preliminary phase of the research presented in this paper explored the extent of this phenomenon through a quantitative analysis of the Auckland case. It showed how the high house type number (exceeding 12,000) corresponds to a very low average repetition (below five occurrences) of realisations for each type [20].

The rich and diverse patrimony embedded in those domestic architectures, in families and communities of their inhabitants and in their physical and cultural landscapes, is an irreplaceable legacy of a foundational milestone of the construction of New Zealand society. As such, it cannot be preserved selectively: its multiplicity and diversity represent one of its highest achievements. It is a living representation of the great modern effort to construct a concrete and effective collective model of social, cultural and physical progress. Having been recognised by Nikolaus Pevsner as "*the most ingratiating chaos one can imagine*" [21], the highest challenge for the preservation of its integrity is to develop appropriate analyses and interpretations of each and every individual condition concerning people, territories and signs.

Figure 1. Glen Innes, Auckland, 12 December 1961. Whites Aviation Ltd.: Photographs. Aerial photograph of the northern part of the Tamaki state house scheme. The pilot project house is featured in the bottom left side of the image. Source: Alexander Turnbull Library, Wellington, New Zealand. Ref: WA-56411-F. http://natlib.govt.nz/records/22856887.

3. New Zealand Housing Retrofit

According to the latest OECD assessment on people's wellbeing [22], New Zealand ranks as the second highest country for housing expenses of households: on average above one quarter of disposable income. Accounting for 16.3% of total household income, the total costs of housing increased rapidly in 2013/14 [23], worsening the problem of affordability. This particularly affects Auckland, the largest city of the country and one of the five most unaffordable major markets worldwide [24]. Without reduction of the housing shortage, this situation cannot be eased, but is expected to worsen: the growing difference between demand and supply cannot be reversed, due to the limited capacity of the construction industry [25]. With this structural scarcity and a market dominated by investors—who in 2014 won more than 45% of all sales [26]—the existing housing stock is subject to intensive exploitation, resulting in an emphasis on profitability through short term capital gains. This draws attention away from running costs and building operating efficiency, further exacerbating problems arising from poor building maintenance—especially critical for those houses already unable to guarantee comfortable and healthy living conditions.

Problems affecting current New Zealand housing stock are well documented, but there is still little information available on their specific extent. The largest source of data currently accessible is a series of House Conditions Surveys conducted on representative samples by BRANZ from 1994 to 2010.

They show that a large portion of the existing stock offers substandard living conditions, due to lack of maintenance: only 40% of the sample assessed in the 2010 survey, including new construction, was considered "well maintained," while 25% had defects to be urgently addressed [27]. The 2010 survey shows a high percentage of homes with some form of insulation, mainly in the roof cavity; but still, 52% of both new and old construction had no wall insulation and 45% of accessible suspended floors remained un-insulated. Lack of insulation appears to be a feature of most pre-1978 houses, despite different policies and programmes implemented within the framework of *National Energy Efficiency and Conservation Strategies* [28,29].

The critical condition of older houses also resulted from the limited effects on existing buildings of the New Zealand Building Code (NZBC), which includes stringent requirements for new houses insulation but offers little guidance for renovations. The provisions for energy performance and comfort of the NZBC are included in *Clause H1-Energy Efficiency* [30] and *Clause G5-Interior Environment* [31]. The latter, despite introducing a minimum internal temperature limit (16 °C) for old people's homes and early childhood centres, has no quantitative requirements for other residential buildings. Conversely, Clause H1 fixes the maximum building heating energy consumption by establishing the *Building Performance Index* (BPI), a thermal performance measure calculated as heating energy divided by the product of heating degrees total, floor area and total wall area. The recently refined BPI, though effectively requiring houses to use less heating energy than before, still allows rather low R-values for building envelopes across all New Zealand regional climates, especially inadequate in the colder areas of the country [32].

3.1. Costs and Benefits of Housing Retrofit

While general awareness about needs and benefits of improving living conditions has increased, it seldom translates into adequate energy retrofit actions. This is mainly because, as capital gain is the primary driver for home renovations, the emphasis on short-term financial return does not appropriately cope with long-term sustainability principles. The fact that New Zealand still enjoys some of the lowest power tariffs in the world often challenges the acceptance of future proof retrofitting plans. Page [33] found that basic efficiency measures, including ceiling and under-floor insulation, are considered worthwhile from a cost-benefit viewpoint in all locations in New Zealand; enhanced solutions though, such as wall insulation and double glazing, are not considered cost-effective when compared to the installation of heat-pumps and solid fuel burners, especially in the milder Auckland region. This approach, disregarding environmental concerns and predicted energy shortage, has supported current retrofitting practice and policies, which make basic interventions on the building envelope and heating devices, instead of addressing a comprehensive requalification of the older building stock. However, it does not consider the "non-energy benefits" of efficiency upgrade measures: comfort levels are not given as an "invariable" but included in the cost-benefit equation with the same value of energy savings. This economically driven approach has been criticised by the social scientist Philippa Howden-Chapman: focusing on positive effects of adequate house insulation on reduction of health inequality, she highlighted the relation between cold and damp houses, and mould growth, which is a trigger for respiratory illnesses [13]. Her findings showed how health costs of substandard living conditions are often paid by the weakest social groups. This study, considering predominantly low-income communities in New Zealand, concluded that retrofitting insulation in existing homes led to significantly warmer and drier living environments, which resulted in improved health and reduced days off school and work. She also proposed a cost-benefit analysis [34] that included health gains, energy and environmental savings, and suggested that the monetary value of improving housing quality is "compelling", as its total benefits equal to "one and a half to two times the magnitude of the cost of retrofitting insulation". She found that 61% of the total benefits were health sector gains, so her research confirms the social relevance of retrofitting and the need of filling the gap of research on cost-effectiveness of regulatory requirements to eventually guide their improvement.

3.2. Retrofit Programmes

Pre-1978 housing stock in New Zealand does not present significant typological differences between public and privately owned dwellings. Although quite varied, state houses and most private houses were mainly single or double-storey detached buildings, generally with small square or rectangular floor plans, with hipped or gabled roofs covered with tiles, asbestos-cement shingles or corrugated sheets. The envelope, largely timber-framed construction, was realised using a limited selection of cladding materials (timber weatherboard, brick veneer, stucco or asbestos-cement cladding) and no insulation [35]. These houses though, were very progressive compared to the bungalows and villas that preceded them. Instead of the villa layout, strictly designed in relation to the main street and disregarding cardinal directions, the new types were usually orientated to the north to allow maximum sun penetration into the main interior living areas in winter, and fitted with eaves for protection from the summer heat. Windows were larger to increase sunlight and fresh air, which were deemed very important to avoid the recent outbreak of tuberculosis and diphtheria: the window area was indeed sized to be 15% of the floor area of the room, with half of the windows openable for ventilation [36].

The innovative planning and robust construction of early state housing types, however, did not prevent them being cold and draughty: within a few years from construction, mould was discovered in 57% of the houses built from the early 1940s [37,38]. Since these and successive findings of research to investigate the correlation between insulation, ventilation, and healthy interiors did not substantiate in any housing improvement, most of those houses that have not been renovated still present the same issues. Indeed, compared to privately owned houses, the maintenance level of state-owned rental properties has often been lower for basic budget constraints, and included minimal upgrade interventions over the time.

After decades without maintenance, in 2001 the government's National Energy Efficiency and Conservation Strategy [28] set up the political and financial framework to support a major energy retrofit programme, providing a platform to extend funding for pre-1978 home insulation retrofits, including state housing stock, around the country. The programme aimed to upgrade two thirds of the stock with ceiling insulation, aluminium foil below suspended floors, hot water cylinder insulation and draught stoppers. Given the magnitude of the programme, the University of Otago was commissioned to monitor results of the upgrade process by measuring energy consumption and indoor thermal environment of a sample of 100 state houses in Dunedin [39]. Findings of this study seriously challenged the efficacy of the retrofit programme, revealing apparent limits of retrofitting to ceilings and floors only. In the South Island, this produced only small temperature increases that were insufficient to guarantee healthy indoor environments. Budget constraints have downsized public upgrade programmes. These, however, should not be evaluated in terms of simple payback times. Instead, social benefits must become the real driver, as affirmed by the successful initiative started by HNZC in the Auckland region in 2000: the Healthy Housing Programme, a comprehensive scheme aimed to reduce the risk of infectious diseases related to overcrowding. Beside health and social interventions to identify families with health needs and provide social services accordingly, the scheme included housing retrofitting intervention to address cold, mould and ventilation issues [40].

In 2009, the Energy Efficiency and Conservation Authority made a further attempt to overcome the poor energy performance of a still significant part of the housing stock, launching the *Warm Up New Zealand: Heat Smart Programme*. Its scope was to subsidise costs of retrofitting insulation and installation of clean heating for houses built before 2000. Although this programme proved beneficial, in particular to health [41], the subsidies are not subject to the achievement of any specific performance level (under the assumption that any amount of insulation installed represents an appreciable improvement), so its efficacy is difficult to assess and its results are questionable. Furthermore, it has been observed that long-term effects can be challenged, considering that basic retrofit interventions may reduce future energy savings, *i.e.*, "*the potential revenues of an energy efficiency investment [as] future amendments to improve quality are very improbable because they will not pay back, thus impeding future sustainable development*" [42]. This highlights the need for a more holistic retrofitting approach, addressing wall insulation, airtightness, and ventilation.

3.3. From Common Practice to EnerPHit Standard

In recent years, the growing market pressure due to housing shortage and speculative investments in the real estate sector have made old houses in need of renovation a valuable target. In this context, interventions are usually market-driven and mainly concentrated on spatial layout improvements, with bathroom and kitchen upgrades being the most effective way to achieve short-term paybacks; there is very little concern for thermal comfort, indoor air quality and airtightness. Building consents must be lodged only when the improvements require substantial alterations of the existing envelope, and for new additions, which must comply with the criteria and energy performance requirements of the Code. Underfloor and roof insulation retrofits are exempt from building consent according to Schedule 1 of the New Zealand Building Act. However, retrofitting insulation into a wall cavity either requires a building consent or specific approval from the authority, as the impact on the existing building can be potentially problematic (e.g., for moisture accumulation, timber framing and cladding durability, fire and electrical safety). Despite that, thermal resistance of retrofitted walls does not need to comply with the Building Code *Clause H1-Energy Efficiency* [30]. This set of rules accepts that the partial and cosmetic retrofit interventions mentioned above are executed without considering housing performance and indoor comfort matters [43].

Even the most forward-thinking examples of retrofitting simply aim at improving the energy performance of buildings to match minimum requirements of the current NZBC, which do not guarantee acceptable indoor air quality and comfort. Indeed, concerning current Code requirements, a recent study undertaken at the University of Auckland [44] focusing on New Zealand houses built after 2000, showed that thermal comfort in Code-compliant houses is generally below the acceptable threshold defined by international standards [45]. Measured air-leakage rates appeared high compared to both European houses and international guidelines, and yet the resulting indoor air quality was relatively poor. This is also consequent to the fact that neither the NZBC nor the New Zealand ventilation standards set any airtightness target for residential buildings that rely on natural ventilation. NZBC *Clause H1-Energy Efficiency* states that the airtightness of the building envelope has to be taken into account, but does not define any specific threshold. *Clause G4-Ventilation* [46] only requires openable areas (window and doors) to be more than 5% of the floor area for air quality and summer cooling, relying on the New Zealand tradition of satisfying ventilation needs with background infiltration, *i.e.*, uncontrolled access of outdoor air into the building through leaks in the envelope. Without any Code requirement, retrofitting practice is not compelled to address airtightness, especially given the additional cost of achieving acceptable performance using conventional construction methods. However, this exposes existing houses to new risks, as shown by recent research on the effects of retrofitting insulation according to common practice (*i.e.*, without a ventilated air cavity or a vapour control layer), which confirmed the possibility of formation of interstitial condensation and related moisture problems [47]. This not only suggests that the Building Code is in need of more detailed identification of benchmarks to promote healthy and energy efficient houses, but also that retrofitting practice must set higher goals to be effective.

The depicted scenario of retrofitting practice demands a new set of upgrade strategies aiming to more ambitious performance targets. Based on energy efficiency and comfort, the performance-based Passive House Standard [48] seems to offer effective answers to the identified issues. Developed in Germany in the early 1990s by Professors Bo Adamson and Wolfgang Feist, this is the fastest growing energy performance standard in the world, with over 30,000 buildings realised to date. Its strengths lie in the simple, non-prescriptive approach, based on the excellent thermal performance and airtightness of the building envelope. According to the Passivhaus Institut (PHI), "*a Passivhaus is a building for which thermal comfort can be achieved solely by post-heating or post-cooling of the fresh air mass, which is required to achieve sufficient indoor air quality conditions—without the need for additional recirculation of air*" [49]. A traditional heating system is no longer essential, due to reduced heating needs. The other important feature is its excellent indoor air quality, achieved by reducing the air infiltration rates and supplying filtered fresh air that can be post heated by a mechanical ventilation unit with heat recovery (MVHR).

The standard was first officially introduced to New Zealand in 2011, with the launch of Passive House Institute New Zealand (PHINZ) and the construction of the first Passive House in Auckland—certified the following year. Since then Passive Houses have spread throughout the country, especially in the North Island. In the retrofitting sector, however, the standard has been widely disregarded, due to its presumed high costs. Indeed, it can be technically and economically difficult to achieve the requirements of the standard in old residential buildings, due to unavoidable thermal bridges, constraints with insulation installation and air-leakage issues. Considering these aspects, in 2010 PHI introduced the EnerPHit standard for energy retrofits with a certification method based on listed Passive House components. These however, are available for cool-temperate climates only. To overcome this limitation, PHI started a research process to define international component requirements that can serve as the basis for international EnerPHit criteria [50]. The requirements are the result of an economic optimization process applied to a simplified set of seven climate zones around the globe.

PHI also developed an alternative certification method for building retrofit based on performance requirements. These are looser than those for new construction, with maximum space heating and cooling demand of 25 kWh/(m^{2}a) and a limit value of the envelope air permeability rate $n_{50} < 1.0\ h^{-1}$—that is expressed as the ratio between the air leakage rate at 50 Pascal air pressure and the total volume of air in cubic metres. While the airtightness limit will remain valid for all climate zones of the future international EnerPHit standard, heating and cooling demand will be tuned to the different zones: from 30–35 kWh/(m^{2}a) in cold and artic climates to 15–20 kWh/(m^{2}a) in warm and warm-temperate climates.

4. Research Methodology

This paper presents the most recent findings of a long-term study carried out at the University of Auckland on eco-retrofitting of pre-1978 state housing in Auckland. This study proposes and tests a new holistic approach to improve upon the sheer economic cost-benefit path driving the majority of common retrofit practices, which disregard social costs (in particular the ones related to health and environment issues highlighted above) and socio-cultural matters. Consistent with sustainable development principles and aiming to preserve the social, cultural and material integrity of the mid-twentieth century housing stock in New Zealand, the proposed approach sets specific criteria for its upgrade. It draws upon a comprehensive investigation of social, environmental and architectural aspects, as well as local and international retrofitting best practices. On the basis of the successful implementation in New Zealand of the Passive House standard for new construction, in the Auckland region, the research proposes to adopt the EnerPHit energy efficiency and comfort targets for retrofitting in warm-temperate climates. The applicability of the standard is tested through the development of a pilot project that includes an intervention strategy articulated on four incremental phases.

Overall, the research included the following consecutive phases: (1) Collection, systematization and analysis of documents and information, and subsequent definition of a type stock pool of pre-1978 low density state houses; (2) Selection and analysis of representative case-studies in Auckland; (3) Comparative evaluation and adaptation to New Zealand of internationally recognized diagnostic and measurement protocols, and building performance standards; (4) Identification and documentation of a pilot project and two control cases including measured drawings, stratigraphic and 3D models, diagnostic survey (building pathologies, Blower Door test, thermography, stratigraphic inspection and analysis of structural and functional inadequacies) and occupants' interviews; (5) Environmental parameters' measurement campaign of the pilot project and the two control cases; (6) Development of an intervention strategy and related retrofit packages with assessment through thermal simulation (comparing annual heating requirements of various insulation and ventilation options) for a pilot-project in Auckland.

While the results of phases 1–5 have already been published [20,51], this paper focuses on the last research phase concerning the definition and evaluation of the intervention strategy and its phased incremental retrofit packages.

4.1. Pilot Project and Measurement Campaign

The pilot project concerns a representative three-bedroom state house. It was built according to one of the most common types (6/1163) developed by the Department of Housing Construction in 1940s, as part of the large Tamaki state house scheme in the outer Auckland suburb of Glen Innes (Figure 2).

It is a single story building of 93 m^2 on sloping ground with a west-oriented living room. Although this house passed into private ownership, it has retained the original layout and features. The typical timber frame construction, enclosed by weatherboard cladding and internal plasterboard lining, has a tiled roof and a suspended (carpeted) timber floor. It sits on perimeter concrete walls that define a basement space partially accessible due to the sloping terrain (Figure 3). Originally without insulation, a few years ago the house was refurbished with the addition of polyester blankets over the ceiling and under the floor, between joists. Lying in a prominent position that benefits from good solar access, it has a well-ventilated basement space, which makes it very dry in comparison to the majority of similar residential buildings of the neighbourhood (Figure 4).

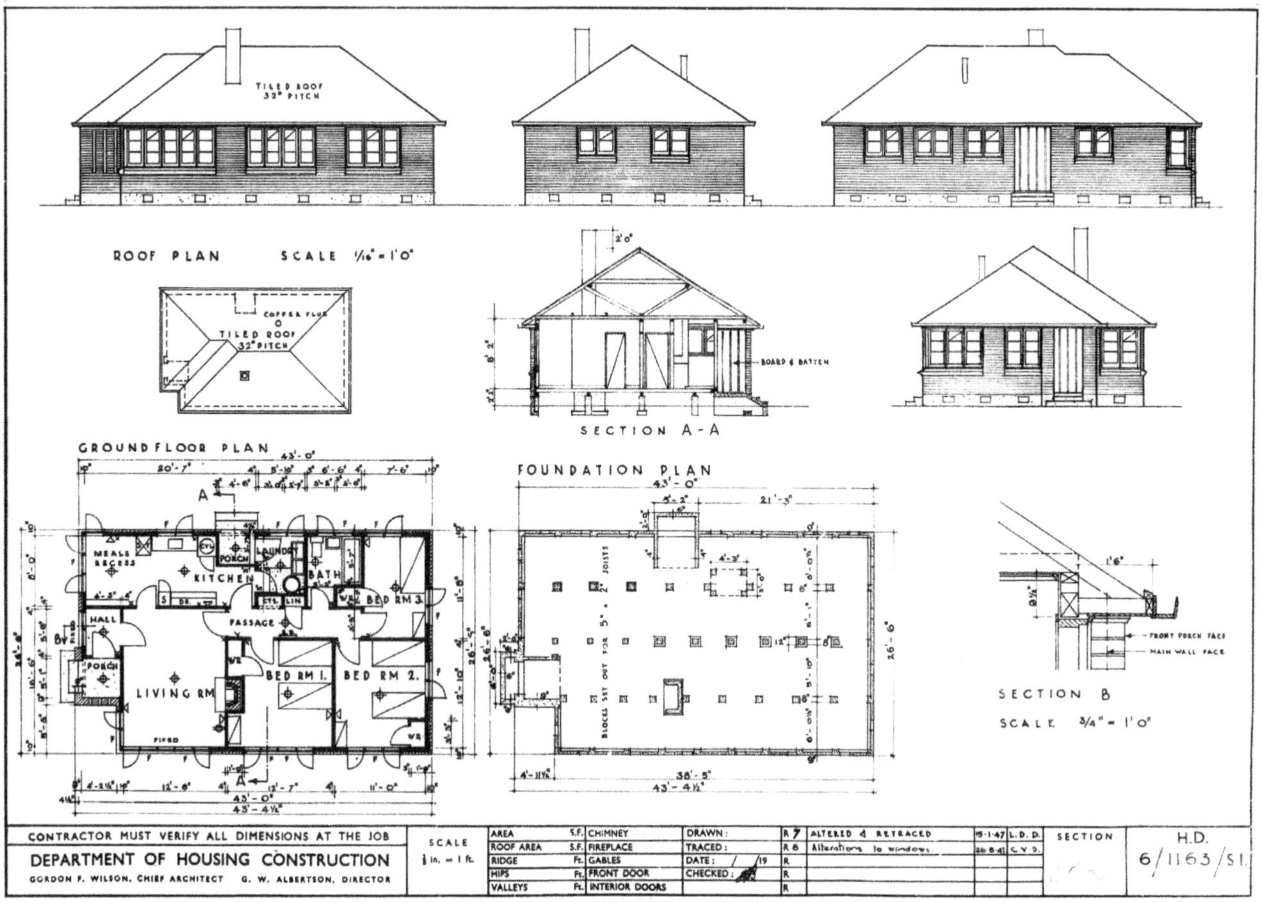

Figure 2. Original documentation of the Department of Housing Construction for the house type 6/1163 of 1947. The house selected for the pilot project is its wooden variant for sloping ground. Ref: Source: Housing New Zealand Auckland Regional Office, Floor plans and details for standard houses, naval housing, pre-cuts, demonstration flats, star flats and point flats. Ref: Q12_47, BBKD 10559 Box1 (Archives New Zealand/Te Rua Mahara o te Kāwanatanga, Auckland Regional Office).

Figure 3. South-west (street side) and north-east (backyard side) views of the pilot project house in 2012.

Figure 4. North-west facades , hipped roof cavity (timber rafters with concrete tiles and polyester insulation) and basement cavity (jack studs, sleepers and bracing struts on concrete blocks supporting floor joists with polyester insulation in-between) of the pilot project house in 2012.

In 2012, a measurement campaign of pre-retrofitting indoor and outdoor environmental parameters was carried out to provide empirical data. It was used to validate the simulation model of the various energy upgrade packages and provide a reliable basis to the quantification of potential energy savings and indoor comfort improvements. It included one year of continuous data collection (from January to December) of indoor and outdoor environmental parameters using data loggers (iButton and HOBO) for measuring air temperature, relative humidity and CO_2, and energy metering systems (smart meters). Weather conditions (air temperature, relative humidity, wind speed and direction, and global solar radiation) were recorded using hourly data provided by the National Institute of Water and Atmospheric Research.

4.2. Intervention Strategy

The strategy adopted to achieve the defined primary social and cultural objectives of the research proposes the conservation and sustainable development of peoples' habitats with particular regard to the integrity of the existing buildings and their physical environment. It includes the definition of a set of actions, framed in participatory processes both in the planning and implementation phases, to allow residents to remain in place, consolidating their communities and sense of ownership (only strictly necessary short-term relocations are foreseen). These actions aim to preserve the material patrimony and aesthetic character of the houses, including their overall building structure, spatial organisation and decorative framework.

Economic concerns have also been fundamental drivers of the strategy and its implementation through incremental retrofit packages, particularly aiming to provide flexible investment options and minimise costs of components, material, procurement and labour. The proposed phasing comprises a four-step series of coherent, effective, discrete and independent retrofit packages, respectively concerning the improvement of thermal performance and airtightness of ceiling (PH-S1), flooring (PH-S2), external walls (PH-S3), as well as the installation of new windows and mechanical heat-recovery ventilation system (PH-S4). All employed building components and construction materials are cost-effective and already available in the New Zealand market. Construction and installation techniques are simple, rapid (mainly add-on non-structural interventions) and common practice in this country.

To further sustain the cost-effectiveness and practicality of the proposed model, the packages are designed either to not require building consent (PH-S1 and PH-S2) or to ease the consenting process (PH-S3 and PH-S4) through the use of standardised components, details and installation methods that may be complemented with pre-compiled documentation. This is of particular relevance in New Zealand as it is aligned with the principles behind the peculiar regulatory framework that, for the consenting process, includes owner-builder (DIYers) exemptions from the 2004 introduced Licensed Building Practitioners Scheme [52].

4.3. Thermal Modelling and Calculation with Passive House Planning Package

The retrofit packages are designed to achieve progressive improvements of the envelope's thermal performance and airtightness up to the EnerPHit standard for warm-temperate climates (heating demand = 20 kWh/(m^{2}a) and air permeability rate $n_{50} < 1.0\ h^{-1}$). For thermal simulation, a "base case" was defined to represent the condition of the house as originally built, without insulation. Four cumulative upgrade levels were then modelled, each including all the interventions of the levels before it. In addition the house was modelled for the current situation (*i.e.*, with the insulation Basic Package, thus labelled SH-BP) and to comply with the current Building Code (SH-NZBC). Simulation models were validated using data from environmental parameter measurements. Air infiltration levels were estimated according to previous studies undertaken by BRANZ [53,54] and to match EnerPHit requirements. To meet the standard, mechanical ventilation with heat recovery (MHRV) was introduced only in the final package (PH-S4).

Preliminary estimates of heat losses and space heating demands were made for the various upgrade options using the computer thermal simulation program Virtual Environment [51]. This helped verify the technical and economic feasibility of the proposed interventions, and the magnitude of associated energy savings. On this basis, in the next phase of the study, refined retrofit packages were tested using the Passive House Planning Package (PHPP), the Excel spreadsheet-based design tool specifically developed by PHI to assist architects and designers plan and verify Passive Houses towards certification. It calculates building components' U-values, heating, cooling and primary energy demand, ventilation rates for comfort as well as the risk of overheating in the warmer season. Furthermore, it compiles climate data from many locations worldwide, including Auckland and other main centres in New Zealand. Validated with dynamic simulation tools as well as with measured data, the PHPP energy balance module has proven to be surprisingly precise, calculating the energy balances of a building to an accuracy of ±0.5 kWh.

The specific tool embedded in PHPP for calculation of *U*-values allowed a more detailed exploration of individual building component assemblies, taking into account specific construction details and locally available products. To achieve the thermal performance required by the EnerPHit standard (PH-S4), the following insulation options were selected: 190 mm polyester insulation in the attic space between and above ceiling joists (*R*-value = 4.55 m^2K/W); 125 mm glasswool insulation under the floor between joists and 45 mm EPS between bearers (*R*-value = 4.18 m^2K/W); 100 mm glasswool insulation between studs plus 45 mm of the same material between battens in the service cavity (*R*-value = 3.20 m^2K/W); low-e, double-glazing timber windows filled with argon gas (*R*-value = 0.73 m^2K/W) (Table 1).

Table 1. Upgrade levels, including a summary of *R*-values (m^2K/W) used for each element of the building fabric for PHPP calculation.

Upgrade	Insulation	Ceiling m^2K/W	Floor m^2K/W	Walls m^2K/W	Window m^2K/W	Airtightness (n_{50}) h^{-1}
Base Case	House as built (no insulation)	0.27	0.74	0.60	0.22	20
PH-S1	Ceiling insulation 100 + 90 mm polyester between and above joists	4.55	0.74	0.60	0.22	20
PH-S2	As PH-S1 + Floor insulation 125 mm glasswool between joists + 40 EPS between bearers	4.55	4.18	0.60	0.22	20
PH-S3	As PH-S2 + Wall insulation 145 mm glasswool between studs and battens	4.55	4.18	3.20	0.22	6.5
PH-S4	As PH-S3 + Windows upgrade and airtightness low-e, argon filled double glazing timber windows	4.55	4.18	3.20	0.73	1
SH-BP	SH insulated with basic package (as-it-is) 50 mm polyester under floor +150 mm above ceiling	2.42	1.73	0.60	0.22	20
SH-NZBC	House to comply with NZBC	2.90	1.30	1.90	0.26	6.5

For this study the new PHPP 9 was used, which allows the direct comparison of different variants, together with their economic evaluation. Additional retrofit options were then tested to verify the specific impact of airtightness, mechanical *versus* natural ventilation and window performance. The aim of this investigation was to assess under what insulation, ventilation and airtightness conditions the EnerPHit requirements could be met, thus, exploring alternative and affordable solutions to overcome constraints imposed by the existing construction.

5. Results and Discussion

Overall results of PHPP calculation confirm the impact on energy savings of different insulation (ceiling, floor, walls and windows) and airtightness improvements that was found with VE thermal modelling—assuming constant indoor thermal comfort conditions associated with air temperature equal to 20 °C. However, the heating demand of the various options tested is higher, due to the different nature of the two tools and slightly different boundaries conditions (e.g., climate data, treated floor area and ventilation schedule). Furthermore, PHPP provides more detailed calculations of heat losses, including the effect of thermal bridges, natural ventilation and solar gains according to surrounding shading elements.

Figure 5 shows specific annual heating demand [kWh/(m^{2}a)] and overheating risk [% of time] for each of the upgrade scenarios calculated with PHPP. The graph illustrates how improvements from the uninsulated house (Base Case) to one that complies with NZBC reduce heating demand by 74%. This result however can be used as a reference only, because old, draughty houses are not usually evenly heated to reach comfortable indoor temperatures—while new Code-compliant houses should be. A more significant result is that, by improving insulation of ceiling, floor and external walls in the path to meet the EnerPHit standard (PH-S3) [50], annual heating demand drops below that of the Code-compliant house SH-NZBC (78 *versus* 81.4 kWh/(m^{2}a))—using for calculation the same airtightness (n_{50}) of 6.5 h^{-1}, the average value for new construction according to BRANZ [54].

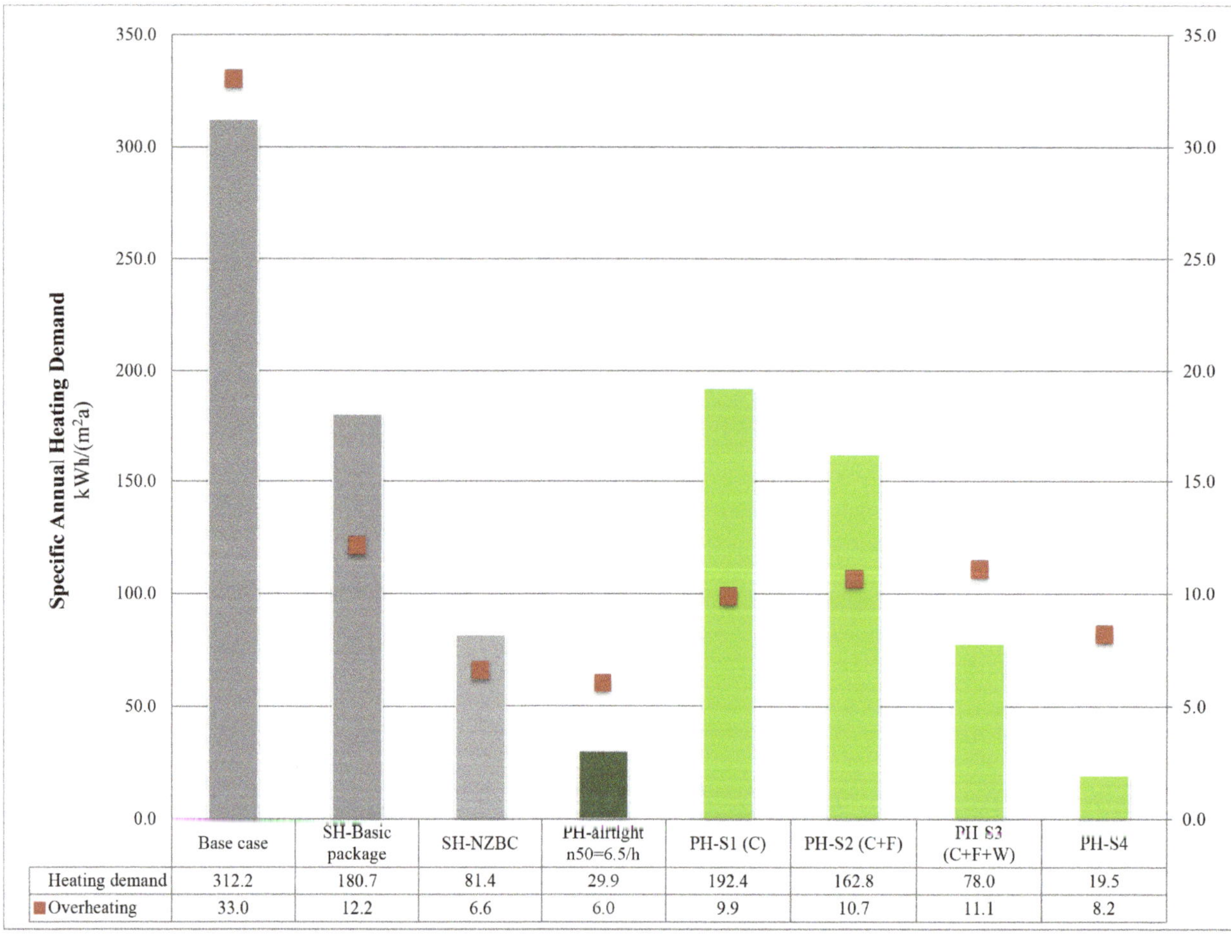

	Base case	SH-Basic package	SH-NZBC	PH-airtight n50=6.5/h	PH-S1 (C)	PH-S2 (C+F)	PH S3 (C+F+W)	PH-S4
Heating demand	312.2	180.7	81.4	29.9	192.4	162.8	78.0	19.5
■Overheating	33.0	12.2	6.6	6.0	9.9	10.7	11.1	8.2

Figure 5. Comparison of specific annual heating demands (kWh/(m^{2}a)) and overheating time (%) of the base case and different retrofitting packages.

The successive retrofit package PH-S4, with additional interventions to achieve the EnerPHit standard, is essential for reducing heating demand, enhancing indoor comfort and moisture control. New windows (R-value = 0.73 m^2K/W) and overall air permeability rate n_{50} equal to 1.0 h^{-1}, reduce heating demand further, to one quarter (from 78 to 19.5 kWh/(m^2a)), while assuring a constant comfortable indoor temperature of 20 °C. Compared to the Code-compliant house (SH-NZBC), this package produces a marginal increase in overheating risk (due to improved airtightness and reduced heat losses through the envelope), although still below the 10% limit. The residual discomfort issues could be easily avoided by employing more effective window shading (such as external shutters or blinds) or selective glazing.

Other relevant results emerge from the evaluation of the initial packages' implementation: the first step (PH-S1) confirms the importance of ceiling insulation not only in terms of energy savings (with 38% heat demand reduction) but also for reducing the risk of overheating. The additional floor insulation in step 2 (PH-S2) has a lower impact on heating demand (15% reduction) and, decreasing heat loss through the floor in summer, slightly increases the overheating risk. The amount of insulation installed under the floor is excessive for achieving comfortable surface temperatures, but is indispensable to attain the envelope's performance required to meet the standard, as it compensates for the limited amount of insulation that can be accommodated in the wall construction.

Overall, results also prove the positive impact on heating demand reduction of two factors often neglected by the common retrofit practice in New Zealand: wall insulation and air infiltration. This becomes evident through the analysis of outward and inward flows (losses and gains) through the envelope, shown in Figures 6 and 7, which provide details of benefits from each intervention. The comparison of the heat flow break-down of the existing house (SH-BP) and that meeting the NZBC requirements (SH-NZBC) shows a significant reduction of heating demand due to new windows, improved airtightness (n_{50} drops from 20 to 6.5 h^{-1}) and wall insulation; the latter proves to have the greatest impact (54% of the reduction).

U-values input in the calculation can be achieved using different building components and mechanical ventilation solutions available in New Zealand, taking into consideration technologies compatible with the country's construction practice; results show that the heating demand requirements of the EnerPHit standard can theoretically be achieved. However, given the extremely low energy consumption involved, performance could be affected by thermal bridging and airtightness issues. Further research is required to develop thermal bridge-free construction details and effective solutions for a continuous airtight envelope. Indeed, while insulation can easily be installed in existing houses to guarantee the necessary thermal performance of the envelope, achieving levels of airtightness required by the EnerPHit standard can be challenging, due to construction and preservation issues. Deep retrofit interventions, including floor and ceiling replacement as well as new internal partitions, offer the best opportunity to realise a continuous airtight layer inside the timber structure of the existing envelope. Airtight layers (oriented strand board, plywood boards or vapour control membranes) can be installed in the flooring between joists and finishing, in the external wall between the existing timber structure and the new service cavity (to be added inside for accommodating additional insulation and reducing thermal bridges), and in the ceiling below the existing joists. All joints of floor, wall and ceiling airtight layers must be tightly sealed. Achieving the required airtightness levels can be challenging in retrofit interventions where existing floor, ceiling, internal load bearing walls and partitions are maintained in place, as the additional airtight layer added on top of existing floor, wall and ceiling surfaces is interrupted by the structure of each internal partition, and so is difficult to seal. Airtightness though, is key to reducing heating demand, as PHPP calculations show that, in the final package (PH-S4) shifting n_{50} from 1 h^{-1} (EnerPHit limit) to 8 h^{-1} (value estimated for contemporary retrofitted Auckland houses), results in 60% heating demand increase. As shown in Figure 7, even increasing n_{50} just to 6.5 h^{-1} (PH-S4*), which is the same airtightness of SH-NZBC, increases specific ventilation losses from 5.1 to 19.7 kWh/(m^{2}a)—higher than the specific annual heating demand of PH-S4—and heating demand to 29.9 kWh/(m^{2}a).

Experimental research is required to evaluate the achievability of the desired infiltration rate in either intervention (also considering cost issues) since there are no examples in New Zealand of retrofitted houses to EnerPHit standard. This should also include studies of technological and economic feasibility to identify the achievable airtight levels and, if lower than required by the standard, to evaluate how to compensate for the increased heating demand with either additional insulation or higher performance windows (without compromising indoor comfort and building integrity).

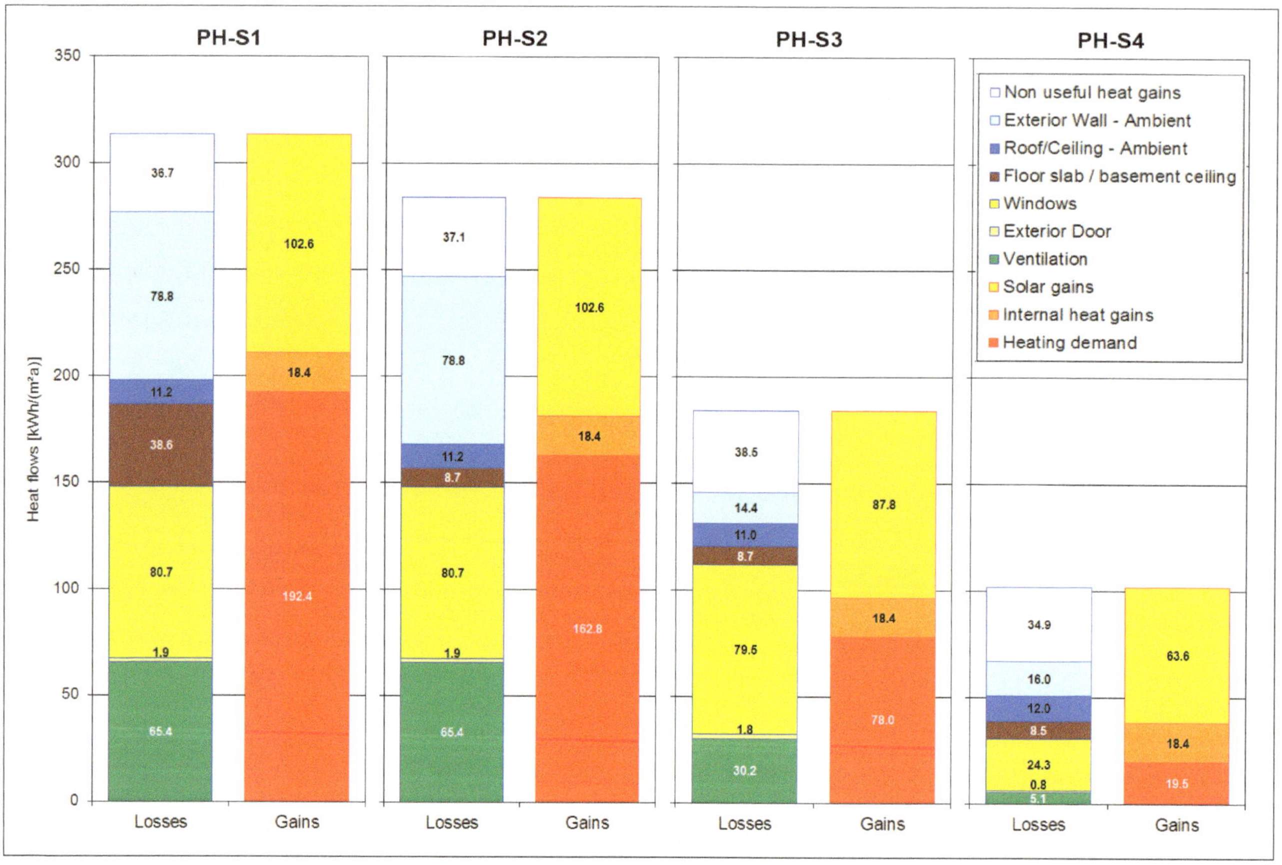

Figure 6. Comparison of heat flows (losses and gains) of the four incremental retrofit packages to meet the EnerPHit standard calculated with PHPP.

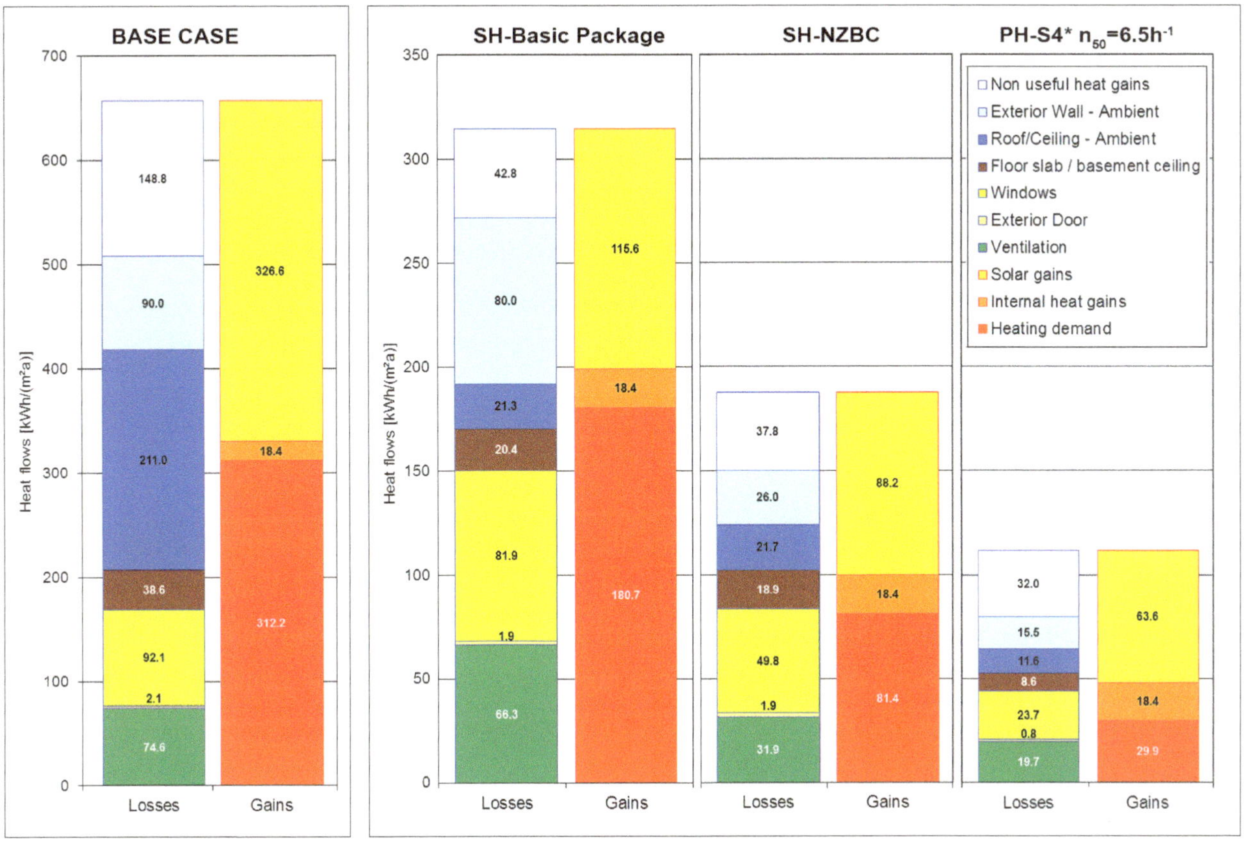

Figure 7. Heat flows (losses and gains) of the Base Case (axis maximum value: 700 kWh/(m^{2}a)) and comparison of heat flows (losses and gains) of SH-BP, SH-NZBC and PH-S4* (axis maximum value: 350 kWh/(m^{2}a)) calculated with PHPP.

The comparative analysis of different ventilation systems showed the effect of MHRV on heat demand and summer comfort. The impact of airtightness and ventilation is evident when comparing the heat flow breakdown of PH-S3 and PH-S4 (Figure 6), where the installation of MHRV and new double-glazed timber windows halves ventilation losses. In winter, a balanced MHRV proves to be key to maintaining heating demand below the limit of 20 kWh/(m^{2}a), as with natural ventilation it would increase from 19.5 to 28.1 kWh/(m^{2}a). Summer overheating was identified as a potential issue for the existing house, which has its largest windows exposed to the western afternoon sun. PHPP calculations show that heat recovery only provides a small reduction of the overheating percentage. Thus, overall energy consumption could be reduced (and the New Zealanders' propensity for indoor/outdoor flow favoured) by coupling winter use of the balanced heat recovery ventilation system with increased (when compared to the actual schedule provided by the occupants) summer natural ventilation. Increasing natural ventilation in summer only reduces the percentage of overheating during the warmest months, given the minimal air temperature difference between inside and outside, but is insufficient for night cooling of lightweight building structures. To achieve better indoor comfort conditions, a suitable option would be to utilise mechanical ventilation in summer (activated by temperature or humidity sensors) without heat recovery.

Costing the upgrades and undertaking a cost-benefit analysis of the investments over a long-term period would enable a full assessment to be made for the upgrade of the national stock to EnerPHit standard. The complete retrofit intervention in the cost-benefit analysis should include a fully insulated and airtight envelope, high performance windows with external shadings to prevent overheating in summer, and supplemental space heating (normally distributed through the low-volume MHRV system). As 1940–1960s standalone houses represent an important portion of the national housing stock, results of this analysis can be extrapolated to different climate zones to estimate levels of insulation required and associated costs. Previous research [55] has provided some indicative costing for upgrading state houses to NZBC requirements in different climate zones: as Passive Houses require extra levels of insulation, airtightness and higher performance windows, this initial estimate needs to be increased. Design optimization/standardization and product availability on the New Zealand market are possible ways to overcome expected barriers to retrofitting to EnerPhit standard.

6. Conclusions

The research discussed in this paper challenges the current New Zealand retrofit practice that, driven mainly by financial return, disregards social, cultural and environmental matters lying beyond regulatory prescriptions and engenders severe problems of health, energy waste and building decay. These issues are of primary importance when dealing with pre-1978 state housing, still largely occupied by the most economically vulnerable part of the population.

In consideration of the World Health Organisation [9] suggestion that, for long-term solutions, financial support to households for fuel expenses needs to be subordinated to housing energy upgrade, this study argues that subsidies should be streamed towards achieving economically feasible nearly zero-energy buildings, to detach living conditions from the risks presented by the use of unreliable heating sources and exacerbation of financial constraints. This goal, for the existing Auckland building stock, can be achieved through retrofitting interventions that meet the EnerPHit standard.

This study uses a holistic retrofit approach to develop a retrofitting strategy based on incremental interventions that are defined, prioritised, evaluated and optimised, using a representative case-study; a 1950s state house located in Auckland. The use of thermal modelling and PHPP calculation allows assessing the impact of different factors—such as wall insulation, ventilation mode, and air infiltration—on the building energy performance.

The assessment of a basic retrofit package—only including ceiling and floor insulation (SH-BP and PH-S2)—shows that the common interventions supported by governmental subsidies are not sufficient to effectively reduce energy consumption. Through the comparative evaluation of the results of different retrofit interventions modelled with the PHPP software, the study illustrates how effective results can be achieved with comprehensive retrofit interventions that combine insulation and airtightness of the whole building envelope with controlled ventilation.

Detailed calculations with PHPP show that, given the mild climate of the region, the specific insulation requirements of the EnerPHit standard can be achieved using existing technology and construction materials available on the local market. However, airtightness represents the main challenge and further empirical research is needed to develop technologically and economically viable retrofit details for a thermal bridge-free and airtight envelope.

Overall, the approach adopted in this study is also consistent with the on-going research project EuroPHit [56], which aims to establish a certification scheme for stepwise retrofit with Passive House components. In cases of step-by-step deep retrofit, a comprehensive renovation plan, covering both present and future steps, can help ensure the integrity of the building throughout the whole process. This scheme though, and the European EnerPHit standard, were developed for application in countries with cool-temperate climates. International criteria for Passive House components and the associated EnerPHit certification, currently under development, are likely to make certification in New Zealand more challenging by decreasing heating demand limits according to warmer climates. However, as the retrofit packages proposed in the study were developed by selecting the most cost effective components, there seems to be scope for further improvements of envelope performance in order to match future requirements.

Lastly, it is important to remark that New Zealand state housing retrofit is case-specific, despite the standardised nature of these buildings, and evolutive, since the achievement of high performance-based standards, such as EnerPHit, benefits from the steady improvement of construction technology. This will be taken into account in the future phases of this research, which will shift from theory and calculation to practice: with the implementation of the identified retrofit strategy, this study aims to prove that this approach to retrofitting is not only technically feasible and economically viable, but also socially and environmentally desirable, thus encouraging essential changes in New Zealand practice.

Acknowledgments

The authors acknowledge the financial support from the University of Auckland. They also acknowledge the support of Housing New Zealand Corporation. Special thanks go to Passivhaus Institut for granting access to their databases, as well as to unpublished documents and tools.

Author Contributions

The co-authors both contributed equally to this research project and the writing and review of this article.

Conflicts of Interest

The authors declare no conflict of interest.

References

1. International Energy Agency. Available online: https://www.iea.org/aboutus/faqs/energyefficiency/ (accessed on 28 April 2014).
2. Preservation Green Lab. *The Greenest Building: Quantifying the Environmental Value of Building Reuse*; National Trust for Historic Preservation: Seattle, WA, USA, 2011.
3. Poot, J. Building activity brings many benefits for NZ. *Build Magazine* **2011**, *125*, 30–31.
4. Page, I.C.; Fung, J. *Housing Life Cycle and Sustainability. BRANZ Study Report 214*; Building Research Association New Zealand: Porirua, New Zealand, 2008.
5. Department of Building and Housing. *New Zealand Housing Report 2009/2010: Structure, Pressures and Issues*; Department of Building and Housing: Wellington, New Zealand, 2010.
6. Auckland Council. *The Draft Auckland Plan*; Auckland Council: Auckland, New Zealand, 2011.

7. Building Research Association New Zealand. *Building a Better New Zealand. Annual Review 2014*; Building Research Association New Zealand: Porirua, New Zealand, 2014.
8. Isaacs, N.; Camilleri, M.; French, L.; Pollard, A.; Saville-Smith, K.; Fraser, R.; Rossouw, P.; Jowett, J. *Energy Use in New Zealand Households: Report on the Year 10 Analysis for the Household Energy End-use Project (HEEP). BRANZ Study Report 155*; Building Research Association New Zealand: Porirua, New Zealand, 2006.
9. World Health Organization. Regional Office for Europe. *Housing, Energy and Thermal Comfort*; World Health Organization: Copenhagen, Denmark, 2007.
10. Pattemore, P.K.; Ellison-Loschmann, L.; Asher, M.I.; Barry, D.M.J.; Clayton, T.O.; Crane, J.; D'Souza, W.J.; Ellwood, P.; Ford, R.P.K.; Mackay, R.J.; *et al.* Asthma prevalence in European, Maori, and Pacific children in New Zealand: ISAAC study. *PediatrPulmonol* **2004**, *37*, 433–442.
11. Regional Growth Forum. *Auckland Regional Affordable Housing Strategy*; Auckland Regional Council: Auckland, New Zealand, 2003.
12. Olssen, A.; McDonald, H.; Grimes, A.; Stillman, S. *A State Housing Database: 1993–2009*; Motu Working Paper; Motu Economic and Public Policy Research: Wellington, New Zealand, 2010.
13. Howden-Chapman, P.L.; Carroll, P. *Housing & Health: Research, Policy and Innovation*; Steele Roberts: Wellington, New Zealand, 2004.
14. Page, I.; Ryan, V. It Takes All Types: A Typology of New Zealand Housing Stock Retrofits. In Innovation and Transformation, Proceeding of the SB10 Conference, Wellington, New Zealand, 26–28 May 2010.
15. Passive House Institute. *EnerPHit: Certification Criteria for Retrofits*; Passive House Institute: Darmstadt, Germany, 2012.
16. Schrader, B. *We Call It Home: A History of State Housing in New Zealand*; Reed: Auckland, New Zealand, 2005.
17. Ferguson, G. *Building the New Zealand Dream*; Dunmore: Palmerston North, New Zealand, 1994.
18. Wakefield, E.G. *A View of the Art of Colonization, with Present Reference to the British Empire*; J. Parker: London, UK, 1849.
19. Holden, H.C. *Summary*; Department of Internal Affairs: Wellington, New Zealand, 1959.
20. Manfredini, M.; Leardini, P. Eco-Retrofitting Analysis, Strategies and Pilot Project for State Housing in Auckland, New Zealand. In Proceedings of the Retrofit 2012 Academic Conference, Manchester, UK, 24–26 January 2012.
21. Pevsner, N. The Ingratiating Chaos. In *Pevsner on Art and Architecture: The Radio Talks*; Games, S., Ed.; Methuen: London, UK, 2002; pp. 259–264.
22. Organisation for Economic Co-Operation and Development (OECD). *How's Life? 2013: Measuring Well-Being*; OECD Publishing: Paris, France, 2013.
23. Statistics New Zealand. Household Economic Survey (Income): Year Ended June 2014. Available online: http://www.stats.govt.nz/browse_for_stats/people and communities/Households/Household EconomicSurvey_HOTPYeJun14/Commentary.aspx#housing (accessed on 10 December 2014).
24. Demographia. *10th Annual, Demographia International Housing Affordability Survey: 2014. Ratings for Metropolitan Markets*; Demographia: Belleville, IL, USA, 2013.
25. Building Research Association New Zealand. *Auckland Housing Forecasts and the Unitary Plan*; Building Research Association New Zealand: Porirua, New Zealand, 2013.

26. Ingerson, J. Where and What Are Investors Buying? *Westpac REDnews*, 6 October 2014. Available online: http://www.westpac.co.nz/rednews/property/where-and-what-are-investors-buying/ (accessed on 10 December 2014).
27. Buckett, N.R.; Marston, N.J.; Saville-Smith, K.; Jowett, J.H.; Jones, M.S. *BRANZ Study Report SR 240—Preliminary BRANZ 2010 House Condition Survey Report—Second Edition*; Building Research Association New Zealand: Porirua, New Zealand, 2011.
28. Energy Efficiency and Conservation Authority and Ministry of the Environment. *National Energy Efficiency and Conservation Strategy. Towards a Sustainable Energy Future*; Energy Efficiency and Conservation Authority and Ministry of the Environment: Wellington, New Zealand, 2001.
29. New Zealand Government. *National Energy Efficiency and Conservation Strategy. Making it Happen*; New Zealand Government: Wellington, New Zealand, 2007.
30. Department of Building and Housing. *Compliance Document for New Zealand Building Code Clause H1 Energy Efficiency*, 3rd ed.; New Zealand Government: Wellington, New Zealand, 2011.
31. Department of Building and Housing. *Compliance Document for New Zealand Building Code Clause G5 Interior Environment*; New Zealand Government: Wellington, New Zealand, 2011.
32. McChesney, I.; Cox-Smith, I.; Amitrano, L. *Thermal Insulation in New Zealand Homes: A Status Report*; Beacon Pathway Limited: Auckland, New Zealand, 2008; pp. 52–53.
33. Page, I. Cost Benefits of Housing Retrofits. In Proceeding of the SB10 Conference Innovation and Transformation, Wellington, New Zealand, 26–28 May 2010.
34. Howden-Chapman, P.L.; Chapman, R.; Viggers, H.; O'Dea, D.; Kennedy, M. Retrofitting houses with insulation: A cost-benefit analysis of a randomized community trial. *J. Epidemiol. Community Health* **2009**, *63*, 271–277.
35. Elkink, A. *Renovate 1940–1960s*; Building Research Association New Zealand: Porirua, New Zealand, 2011.
36. Firth, C. *State Housing in New Zealand*; Ministry of Work: Wellington, New Zealand, 1949.
37. Brien, R.M.; Winsome, D.R. *Investigation into Causes and Control of Moulds in State Houses*; Department of Scientific and Industrial Research: Wellington, New Zealand, 1944.
38. Marsden, E. *State House—Erection at Dominion Physical Laboratory*; Department of Scientific and Industrial Research: Wellington, New Zealand, 1946.
39. Lloyd, C.R. Fuel poverty in New Zealand. *Soc. Policy J. N. Z.* **2006**, *27*, 142–155.
40. Clinton, J.; Mahony, F.; Irvine, R.; Bullen, C.; Kearns, R. *The Healthy Housing Programme: Report of the Outcomes Evaluation*; Housing New Zealand Corporation: Wellington, New Zealand, 2006.
41. Grimes, A.; Denne, T.; Howden-Chapman, P.; Arnold, R.; Telfar-Barnard, L.; Preval, N.; Young, C. *Cost Benefit Analysis of the Warm Up New Zealand: Heat Smart Programme*; Ministry of Economic Development: Wellington, New Zealand, 2012.
42. Ebel, W.; Feist, W.; Kaufmann, B. Economy and Financing of Efficiency: New Buildings, Renovation and Step by Step Renovation. In Proceedings of the 18th International Passive House Conference 2014, Aachen, Germany, 25–26 April 2014; Passive House Institute: Darmstadt, Germany, 2014; pp. 145–154.
43. Guidance on Building Code Compliance for Retrofitting Insulation in External Walls. Available online: http://www.dbh.govt.nz/retrofitting-insulation-guidance#fid20 (accessed on 10 December 2014).

44. Leardini, P.; Rosemeier, K.; Ong, A. Ventilation's Pivotal Role for Indoor Air Quality of Houses in New Zealand. In Proceedings of the Healthy Buildings 2012, Brisbane, Australia, 8–12 July 2012.
45. American Society of Heating, Refrigerating, and Air-Conditioning Engineers. *ASHRAE Standard 55–2010: Thermal Environmental Conditions for Human Occupancy*; American Society of Heating, Refrigerating, and Air-Conditioning Engineers: Atlanta, GA, USA, 2010.
46. Ministry of Business, Innovation and Employment. *Acceptable Solutions and Verification Methods for New Zealand Building Code Clause G4 Ventilation*, 3rd ed.; New Zealand Government: Wellington, New Zealand, 2014.
47. Leardini, P.; de Groot, H. Indoor Air Quality and Health in New Zealand's Traditional Homes. In Proceedings of the 44th Annual Conference of the Australian and NZ Architectural Science Association, Auckland, New Zealand, 24–26 November 2010.
48. Passive House Institute. *Certified Passive House—Certification Criteria for Residential Passive House Buildings*; Passive House Institute: Darmstadt, Germany, 2013.
49. The Passive House—Definition. Available online: http://passipedia.passiv.de/passipedia_en/basics/the_passive_house_-_definition (accessed on 14 April 2014).
50. Bastian, Z. International EnerPHit Certification Criteria for Energy Retrofit with Passive House Components. In Proceedings of the 18th International Passive House Conference 2014, Aachen, Germany, 25–26 April 2014; Passive House Institute: Darmstadt, Germany, 2014; pp. 615–624.
51. Leardini, P.M.; Manfredini, M.; Callau, M. Energy Upgrade to Passive House Standard for Historic Public Housing in New Zealand. In Proceedings of the 49th AICARR International Conference, Rome, Italy, 26–28 February 2014.
52. *The Building Act 2004*; Brookers: Wellington, New Zealand, 2004.
53. Bassett, M.R. Building Site Measurements for Predicting Air Infiltration Rates. In *ASTM Special Technical Publication 904*; ASTM: Philadelphia, PA, USA, 1986; pp. 365–383.
54. McNeil, S.; Quaglia, L.; Bassett, M.; Cunningham, M.; Overton, G.; Plagmann, M.; Boulic, M.; Leardini, P.; Rosemeier, K.; Smith, I. Trends in Airtightness of New Zealand Homes. In Proceedings of the 19th International Congress of Biometeorology Conference, Auckland, New Zealand, 5–9 December 2011.
55. Lloyd, C.R.; Callau, M.F.; Bishop, T.; Smith, I.J. The efficacy of an energy efficient upgrade program in New Zealand. *Energy and Buildings* **2008**, *40*, 1228–1239.
56. EuroPHit. Available online: http://europhit.eu/project (accessed on 10 December 2014).

8

Marginal Productivity Gained Through Prefabrication: Case Studies of Building Projects in Auckland

Wajiha Shahzad *, Jasper Mbachu and Niluka Domingo

School of Engineering and Advanced Technology, Massey University, P/Bag 102904, Auckland 0745, New Zealand; E-Mails: j.i.mbachu@massey.ac.nz (J.M.); n.d.domingo@massey.ac.nz (N.D.)

* Author to whom correspondence should be addressed; E-Mail: w.m.shahzad@massey.ac.nz

Academic Editor: George Baird

Abstract: Several studies have documented benefits of prefabrication system (prefab) compared to the traditional building system (TBS). However, the documented benefits have been anecdotal or fragmented with reports of isolated case study projects. Few studies have looked at the objectively quantified benefits from statistical significance point of view and across building types in New Zealand. This study contributes to filling this knowledge gap by analyzing cost and time-savings, and productivity improvement achievable by the use of prefab in place of the TBS. Records of completion times and final contract values of 66 building projects implemented using prefab in Auckland were collected. The building types included commercial, houses, apartments, educational, and community buildings. The project details included final contract sums, completion dates, gross floor areas, and number of floors. Based on these details, the equivalent completion times and the final cost estimates for similar buildings implemented using the TBS were obtained from the Rawlinsons construction data handbook and feedback from some designers and contractors. Marginal productivity outcome for each building project was computed as the product of the cost and time-savings achieved using the prefab. Results showed that the use of prefab in place of TBS resulted in 34% and 19% average reductions in the completion times and costs, respectively. This also translated to overall 7% average improvement in the productivity outcomes in the building projects. Univariate ANOVA-based hypothesis test results showed that 'building type' had no significant effects on the cost and productivity improvement outcomes, but had significant effect on the time savings analyzed in the case study projects. The greatest productivity gain of 11% was achieved in house projects. These evidence-based results could guide optimized use of prefab for specific building application. The hypothesis-testing outcome provides insights on one of several potential influences on prefab improvements, which will be analyzed in subsequent research.

Keywords: building systems; efficiency; off-site manufacturing; prefabrication; productivity

1. Introduction

Offsite manufacture or prefabrication of building components and systems has been acknowledged globally by many industry-driven commissions of enquiry as an effective solution to the several problems faced by the construction industry [1]. The technique readily supports standardization and rapid prototyping (including 3D printing/additive manufacturing technologies), which are expected to re-engineer the future of the construction industry [2].

Numerous benefits that prefab offers compared to the traditional building system (TBS) have been documented in several studies. These included better quality of work [3,4], better environmental performance and safety records [5,6], significant cost and time savings [2,7], and enhanced productivity and efficiency [4]. However, the benefits have been anecdotal or fragmented, and involved reports of isolated case studies [5,8]. Few studies have looked at the evidence-based and quantified benefits from statistical significance point of view and across building types. The low level of industry uptake of prefab hinges in part on the lack of quantifiable evidence to support claims about the numerous benefits credited to the technology [1]. This research contributes to filling this knowledge gap by seeking and analyzing hard data on the marginal benefits of the technology.

1.1. Prefabrication in Context

The Modular Building Institute [9] defines "prefabrication" as the process of manufacturing and assembling the major building components at remote offsite locations for their subsequent installation on construction site. Operationally, prefabrication is a construction innovation, which aims to take as much as possible the construction activities away from the project site to the factory to ensure better quality and safer production under controlled working conditions. This construction approach is seen as being more environmentally friendly, safer and productive than the conventional stick-built approach [10,11]. Shahzad and Mbachu [1] argue that there is increasingly little differentiation between the conventional building types and the "componentized" and the "panelized" prefabrication types. This is because conventional buildings involve some form of "componentized" and "panelized" prefab units or the other. In this context, the differentiation is made by checking the value or proportion of the prefabrication components compared to the onsite manufactured components. On this basis, a building is classified as a prefab building where the prefab component is more than 50% of the total building value, or *vice versa*.

1.2. Benefits of Prefabrication Compared to Conventional Construction Methods

Several studies have explored benefits of prefabrication technology. These benefits included reduced project cost, shorter project duration and on time delivery of projects [12], enhanced quality of construction [12,13] improved onsite health and safety [14], reduced onsite wastage and environmental impact, and reduced whole life cycle costs [1]. Lu [14] observes that prefab does not only save construction costs but also offers more reliable estimates of the upfront costs, total investment outlay and overall returns on investments. Other advantages of the modular prefabrication over and above conventional construction methods include better compliance with the Building Codes, quicker processing of the building consents/permits, and fewer building inspections [1]. In addition, Modular Building Institute [9] maintains that prefabrication optimizes the use of construction materials, resulting in less amount of waste generation.

1.3. Issues with Prefabrication System of Construction

Despite the benefits of prefabrication, there is still low-uptake of the technology in many countries [1]. Several studies explored the issues with the prefabrication technology in general. Some of the issues relate to the need for expensive haulage and cranage for handling large prefab components [9,13], the long lead time required for ordering and supply of prefab components, and the lack of flexibility to permit bespoke designs at the design phase [15].

From a socio-cultural perspective, Bell [3] notes the misperceptions about the technology based on cultural issues and social stigma attached to the technology due to negative experiences in the past about the quality of prefab buildings, especially, during the post-world war reconstruction era, as well as the temporary nature and poor aesthetics of prefab buildings. Shahzad and Mbachu [1] note the building owner's penchant for bespoke designs which allows them to make changes to suite lifestyle preferences throughout the design and initial construction stages. The conventional building approach offers this flexibility to a large extent and also allows room for more proactive change management, whereas the prefab approach usually limits the extent of the owner's changes to the standard designs. If significant changes are made, especially at the construction phase, the outcomes for the prefab technology in terms of costs, speed and wastage will be less desirable when compared to the corresponding outcomes for the conventional system. Other issues include the logistic challenge (especially for projects located in traffic congested areas), onsite connection or interface problems, and the reality that construction-phase changes are bound to be made due to the variability of site conditions from the initial design assumptions. In these circumstances, the conventional building method proves to be more suitable than the prefab system.

In sum, prefab is still globally viewed as the way of the future in the construction industry. Its benefits certainly outweigh its shortcomings. The question is, why is it then that the technology suffers low industry-wide uptake? Part of the answer to this problem could be lack of empirical evidence with which to support the numerous benefits credited to the technology. To enable a choice to be made between prefab and conventional building systems, owners would want credible and quantified evidence of the marginal benefits the technology offers. Current evidence is still anecdotal or is not robust enough due to hasty conclusions being made on isolated case study results.

1.4. Research Objectives

The main objectives of this study were: (1) To quantify the marginal benefits of the prefab building system in terms of the cost, time and productivity improvements it can offer over and above the corresponding benefits achievable with the TBS; (2) To examine how any observed benefits may be affected by the "building type" as a fixed independent factor.

1.5. Scope and Limitations

The study is limited to the historical records of final contract sums and durations of 66 building projects completed in Auckland. The building types covered included commercial, houses, apartments, educational, and community buildings. The need to limit the project locations to Auckland is to minimize any bias which location as a fixed independent factor could have on the research outcomes. Case study building projects investigated under each building category comprised 33 houses (50%), 16 commercial (24%), 7 educational (11%), 5 community (8%), and 5 apartment (8%) buildings, totalling 66 projects. Majority of the case study building projects were houses.

Proportions of the prefab systems used in the 66 case study projects were as follows: Componentized and panelized (CP) 30 (45%), hybrid (modular and CP) 20 (30%), modular 13 (20%), and whole house 3 (5%). The CP form of prefab therefore constituted the greatest proportion of the projects investigated.

2. Research Methods

Historical cost information on some completed projects provided empirical data for the study. A mixed method of archival research and case study was therefore adopted as the appropriate research method as it permitted the extraction of information from archived records [16]. This method also aligns with the research objective of obtaining multiple sources of evidence from different cases (*i.e.*, building types) [17].

In addition, Cooper and Emory [18] recommend the use of case study research method where data samples are chosen for relevance to the breadth of the issue under investigation rather than on the basis of how well they represent the target population. Investigations were focused on exploring the potential marginal time and cost savings that could be achieved by the use of prefab system over and above the outcomes for conventional building system. With no pre-defined sampling frame for the study, final contract sums and completion times of the prefab buildings covered in the study were obtained through contacts with contracting and consulting firms in Auckland. In total archived records of 66 prefab building projects were examined. For the purpose of comparative analysis, the key historical project information required for each prefab building type was mainly the building characteristics that determined the price information in the Rawlinsons Construction Handbook [19]. Information for each of the five building types included cost at completion, duration, gross floor area (GFA) and number of floors. The need to adjust the effects that differing regimes of inflation and exchange rates had on the historical cost information was obviated by the use of percentage differences as the common denominator for comparison.

To obtain the empirical data, purposive sampling method [20] was used, since there were no databases from which to sample the projects. Through the assistance of the Registered Master Builders

Federation of New Zealand, contractors who were willing to release records of their past projects for the purposes of the research were contacted. They were promised anonymity, assuring that the data would be used solely for academic research purposes with no revealing details about their projects, their companies or their clients. The efforts in this direction yielded 66 prefab project records. The equivalent completion times and the final cost estimates for similar buildings implemented using the TBS were obtained from the Rawlinsons construction data handbook as well as feedback from some designers and contractors.

2.1. Method of Data Analysis

The data required for the first objective were the final costs and completion times of prefab buildings under each building category. The data analysis involved computing for the following parameters for the *j*th project within a set for a particular building type *i*:

2.1.1. Marginal Cost Saving (MCS_{ij})

This was computed as the difference between the final cost of the prefab building and the corresponding cost for a similar building erected using the traditional system, expressed as a percentage of the latter. Equation (1) provides the expression for the MCS_{ij}:

$$MCS_{ij} = \left(\frac{C_{TRADij} - C_{PREFABij}}{C_{TRADij}} \right) \quad (1)$$

where MCS_{ij} is marginal cost saving (MCS) achieved in the *j*th project within the set of buildings for the particular building type *i*; $C_{PREFABij}$ is final cost of the *j*th prefab building; C_{TRADij} is corresponding final cost of a similar building completed using the traditional building system.

The rationale for computing the productivity improvement as a product of marginal time and cost savings draws upon two streams of thoughts:

First, in the construction industry context, productivity performance depends largely on the cost and schedule performance [21]. This strategic perspective of the concept of productivity differs to some extent from the economist's perspective of productivity which is basely solely on output *versus* input resource ratio, featuring variants, such as labour, capital and multi-factor productivity measures. Mbachu and Shahzad [7] clearly made this distinction. The mathematical expression for an integrated productivity measurement based on the two key parameters of cost and time savings draws from the fact that productivity is directly proportional to the cost and schedule performance, *i.e.*:

$$P \propto (S_p, C_P) \quad (2)$$

$$P = \beta(S_P \times C_P) \quad (3)$$

where P is the productivity performance achieved in a project; β is an empirically determinable constant of proportionality that depends on the dynamics of the operational environment. The constant could be taken as unity (*i.e.*, value of 1) for projects executed under normal operating conditions as assumed in the study; S_p is the schedule performance computed using a modified form of Equation (1); C_p is the cost performance computed using Equation (1).

2.1.2. Average Marginal Cost Saving ($AMCS_i$)

This was computed as the average of the MCS for all the case study projects for a particular building type. Equation (4) provides the expression for evaluating the average marginal cost saving ($AMCS_i$):

$$\mathrm{AMCS}_i = \left(\frac{\sum_{j=1}^{n} \mathrm{MCS}_{ij}}{n} \right) \tag{4}$$

where $AMCS_i$ is average marginal cost savings achieved in all the n case study buildings for the building type i.

2.1.3. Marginal Time Saving (MTS_{ij}) and Average Marginal Time Saving ($AMTS_i$)

These were computed as for the MCS_{ij} and $AMCS_i$ in Equations (1) and (4) above, respectively.

2.1.4. Productivity Improvement (PI_i)

This was computed on two levels: At the level of the individual building types, it was computed as the product of the marginal cost and time savings achieved in each case study project j within the set for the building type i, *i.e.*:

$$\mathrm{PI}_{ij} = \mathrm{MCS}_{ij} \times \mathrm{MTS}_{ij} \tag{5}$$

where PI_{ij} is productivity improvement achieved in the jth project for the building type i; MCS_{ij} is marginal cost saving achieved in the jth project for the building type i; MTS_{ij} is marginal time savings achieved in the jth project for the building type i.

At the level of the building type i, the productivity improvement was computed as the product of the average cost and time savings achieved in the case study projects for a particular building type i:

$$\mathrm{PI}_i = \mathrm{AMCS}_i \times \mathrm{AMTS}_i \tag{6}$$

where PI_i is productivity improvement achieved in all the n case study buildings for the building type i; $AMCS_i$ is average marginal cost savings achieved in all the n case study buildings for the building type i; $AMTS_i$ is average marginal time savings achieved in all the n case study buildings for the building type i.

2.1.5. Average Productivity Improvement (API_{av})

This was computed as the average of the productivity improvements achieved across all m building types:

$$\mathrm{API}_{av} = \left(\frac{\sum_{i=1}^{m} PI_i}{m} \right) \tag{7}$$

where API_{av} is Average Productivity Improvement achieved in all the m building types.

2.2. Hypothesis Testing

The statistical tests of significance, which informed the second objective of the study involved a null hypothesis that "building type" as a fixed independent factor did not have significant effect on any marginal benefits achieved by the use of prefab system in the case study projects. The alternative hypothesis assumed otherwise. The hypothesis testing required a single factor analysis of variance (ANOVA) given the single independent factor involved (*i.e.*, "building type") and the interval scale of the empirical data [22]. The hypothesis testing was conducted separately for the observed cost, time and productivity marginal benefits. Typical procedure for the test involving time savings is defined as follows:

H0T: There is no significant difference in the average time-savings achieved in the five building groups. If this were true, it would mean that "building type" would have no significant effect on the time-savings achieved by the use of prefab in place of conventional method.

The alternative hypothesis was formulated as follows:

HaT: The differences in the average time-savings achieved in the five building groups are significant.

Acceptance condition: Accept H0T if the *p*-value or confidence level of the single factor ANOVA test is higher than the 0.05 alpha value for the test.

Rejection condition: Reject H0T if the *p*-value is equal to or lower than the alpha value for the test; accept the alternative hypothesis instead.

The above hypothesis tests were replicated for the cost and productivity improvements.

3. Results and Discussions

Results relating to the two objectives of the study are presented and discussed in the following subsections.

3.1. Cost, Time, and Productivity Improvement Achieved by Use of Prefab

The first research objective was to quantify the marginal benefits of the prefab building system in terms of the cost, time and productivity improvement it delivers over and above the outcomes achievable with the traditional building system (TBS). Table 1 presents typical example of the marginal benefit analysis for the office/commercial building projects. This building category comprised 18 case study projects. Similar computations were done for the other 4 building categories. The results were summarized for all the buildings types and plotted in Figure 1 for better visual appreciation.

Figure 1 shows that the marginal cost saving delivered by the prefab system over the TBS was 19% on average. The highest cost saving (24%) was achieved in the community building projects. Perhaps, the relatively simpler design nature of community buildings might have contributed to this result.

On the other hand, the average marginal time saving achieved by the prefab system in the case study projects was 34%. The highest time saving (*i.e.*, 50%) was achieved in the house projects. This result could be due to majority of the houses being developed off standard plans provided by group home builders. Such houses lend readily to standard components, which make for faster manufacture and installation, thereby enhancing speed of construction. In addition, in terms of costs, this method of housing procurement has been found to be 15% cheaper than houses that were of one-off designs [23]. The overall marginal productivity improvement offered by prefab system was 7%, with the house projects having the highest (11%) productivity gain. This result might suggest that prefab could provide the most effective approach to delivering the 400,000 houses planned to be built over the next 3 years under the Auckland Housing Strategy [24], as well as the $30 billion Canterbury rebuild [25].

Overall, the findings of this study in relation to the prefab marginal cost, time and productivity improvements were in agreement with conclusions reached in previous studies overseas. For instance, Egan [26] argues that the use of prefab system has many benefits including speed of construction, lower cost, reduced need for skilled labour and achievement of zero defects, all of which could have significant impact (about 30%–40% improvement) on on-site productivity level. However, some New Zealand studies did not concur with overseas results on superior economic benefits of prefab system. For instance, Burgess *et al.* [27] found that prefabricated buildings did not yield better economic choice than onsite construction in a case study house investigated in their work. This can be associated to small size of the New Zealand market and its inability to exploit the scale efficiencies offered by mechanized bulk production of similar products. It was not clear whether the New Zealand results were due to inadequate sample size—with the study being based on a few case study projects.

Table 1. Prefab marginal benefits (typical analysis for commercial building projects).

Project	GFA (m^2)	Storey	Date Completed	Prefab Cost ($)	* Equiv TBS Cost ($)	% Cost saving [a]	** Prefab dur (wks)	*** Equi TBS dur (wks)	% Time Saving [b]	Productivity Improvt [c]
1	5,900	6	5 October	6,000,000	14,553,000	58.77	72	84	14.29	8.40
3	1,290	2	8 August	1,500,000	3,160,500	52.54	24	52	53.85	28.29
4	2,850	2	8 April	5,500,000	6,982,500	21.23	32	64	50.00	10.62
6	5,400	1 and 2	13 August	7,000,000	10,665,000	34.36	48	72	33.33	11.45
7	4,082	1	13 November	7,000,000	8,061,950	13.17	32	66	51.52	6.79
8	1,001	1	12 August	2,000,000	3,461,000	42.21	36	50	28.00	11.82
9	9,000	1	11 March	8,000,000	14,085,000	43.20	32	84	61.90	26.74
10	150	2	9 January	160,000	189,000	15.34	4	10	60.00	9.21
11	100	1	13 March	220,000	243,000	9.47	16	18	11.11	1.05
12	10,000	8	12 October	30,000,000	35,400,000	15.25	80	100	20.00	3.05
13	2,100	2 and 3	12 March	3,200,000	3,831,100	16.47	30	52	42.31	6.97
14	10,000	4	8 July	19,000,000	20,488,000	7.26	60	90	33.33	2.42
15	23,000	6	13 February	105,000,000	112,125,000	6.35	112	130	13.85	0.88
16	5,240	4	8 June	9,600,000	12,602,200	23.82	40	76	47.37	11.28
17	2,100	1	12 October	6,500,000	6,615,000	1.74	40	60	33.33	0.58
18	1,547	5	13 September	7,800,000	7,967,050	2.10	44	64	31.25	0.66
–	–	–	–	–	Averages [d,e,f]	22.71	–	–	36.59	8.76

Notes: [a] % Cost saving = Marginal Cost Saving (MCS_{ij}); (see Equation (1)); [b] % Time saving = Marginal Time Saving (MTS_{ij}); (modified Equation (1)); [c] Productivity improvement (PI_{ij}) (see Equation (5)); [d] Average % cost saving = Average Marginal Cost Saving ($AMCS_i$); (see Equation (4)); [e] Average % time saving = Average Marginal Time Saving ($AMTS_{ij}$); (modified Equation (4)); [f] Average productivity improvement (API_{av}) (see Equation (7)). * Equivalent TBS cost in NZ$; ** Prefab duration in weeks; *** Equivalent TBS duration in weeks.

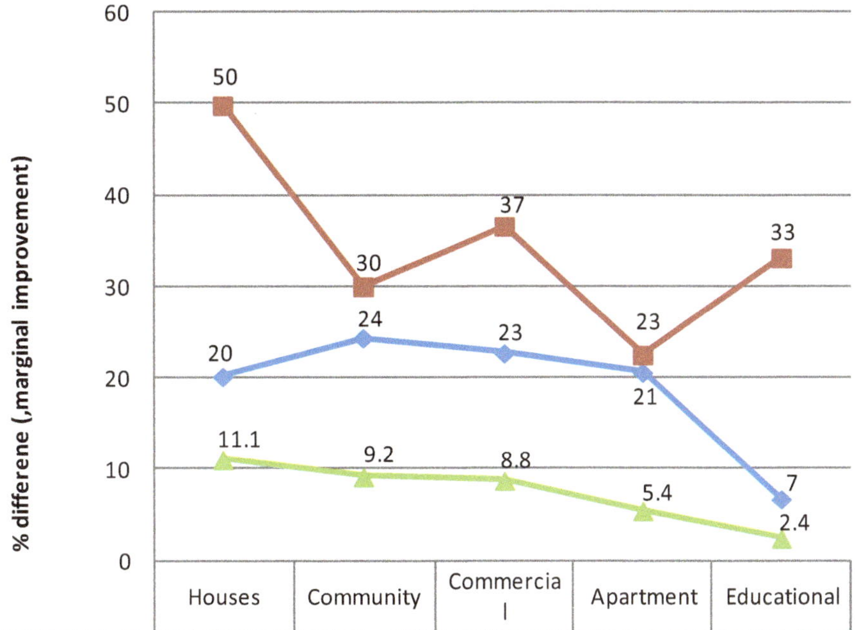

	Houses	Community	Commercial	Apartment	Educational
% Cost saving (19% avg)	20	24	23	21	7
% Time saving (34% avg)	50	30	37	23	33
% Productivity Improvement (7% avg)	11.1	9.2	8.8	5.4	2.4

Figure 1. Time, cost and productivity improvement achieved by the use of prefab in place of traditional building system in the case study projects.

3.2. Influence of Building Type on the Analyzed Prefab Marginal Benefits

The second research objective was to examine how any observed benefits offered by prefab building system could be affected by "building type" as a fixed independent factor. As discussed under the methodology section, achieving this objective required testing for the null hypothesis formulated to imply that "building type" as a fixed independent factor did not have significant effect on any marginal benefits achieved by the use of prefab system in the case study projects. Results of the single factor ANOVA employed in the test are presented in Table 2. The tests were carried out separately for prefab marginal productivity improvement, cost saving and time saving.

Table 2 shows that the *p*-value analyzed in the null hypothesis test for the productivity improvement was greater than the alpha value of 0.05 used in the test (*i.e.*, 0.09 > 0.05). Additionally, the F-ratio of 2.11 was less than the χ^2 *F*-critical value of 2.52. On account of these results, the null hypothesis could not be rejected. It was therefore concluded that "building type" had no significant influence on the marginal productivity improvement benefits delivered by the prefab system.

Similar conclusion was also reached for the marginal cost saving. However, the result was different for the time saving: A *p*-value of 0.005 which is less than the alpha value of 0.05 meant that building type impacted significantly on the observed prefab marginal time saving. This suggests that the use of prefab system may result in shorter completion times of buildings than the TBS depending on the building type, whereas the cost and productivity improvement benefits are not influenced by building type, and are likely to be consistent across all building types. Thus, claims about the superior benefits of prefab can only be generalised in respect of cost and productivity improvements across all building types, whereas claims about time savings should be cautiously applied to specific building types. There is no empirical evidence in this study to conclude that the technology is superior to the traditional building system in terms of time savings across all building types. This latter finding contrasts with conclusions reached in a number of New Zealand and overseas studies which seemed to generalize the benefits across all building types. Burgess *et al.* [27]—Citing Kaufmann and Remick [28], Atkin and Wing [29], and Bell [3]—Stated that "time savings for prefabricated construction for residential applications is between 30% and 60%" (p. 25). The qualified statement (*i.e.*, "residential applications") made by Burgess *et al.* [27] is therefore worthy of emulation when it comes to reporting on the time saving benefits of prefab system. The question that may be worthy of further investigation is why the superior time saving benefits of the technology may not be consistent across all building types. A possible clue might be found in the observations of Shahzad and Mbachu [1] that there is increasingly little differentiation between the conventional building types and the "componentized" and the "panelized" prefabrication types. This is because conventional buildings involve some form of "componentized" and "panelized" prefab units or the other. This is true for most non-residential building constructions. On the other hand, clear differentiation exists in the application of the technology to residential buildings—a clear departure from the traditional stick-built system which dominates residential construction. The large proportion of residential prefab projects (about 50%) might have influenced the hypothesis test result in this regard. However, then, why were the cost and productivity improvement benefits not also influenced by the larger proportion of residential building project samples? These are issues to be resolved in future research.

Table 2. Univariate ANOVA analysis results for hypothesis testing.

Dimension	Variation Source	[a] SS	[b] DF	[c] MS	[d] *F*-ratio	[e] *p*-value	[f] *F*-crit	Result
% productivity improvement	Between groups	505.7	4	126.4	2.11	0.09	2.52	Do not reject H0 (*p*-value > 0.05)
	Within groups	3,651.6	61	59.9	–	–	–	–
	Total	4,157.2	65	–	–	–	–	–
% cost saving	Between groups	1,441.0	4	360.3	1.50	0.21	2.52	Do not reject H0 (*p*-value > 0.05)
	Within groups	14,618.8	61	239.7	–	–	–	–
	Total	16,059.8	65	–	–	–	–	–
% time saving	Between groups	5,617.3	4	1,404.3	5.81	0.0005	2.52	Do not accept H0 (*p*-value < 0.05 a)
	Within groups	14,747.0	61	241.8	–	–	–	–
	Total	20,364.3	65	–	–	–	–	–

Notes: [a] SS (Sum of squares) = sum of squares of variations of data points from the means for the between and within group; [b] DF (degree of freedom) = sum of degrees of freedom for each group computed for each group as number of data points − 1); [c] MS (mean squares) = SS/DF; [d] *F*-ratio = χ^2 test statistic = MS (between groups)/MS (within groups); [e] *p*-value = probability value associated with the *F*-ratio; [f] *F*-crit = critical *F*-ratio value from statistics tables corresponding to the alpha level of test (*i.e.*, 0.05).

4. Conclusions

This study has quantified the marginal benefits of the prefab building system in terms of cost, time and productivity improvements it can offer over and above corresponding benefits achievable with the traditional building system (TBS). Potential influence of building types on the observed benefits was also examined. Results showed that the marginal cost saving delivered by the prefab system over the

TBS was 19% on average. The highest cost saving (24%) was achieved in community building projects. On the other hand, average marginal time saving achieved was 34%, with the highest time saving of 50% achieved in the house projects. The overall marginal productivity improvement offered by prefab system was 7%, with house projects having the highest productivity gain of 11%. This result might suggest that prefab could provide the most effective approach to delivering the 400,000 houses planned to be built under the Auckland Housing Strategy over the next three years [24], as well as the $30 billion Canterbury rebuild [25].

Results of ANOVA tests on how the observed benefits offered by prefab building system could be affected by "building type" were mixed. Whereas cost and productivity improvement benefits offered by the system were consistent across building types, time saving benefits were influenced by the building type. Empirical evidence therefore supports the generalization of the cost and time benefits across building types. This is not applicable to time savings, which require being reported only in relation to specific building types.

Influences of other fixed factors such as location, project size/complexity, procurement and contract strategies, type of prefabrication and the degree of standardization or replication are key limitations of the study which are recommended for further investigations. It should also be noted that variability issues could be associated with the limited dataset of 66 projects used in the study; these could raise some reliability concerns if the findings were to be generalized beyond the scope of this study.

Acknowledgments

This research is funded by Building Research Association of New Zealand (BRANZ) building research levy through the Postgraduate Research Scholarship for which the authors are grateful. In addition to this, the cooperation of construction companies and project managers in providing the requisite data is highly acknowledged.

Author Contributions

The research was part of a thesis undertaken by the author for correspondence Wajiha Shahzad. It was supervised by Jasper Mbachu (principal supervisor) and Niluka Domingo (assistant supervisor).

Conflicts of Interest

The authors declare no conflict of interest.

References

1. Shahzad, W.M.; Mbachu, J. *Productivity Enhancement of Construction Industry Using Prefabrication: Impact Levels of the Underlying Constraints and Improvement Measures*; Lambert Academic Publishing: Saarbrücken, Germany, 2012.
2. National Research Council. *Advancing the Competitiveness and Efficiency of the U.S. Construction Industry*; The National Academies Press: Washigton, DC, USA, 2009.

3. Bell, P. Kiwi Prefab: Prefabricated Housing in New Zealand: An Historical and Contemporary Overview with Recommendations for the Future. Master's Thesis, Victoria University of Wellington, Wellington, New Zealand, 30 June 2009.
4. Department of Building and Housing (DBH). *Report of the Building and Construction Sector Productivity Taskforce*; DBH: Wellingtion, New Zealand, 2009.
5. Cooperative Research Centre (CRC). *Off-Site Manufacture in Australia: Current State and Future Directions*; CRC: Canberra, Australia, 2007.
6. Jaillon, L.; Poon, C.S. Design issues of using prefabrication in Hong Kong building construction. *Constr. Manag. Econ.* **2010**, *28*, 1025–1042.
7. Mbachu, J.; Shahzad, W.M. A Holistic Understanding of the Concept and Measurement of Productivity in the Construction Industry: A New Zealand Perspective. In Proceedings of the WABER 2012 International Conference, Abuja, Nigeria, 24–26 July 2012; pp. 87–98.
8. Davis, N. *Construction Sector Productivity: Scoping Report*; Department of Building and Housing: Wellington, New Zealand, 2007.
9. Modular Building Institute (MBI). *Improving Construction Efficiency & Productivity with Modular Construction Charlottesville*; MBI: Charlottesville, VA, USA, 2010.
10. Arif, M.; Egbu, C. Making a case for offsite costruction in China. *Eng. Constr. Arcitectural Manag.* **2010**, *17*, 536–548.
11. Azman, M.N.A.; Ahamad, M.S.S.; MAjid, T.A.; Hanafi, M.H. The common approach in off-site construction industry. *Aust. J. Basic Appl. Sci.* **2010**, *4*, 4478–4482.
12. Lusby-Taylor, P.M.S.; Ainger, C.; Ogden, R. *Design and Modern Methods of Construction*; The Commission for Architecture and the Built Environment (CABE): London, UK, 2004.
13. Gibb, A. *Off-Site Fabrication*; Whittles Publishing: Scotland, UK, 1999.
14. Lu, N. *The Current Use of Offsite Construction Techniques in the United States Construction Industry*; Construction Research Congress: Seatle, WA, USA, 2009.
15. Bradsher, K. An Ambitious Construction Schedule Using Prefab Modules, with Questions. *N. Y. Times* 2013. Available online: http://www.nytimes.com/2013/08/28/business/global/an-ambitious-construction-schedule-using-prefab-parts.html (accessed on 6 February 2015).
16. Cooper, D.R.; Schindler, P.S. *Business Research Method*; McGraw Hill/Irwin: New York, NY, USA, 2006.
17. Mbachu, J.; Nkado, R. Conceptual framework for assessment of client needs and satisfaction in the building development process. *Constr. Manag. Econ.* **2006**, *24*, 31–44.
18. Cooper, D.R.; Emory, C.W. *Business Research Methods*; Irwin Inc.: Chicago, IL, USA, 1995.
19. Giddens, C. *Rawlinsons New Zealand Construction Handbook*, 28th ed.; Rawlinsons Limited: Auckland, New Zealand, 2014.
20. Saunders, M.; Lewis, P.; Thornhill, A. *Research Methods for Business Students*; Pitman Publishing: London, UK, 1997.
21. Takim, R.; Akintoye, A. Performance Indicators for Successful Construction Project Performance. In the Proceedings of 18th Annual ARCOM Conference, Northumbria, UK, 2–4 September 2002; Greenwood, D, Ed.; Association of Researchers in Construction Management: Edinburgh, UK, 2002; Volume 2, pp. 545–555.
22. *Statistical Product and Services Solution (SPSS)*, version 21.0; IBM: Armonk, NY, USA, 2012.

23. Building Research Association of New Zealand (BRANZ). *BRANZ Response to Ministry of Business, Innovation and Employment: Residential Construction Sector Market Study*; BRANZ: Wellington, New Zealand, 2013.
24. 40,000 Homes to Be Built Under Auckland Housing Strategy. *OneNews*, 13 June 2013. Available online: http://tvnz.co.nz/national-news/40-000-homes-built-under-auckland-housing-strategy-5433767 (accessed on 5 February 2015).
25. Ministry of Business, Innovation and Employment (MBIE). *New Zealand Sectors Report: Featured Sector—Construction*; MBIE: Wellington, New Zealand, 2013.
26. Egan, J. *Rethinking Construction*; Secretary of State for the Environment, Transport and the Regions: London, UK, 1998.
27. Burgess, J.C.; Buckett, N.R.; Page, I.C. *Prefabrication Impacts in the New Zealand Construction Industry*; Building Research Assocition of New Zealand (BRANZ): Wellington, New Zealand, 2013.
28. Kaufmann, M.; Remick, C. *PreFab Green*; Gibbs Smith: Santa Barbara, CA, USA, 2009.
29. Atkin, B.; Wing, R. Future Home—Manufactured Housing for Europe. In Proceedings of the 17th International Symposium on Automation and Robotics in Construction, Madrid, Spain, 22–24 September 1999.

9

Daylight Design of Office Buildings: Optimisation of External Solar Shadings by Using Combined Simulation Methods

Javier González and Francesco Fiorito *

Faculty of Architecture, Design and Planning, The University of Sydney, 148 City Road, Sydney 2006, Australia; E-Mail: jgonzalez.arquitectura@gmail.com

* Author to whom correspondence should be addressed; E-Mail: francesco.fiorito@sydney.edu.au

Academic Editor: Richard Mistrick

Abstract: Integrating daylight and energy performance with optimization into the design process has always been a challenge for designers. Most of the building environmental performance simulation tools require a considerable amount of time and iterations for achieving accurate results. Moreover the combination of daylight and energy performances has always been an issue, as different software packages are needed to perform detailed calculations. A simplified method to overcome both issues using recent advances in software integration is explored here. As a case study; the optimization of external shadings in a typical office space in Australia is presented. Results are compared against common solutions adopted as industry standard practices. Visual comfort and energy efficiency are analysed in an integrated approach. The DIVA (Design, Iterate, Validate and Adapt) plug-in for Rhinoceros/Grasshopper software is used as the main tool, given its ability to effectively calculate daylight metrics (using the Radiance/Daysim engine) and energy consumption (using the EnergyPlus engine). The optimization process is carried out parametrically controlling the shadings' geometries. Genetic Algorithms (GA) embedded in the evolutionary solver Galapagos are adopted in order to achieve close to optimum results by controlling iteration parameters. The optimized result, in comparison with conventional design techniques, reveals significant enhancement of comfort levels and energy efficiency. Benefits and drawbacks of the proposed strategy are then discussed.

Keywords: daylight modelling; daylighting metrics; evolutionary optimization; Genetic Algorithms (GA); external shadings; performance driven design

1. Introduction

Since the introduction of performance simulation in buildings, several software packages have been developed to cover the different needs of the building industry [1]. Although accuracy and levels of complexity vary among them, most accurate software generally require highly specialized knowledge from users. Moreover, an extended process of trial and error is needed for testing different strategies and obtaining satisfactory results. The integration of parametric design with performance simulation tools has opened new possibilities to overcome these issues by giving the user the ability of testing and comparing and selecting the best possible solution for multi-dimensional problems like the ones found in the built environment.

1.1. Building Performance Simulation: Available Combined Tools

Grasshopper©, is a graphical algorithm editor integrated with Rhinoceros© 3D modelling tools [2]. It allows the modelling of simple and complex geometries, also parametrically controlled. It has gained popularity in the recent years among architects and designers due to its versatility in shape generation. Since its creation, several scripts have been developed, aiming at the integration of simulation tools for different aspects of building performance, such as but not limited to geometry, structures, thermal and daylight performance. One example of these is DIVA, a plug-in for Grasshopper, which stands for "Design, Iterate, Validate and Adapt" [3]. This software allows for environmental analyses in Rhinoceros and its Grasshopper components. It integrates Radiance/Daysim for daylight calculations and EnergyPlus for thermal analyses [4]. Both Daysim and EnergyPlus have been extensively used and validated in research and practice, and their integration in Grasshopper allows them to be accessible to a wider range of users.

Genetic Algorithms (GA) are a particular class of evolutionary algorithms that use techniques inspired by evolutionary biology such as inheritance, mutation, selection, and crossover to evolve a solution for general or specific problems. They have been proven to be an effective strategy for addressing multi-objectives design problems and calculate multiple performance criteria, finding close to optimum solutions in a short period of time [5]. However, their application requires extensive mathematical and computer programming knowledge, far beyond the domain of most professionals. This gap has been recently filled by the creation of Galapagos by David Rutten [6]. The tool is a generic evolutionary solver component available in Grasshopper that integrates GA into a highly intuitive solver. Therefore, different optimization problems can be explored without the need for advanced computing or mathematics skills.

The combination of Galapagos with performative plug-ins could be one of the answers for ensuring high levels of environmental performance while reducing the time needed for extensive trial and error processes. Simulations can be performed automatically and the results stored and organized according to their performance. Thus, rather than setting changes for each simulation, users can set a design problem, and then, obtain the solution optimized for that problem. Alternatively, when a set of multiple good solutions—all meeting the selected screening criteria—are found, professionals can use this collected information as a basis for highly informed decisions. An increase of the environmental performance of new and existing buildings would undoubtedly be the final outcome of this process.

Examples of evolutionary solvers applied in buildings are widely available in the scientific literature. However, due to the complexity of using algorithms to solve design problems, they are generally circumscribed to engineering applications. Uses of GAs can be found for example in the optimization of Heating Ventilation and Air Conditioning (HVAC) control systems for temperature and relative humidity regulation, to deliver high levels of thermal comfort with reduced energy loads [7]. The definition of trade-off between users' comfort and expenditures (both capital and annual operational) is another example of application of GAs [8]. At the structural level, GAs have been used for the optimization of different structural components of a building. A comprehensive review of different optimization techniques in use to achieve minimum cost, weight, or other specific goals is provided by Sahab [9]. Design of large span structures is the main focus of most of the optimization techniques adopted in structural engineering problems. An example is represented by ParaGen, a parametric modelling and performance simulation software, which integrates genetic algorithms and a database of solutions [10]. This tool, currently under development, was used to optimise the design of large spam roof structures with positive results. Applications for spatial problems have been found for floor layout distribution applied to planning problems [11] or for assigning vertical and horizontal space distribution in large buildings [12]. Regarding acoustic performance, GAs were found in the optimization of noise barriers with variable cross sections [13]. GAs were also found in defining criteria for sizing windows using façade orientation and climatic conditions as dominant parameters [14].

The mentioned studies represent just a few examples in which GAs have been used, in order to optimize the design of one or more building's sub-systems. However, for most of the cases reported before, the main focus was the solution of a particular problem using specifically developed GAs. In that sense, replicating the proposed logic for different problems would require the development of a new set of specific algorithms and thus, high levels of mathematical knowledge.

Since it has been demonstrated that GAs can play a crucial role in a building's performance optimization, user friendly tools are needed to expand their use beyond the engineering field.

1.2. Application of Building Performance Simulation to Envelope's Design: Combining Parametric Design and Evolutionary Optimization

Complex optimization techniques, also involving the use of GAs, have been found to be useful in several problems involving buildings' envelope designs. Energy and daylight-based optimization functions are usually adopted to find the best solution for a particular design problem.

Trubiano *et al.* in 2013 [15] developed an evolutionary optimization script, integrating Grasshopper with Radiance and EnergyPlus through Matlab. They used the script to demonstrate the possibility of generating the optimal shape of atriums with the adoption of GAs and a single objective function. A similar approach was used by Lobaccaro *et al.* [16] for the optimization of the geometry of a building in order to maximise the annual envelope's exposure to solar radiation.

David *et al.* [17] applied the combination of daylight and thermal analyses for the assessment of the efficiency of solar shades. They used the shading coefficient, cooling energy demand, daylight autonomy, sun patch index on work plane, and useful daylight illuminance as indexes for rating the performance of different typologies of external overhangs.

Some other research groups have worked on the topic of the optimization of configuration of façade's modules for office buildings using integrated thermal and lighting simulations. Goia *et al.* in 2013 [18] carried out the optimization of the window-to-wall ratio of a typical office using total energy consumption as optimization function. They found that, for continental climates (Koppen Geiger Cfb climatic zone), the optimal solution could be obtained for a transparent percentage between 35% and 45%, regardless of the façade's orientation. Previously, in 2012, Ochoa *et al.* [19] identified four criteria for the optimization of window's size in an office building. Together with the minimization of the total yearly energy consumption, they considered illuminance, uniformity, and glare criteria. Instead of defining a univocal solution, they proposed a range of possible solutions, meeting more than one criterion. An open issue was, however, the weight of each criterion in the decision making process. Favoino *et al.* [20] worked on the definition of an ideal glazed façade for office buildings using thermal transmittance, Solar Heat Gain Coefficient (SHGC) and visible light transmittance as control variables. Also in this case the total primary energy consumption was used as the optimization function.

Since the creation of Galapagos, several researches have explored the implication of this methodology on addressing single or multi objective problems. Although its creator points out that the reliability of the solver depends on the nature and complexity of the studied problem, the results can be better than the ones achieved by traditional approaches.

In 2011, a study performed by El Sheikh & Gerber [21] focused on the parametric design of external shading systems using evolutionary optimization principles. Galapagos was used as the main GA solver, while the optimal solution was searched among the ones able to maximize daylight performance of indoor space. Indicators such as minimum levels of indoor illuminance and luminous distribution of light in the selected space were used.

A similar approach was used by Gadelhak [22] in the form-finding and shading optimization of high performance façades. Similarly to the previous research, daylight indicators such as the percentage of daylit area were used.

Furthermore, the full potential of parametric design integrating simulation tools and genetic optimization using Galapagos has been studied by Shi and Yang [23]. Three simulation software packages were integrated with Grasshopper by writing a customised computer script for each one. Ecotect was used to predict solar radiation availability in complex roof geometry. Radiance was used to define the best window proportion from a given size (height and width) for achieving a specific daylight factor; and EnergyPlus was used for determining the optimal distribution of a given window area for four façades's orientations (N, S, E and W), with the aim of reducing energy consumption. The research demonstrated that generic evolutionary solvers, such as Galapagos could successfully achieve specific optimization goals. However, two important limitations of the study were pointed out by the authors. Firstly, the studied problems are general and single objectives, whereas, problems in architecture are generally complex and multi-objective. Secondly, coding capabilities are needed to customise computer scripts for integrating different software packages into Grasshopper. This poses a limitation, as designers and architects generally do not have programming skills.

In 2013, Brotas & Rusovan [24] performed an overall energy simulation (including both lighting and energy calculations) of four different envelope solutions for continental climates (simulations were based in London, UK). One of these presented a shading screen previously designed to optimize the daylight performance of internal spaces. The optimization was performed carrying out an evolutionary definition of the shape of hexagonal shadings based on daylight indicators. They found that in such as those climatic conditions, envelope systems adopting optimized shadings could overall achieve better results in terms of CO_2 emissions than traditional solutions.

1.3. Findings from Literature and Aims of Current Research

Overall it can be concluded that the adoption of GA and advanced optimization techniques applied to building envelope design can be highly beneficial. In the selected studies, summarized in the previous paragraph, more accurate solutions than traditional ones were found. Moreover, benefits under visual and thermal comfort viewpoints, as well as substantial energy savings have been predicted.

However, as the literature review indicates, further studies are needed in the area of evolutionary optimization, especially in the development of tools and methodologies that can be used by a wider range of professionals. Explorations on new design methods for integrating parametric design with evolutionary optimization and making use of the existing tools without the need of specific programming knowledge is needed. In addition, the capabilities of generic solvers, such as Galapagos, need to be tested in multi-objective design problems, where the result of one variable has implications on others.

A methodology that integrates the use of genetic algorithms with parametric design for achieving optimised solutions in a highly familiar design software environment is explored here though the combination of the tools mentioned before. The proposed process is tested in a highly recurring design task, such as the optimization of external shadings of office buildings.

2. Materials and Methods

The external shading of a typical office building was set as a case of study, in order to compare traditional design approaches with optimised scenarios. Traditional approaches are defined by experts' recommendations and by the adoption of deemed-to-satisfy provisions included in energy codes.

Setting external shading as a case study allows for understanding the implication of the shading geometry in both the daylight and thermal performance of the office.

The office is a large rectangular space, with an indoor area of 50 m^2 and is representative of a multi-user office in large Australian commercial buildings. An overall length of the space of 10 m has been selected, in order to maximize daylight needs typical of open-plan offices. Moreover, the resulting space represents well the case of a perimeter occupied zone of a large-plate (>30 m) multistorey office.

In order to simulate the presence of adjoining spaces, thus considering an office fully included in a whole building, only one wall has been modelled as fully exposed to outdoor. All the other surfaces (back wall, side walls, floor and ceiling) are adiabatic. A geometric representation of the model can be found in Figure 1.

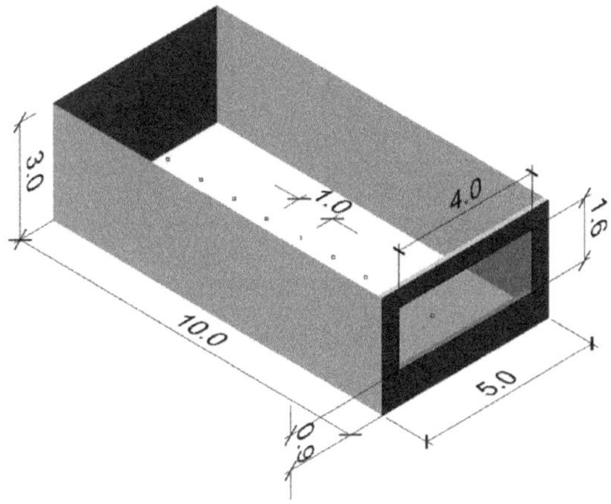

Figure 1. Geometry of the case study and control grid.

A north orientation (for the southern hemisphere) of the exposed façade has been chosen in order to maximize issues related to the control of incoming solar radiation. For the same reason, the size of the transparent portion of the wall has been maximized. Except a spandrel portion of 1.6 m (0.9 m below and 0.5 m above the window) and a side portion of 1 m (0.5 m on each side of the window), the wall is modelled as entirely transparent. This results in a window-to-wall ratio of about 43%.

The office has been located in Sydney (New South Wales, Australia, 33°57' S 151°10' E, Koppen-Geiger climatic zone Cfa). The weather file of the Typical Meteorological Year (TMY) for this location has been collected from the database of the United States Department of Energy (DOE).

Properties of building fabric, internal loads, occupancy, and equipment operational schedules are based on minimum standards (deemed-to-satisfy provisions) of the Australian National Construction Code (NCC), section J-energy efficiency [25]. Standards for class 5 (commercial) buildings have been adopted. A detailed list of the provisions is included in the following Table 1.

Table 1. Input parameters for thermal dynamic simulations.

Data	Units of Measure	Value
External wall thermal transmittance (*U*-value)	W/m^2K	0.36
External window thermal transmittance (*U*-value)	W/m^2K	3.80
External window Solar Heat Gain Coefficient (SHGC)	-	0.75
Air infiltration	ACH	1
INTERNAL HEAT GAIN RATE (OCCUPANTS)	W/m^2	12
Internal heat gain rate (equipments)	W/m^2	15
Artificial lighting power rate (at 320 Lx)	W/ m^2	9
Heating set point temperature	°C	20
Cooling set point temperature	°C	24
Heating C.O.P.	-	0.8
Cooling C.O.P.	-	2.5

In order to maximise solar gain, therefore maximizing the need for its control by means of external solar shadings, an ideal clear Double-Glazed Unit (DGU) has been selected as window's panel. The overall Solar Heat Gain coefficient (SHGC) of 0.75 and a Visible Transmittance of 0.81 reflect the use of two clear 3 mm glasses.

Furthermore, the room has been modelled as fully air-conditioned. Heating and cooling are provided by fan-coil units, and energy is generated by a natural gas boiler and a chiller powered by electricity. Heating and cooling set point temperatures have been defined following currently adopted industry standards in Australian property market [26].

With the aim of simulating the real behaviour of users working in mixed light conditions (*i.e.*, natural and artificial), artificial lighting has been modelled as linearly dimmable and controlled by means of illuminance sensors. Nine sensors, located at 0.8 m above the floor's level and spaced 1 m each, have been identified. They continuously record horizontal illuminance levels due to natural and artificial light. The combination of natural and artificial light is designed to match a target illuminance value on the workplane of at least 320 Lx across the entire room. This value corresponds to the minimum maintained illuminance required for normal tasks in office buildings in Australia [27].

In order to achieve a good accuracy of daylight modelling results, the following input values have been used for Radiance simulation parameters:

- Ambient bounces (-ab). This parameter represents the maximum number of diffuse bounces considered. As demonstrated by Mardaljevic [28], -ab values higher than 4 produce high accuracy in results. For the scope of current research, the following previous studies carried out on similar geometries [29] an -ab value of 5 has been adopted.
- Ambient division (-ad) represents the number of sampling rays projected from each point into the sky for the calculation of indirect radiation. The higher the number, the lower the error in predicting indirect solar radiation. An -ad value of 1024 has been adopted according to [29].
- Ambient Accuracy (-aa) represents the error—in percentage—of ambient interpolations. An -aa value of 0.1, representative of an error not higher than 10%, has been adopted according to [29].
- Ambient Supersamples (-as) represent the number of extra rays created between two neighbouring samples when a significant difference (specifically 10% in the current set-up) is found among them. An -as value of 16 has been adopted according to [29].
- Ambient Resolution (-ar) controls the density of ambient values. An -ar value of 256 has been adopted according to [29].

Table 2 contains materials' properties used for Radiance daylight models. Materials' reflectivity and visual transmissivity (*i.e.*, transmission of light at normal incidence) assume the same values for red, green, and blue bandwidths as Daysim scene is simulated in grey colours. In order to consider perfectly diffusive materials, specularity and roughness values of 0 have been used for all opaque materials.

Five different traditional shading settings, grouped in three main scenarios, have been tested. As a base case, an unshaded and unobstructed window has been considered. Then 4 possible shading scenarios have been analysed. The first two involve the use of external multiple metallic louvres sized to shield the window from summer solar rays. The other two represent shading solutions satisfying the requirements of Australian National Construction Code [25] and involving, respectively, the use of an horizontal overhang placed above the window and of multiple external metallic louvres. The 5 traditional settings have been, then, compared with an optimized shading solution, involving the adoption of multiple external metallic louvres. Depth, spacing, and slope of the louvres have been selected as controlled variables for the optimization process.

Table 2. Materials' properties for daylight simulations.

Building Component	Reflectivity	Visual Transmissivity
Walls and partitions	0.5	-
Ceiling	0.8	-
Floor	0.2	-
Shadings	0.9	-
External ground	0.2	-
Glass	-	0.87

2.1. Performance Indicators

Results are then analysed comparing the following indicators:

- Daylight Autonomy (DA). DA represents the percentage of time in which the indoor illuminance levels at the selected reference point exceed a defined threshold value. For this aim, the value of 320 Lx has been considered as the set point for the calculation of DA.
- Useful Daylight Illuminance (UDI). UDI is a parameter firstly introduced by Nabil & Mardaljevic in 2005 [30] and accounts for the overall percentage of time during a statistical year in which the indoor illuminance at the selected reference point falls within a defined range. In the current paper, the 4 UDI indicators introduced by Mardalievic *et al.* in 2012 [31] have been used:
  - o UDI fell-short (UDI-f). It represents the occurrence of daylight illuminance levels lower than 100 Lx.
  - o UDI supplementary (UDI-s). It represents the occurrence of daylight illuminance levels in the range 100–300 Lx.
  - o UDI autonomous (UDI-a). It represents the occurrence of daylight illuminance levels in the range 300–3000 Lx.
  - o UDI exceeded (UDI-e). It represents the occurrence of daylight illuminance levels greater than 3000 Lx.
- Energy consumption for Heating. This indicator accounts for the total energy used yearly on site for feeding the gas heating system and is measured in kWh consumed per year.
- Energy consumption for Cooling. This indicator accounts for the total energy used yearly on site for feeding the electric cooling system and is measured in kWh consumed per year.
- Energy consumption for Lighting. This indicator accounts for the total energy used yearly on site for feeding the electric lighting system and is measured in kWh consumed per year.
- Total Energy Consumption. Sum of the site energy consumed for heating, cooling and artificial lighting.
- Total Energy Savings. This indicator compares each scenario with the base case on a yearly base.
- Total CO_2 emissions. This indicator accounts for the total emissions of greenhouse gasses during the process of production of the energy consumed on site for feeding HVAC and lighting systems. The unit of measure is kg of CO_2-e emissions per year. According to [32],

emission factors of 0.184 kg CO_2-eq per kWh for natural gas and of 0.86 kg CO_2-eq per kWh for electricity have been used.

Some of the other visual comfort indicators, such as external view and glare probability are here not considered. As a matter of fact the optimized solution, involving the use of external metallic louvres, always offers an obstructed external view. Moreover, as demonstrated in previous researches [33], in the current scenario glare probability is limited.

2.2. Traditional Approaches of Shadings' Designs

The most basic scenario (*no shadings* hereafter) is the one in which the external window is not provided with any shading device. The control of solar gains and indoor illuminance levels is simply managed by the transmittance of the glass (in this case fixed throughout all the year). This usually constitutes the starting point of designers for optimizing envelope's characteristics.

A first improvement is described in Figure 2, where lamellar shadings have been implemented on the external side of the window. Their design follows experts' recommendations for Sydney's weather. The base assumption is to avoid solar gains in mid-summer, while gradually allowing them for the rest of the year. This approach is generally adopted by designers who want to go beyond the minimum standard, without engaging with detailed thermal modelling. Two types of shading have been selected. The first one (T1 hereafter) is composed of 4 external lamellas with a width of 150 mm and an inclination of −31° on horizontal plane, allowing a full penetration of winter solar rays and a full blockage of the summer ones. The second type (T2 hereafter) is composed of 4 horizontal louvres with a width of 150 mm. In this case, summer solar rays (calculated at 12 noon of solstice day) are also entirely blocked, with the additional benefit of maximising external views.

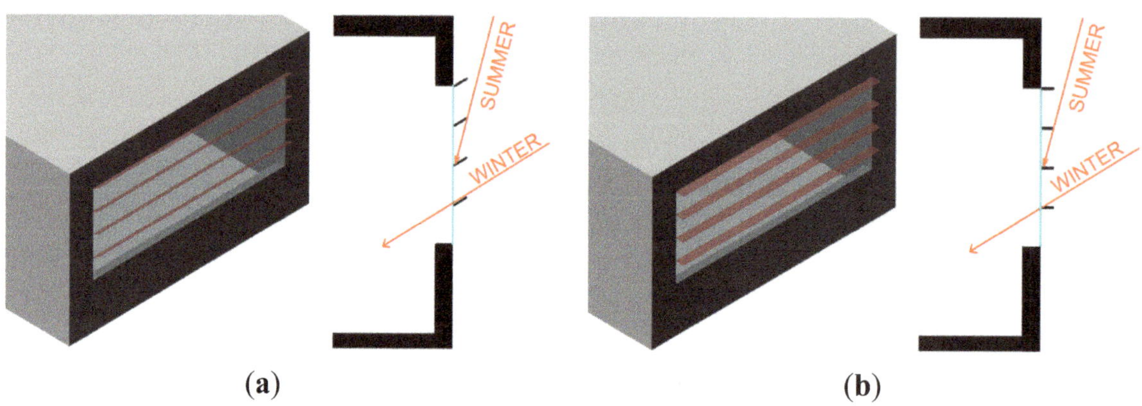

Figure 2. Traditional shading regarding building latitude. (**a**) T1; (**b**) T2.

The second scenario considers a façade that meets the minimum requirements of the Australian National Construction Code (NCC). The dimension of the shading was defined in accordance with deemed-to-satisfy requirement for transparent surfaces. As a result, a first case (NCC1, thereafter) with an external shading 0.95 m wide added to the upper portion of the window is considered (Figure 3a). The second case (NCC2 thereafter) explores the solution of splitting the external overhang into six horizontal louvres but maintaining the same sun-blocking angle of the NCC1 scenario (Figure 3b).

This approach is usually adopted by a designer who wants to maximise daylight through the reflectance of the louvres while still meeting code standards.

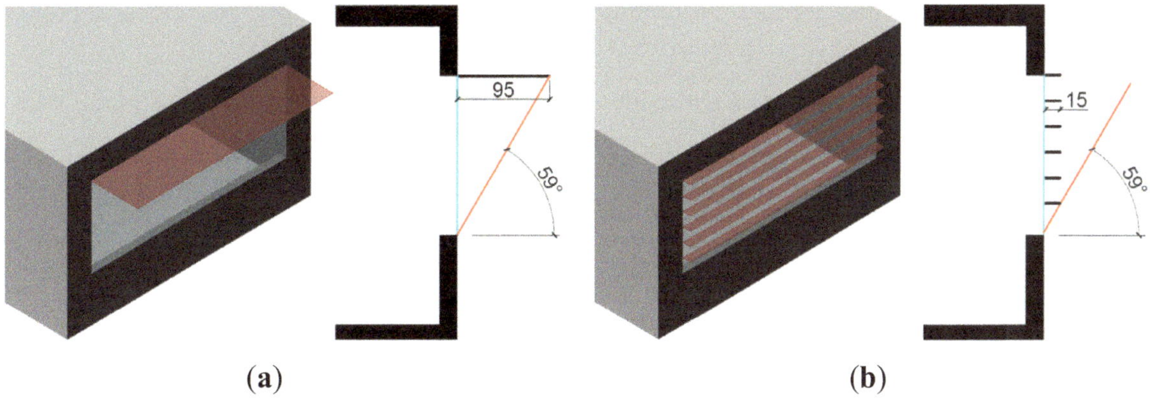

Figure 3. (**a**) Scenario NCC1; (**b**) scenario NCC2.

2.3. Optimised Shading

Several variables influence performance of external shading systems and affect both HVAC and lighting energy consumptions. Thus, the optimization of solar shadings is a multi-variable problem. As mentioned before, as the variables cannot be controlled independently, an approach based on GAs for evolutionary optimization has been adopted. The overall geometry of the model has been generated in Rhinoceros and the variation of a limited number of selected variables (*genes* thereafter) has been managed through Grasshopper. Each gene can assume specific values (*genomes* thereafter). In the case of the selected optimization, the following genes have been selected. Specific genomes are included in round parenthesis.

- Shading depth (linearly varying from 0.1 to 0.3 m with a progression of 0.01 m);
- Shading angle on horizontal plane (linearly varying from −20° to 0° with a progression of 1°);
- Shading number (positive integer varying from 10 to 20).

The following Figure 4 shows genes and genomes adopted.

Optimization algorithms embedded in Galapagos have been used and total CO_2 emission has been selected as the optimization function (*fitness function* thereafter).

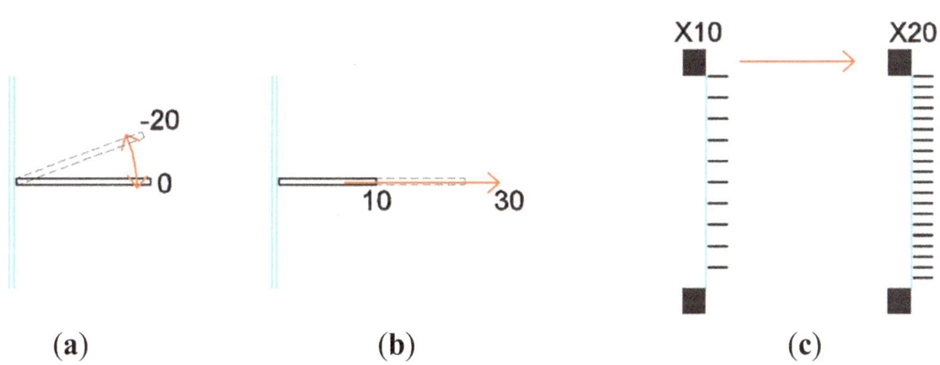

Figure 4. External shading controllers and function associated: (**a**) Angle; (**b**) Depth; (**c**) Number.

The workflow for searching for the optimized solution is summarised in Figure 5. It consists of the following steps:

(a) Genes are associated with Galapagos (though Grasshopper) that randomly defines a first set of triads of genomes (*individuals* hereafter).

(b) DIVA daylight runs the initial daylight calculation based on the first set of individuals. Indoor illuminance levels for each of the 9 sensors are then calculated for every hour of the 365 days of the Typical Meteorological Year. Therefore, the yearly profile of DA is determined for each reference point and, consequently, schedules of artificial lighting are generated.

(c) DIVA thermal component collects artificial lighting schedule and performs a full yearly thermal dynamic simulation. As a result, final energy consumptions for heating, cooling and artificial lighting are provided.

(d) Total source CO_2 emissions are calculated, adding components due to heating, cooling and artificial lighting. Total CO_2 emission rate is then defined as the fitness function, and the Genetic Algorithm is set up in order to optimize it (*i.e.*, to find the minimum value).

(e) The GA solver Galapagos collects and processes the data, commanding the next generation of sets of individuals.

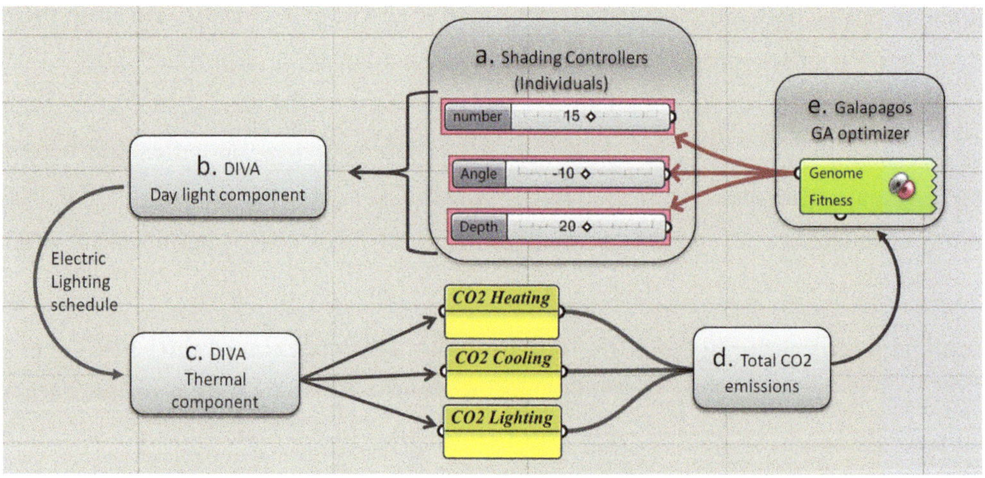

Figure 5. Workflow using DIVA daylight and thermal components, and Galapagos evolutionary solver.

Given the time required for performing daylight calculations, the total search space has been limited to a maximum number of individuals, reducing the number of genomes that each gene could assume. The selection of genomes was based on design constrains, such as the availability of an external unobstructed view and the maximum projection of the shading towards the outside.

The population was set to 10 individuals per generation with an initial population (initial boost) of 20 individuals. Ten percent of individuals passing on at the next generation and a maximum inbreeding factor of 50% insure reasonable variability among generations. Maximum stagnant was set to stop the solver if no improvement of the fitness function was reached after 10 generations. These settings have been defined in order to balance simulation time and accuracy of the optimization process. A similar approach has been previously used by Caldas & Norford [14], Trubiano *et al.* [15] and Gadelhak [22], adopting respectively 5, 15, and 20 individuals per each generation.

3. Results

The predicted overall annual energy consumption of the typical office in its base case configuration (no shading) is equal to about 2500 kWh, with atmospheric emissions of about 1800 kg CO_2-eq. Adopting traditional shading strategies (T1 and T2), annual overall energy consumptions drop down to about 2000 kWh, and the predicted emissions consequently decrease to about 1400 kg CO_2-eq. These results mostly depend on a substantial decrease of the cooling energy needs due to the reduction of summer solar heat gains. Traditional shading techniques based on solutions compliant to Australian energy code show similar results, with an average overall annual energy consumption of about 1900 kWh and average predicted atmospheric emissions of about 1300 kg CO_2-eq.

The optimization process has been demonstrated to be beneficial in reducing consistently both energy demand and carbon dioxide emissions. Annual energy demand of less than 1600 kWh and annual energy-related atmospheric emissions of less than 1000 kg CO_2-eq have been obtained. Therefore, annual energy savings of about 35%—if compared to the base case scenario—and of 15% if compared to the best non-optimized scenario—have been achieved.

The optimized solution was found with an external system composed of 14 louvres with a slope of −10° on horizontal plane, and a depth of 20 cm (Figure 6). Therefore, the optimized system is able to block completely solar radiation with an incident angle higher than 42°. In addition, the negative inclination of louvres increases the sky view factor of inside points and could potentially increase the reflection of solar rays towards the ceiling. The combination of these two effects could augment daylighting illuminance levels.

Figure 7 shows the outcomes of the optimization process. The upper curve (in blue) represents the progress in reaching minimum values of energy consumptions after each generation. As can be seen from the figure, a minimum value was already reached within the initial boost (area highlighted in yellow) and was matched in the last generation. The lower curve (in red) represents the progress in reaching minimum values of the fitness function (CO_2 emissions). As can be seen from the picture, in this case, the optimization was also already reached during the initial boost. No better result was found then after the 10 subsequent generations, even though a nearly optimized value was reached in the last generation. For this reason the optimization process was stopped after processing and analysing the 10th generation of values.

Figure 6. Best case scenario.

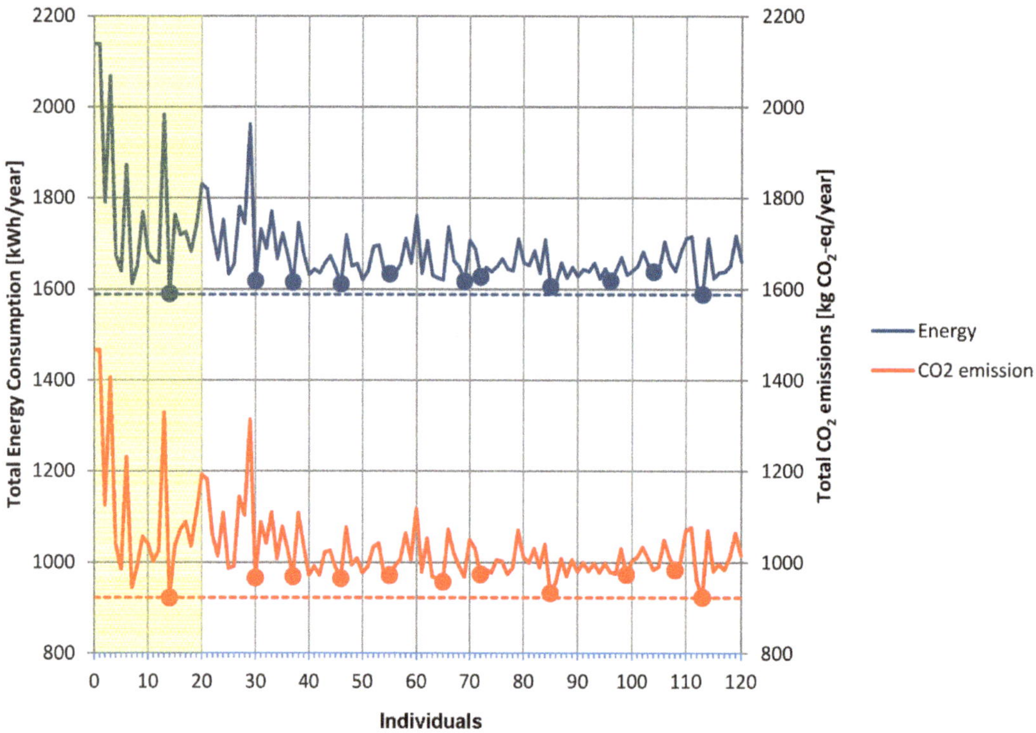

Figure 7. Outcomes of the optimization process.

3.1. Energy Efficiency and CO_2 Emissions

Tables 3 and 4 summarise results obtained during the simulation process and compare them in terms of energy consumptions and CO_2 emissions.

As can be noticed from the tables, energy savings increase progressively with the adoption of more sophisticated design techniques. This increase in energy efficiency is mainly due to a substantial reduction of cooling loads (47%, if NCC2 scenario is compared to the base case). Heating and lighting consumptions, on the contrary, do not increase significantly: for the NCC2 scenario, increases of respectively 3.6% and 13.9% were recorded.

As it was demonstrated, avoiding only summer solar radiation (*i.e.*, blocking solar rays sloped more than 75° on horizontal plane) is not a good strategy for offices that generally present high internal heat gains. National Construction Code minimum standards (represented by NCC1) produce good results, avoiding penetration of solar radiation with an angle higher than 59°. However, maintaining the same sun blocking angle but distributing the overhang in a larger number of louvres (NCC2), determines further benefits.

The optimised solution, indeed, experiences significant benefits from the reduction of cooling loads, with a further 47% reduction in comparison to the NCC2 scenario and an overall reduction of 71.9% in comparison to base case. However, due to the reduction of direct solar gains beneficial in the winter period, heating consumption shows an increase of 16.5% in comparison to the NCC2 case and of 20.8% in comparison to base case.

Overall, it can be seen that the optimization process is successful in the reduction of energy consumption, due to the definitions of shading solutions able to better control solar radiation. The contemporaneous reduction of artificial lighting consumption is another proof of this ability.

Table 3. Result summary and comparison for all simulated scenarios. Energy consumption.

Scenario	Shading Setting			Energy Consumptions (kWh/year)				Energy Savings	
	Nr.	Angle	Depth	Heating	Cooling	Lighting	Total	(kWh/year)	(%)
No shading	-	-	-	546.33	1458.01	472.73	2477.07	0.00	0.0
T1	4	−31	15	551.28	1205.73	489.68	2246.69	230.38	9.3
T2	4	0	15	559.73	966.05	521.83	2047.61	429.46	17.3
NCC1	1	0	95	556.41	844.27	625.74	2026.42	450.65	18.2
NCC2	6	0	15	566.29	771.32	538.70	1876.30	600.77	24.3
Optimised	14	−10	20	659.73	408.59	521.91	1590.23	886.84	35.8

When it comes to the reduction of CO_2 emissions, the different emission rates for electricity and natural gas slightly change final results. In particular, the CO_2 emission rate for purchased electricity is 4.67 times higher than the one for purchased natural gas. Therefore, cooling and lighting loads have a higher impact on CO_2 emissions than heating loads. As a result, the adoption of optimized external shadings leads to a reduction of annual CO_2 emissions of 47.7%.

Table 4. Results summary and comparison for all simulation scenarios. CO_2 emissions.

Scenario	Shading Setting			CO_2 Emissions (kg CO_2-eq/year)				Reduction of CO_2 Emissions	
	Nr.	Angle	Depth	Heating	Cooling	Lighting	Total	(kg CO_2-eq/year)	(%)
No shading	-	-	-	100.52	1253.89	406.55	1760.96	0.00	0.0
T1	4	-31	15	101.44	1036.93	421.12	1559.49	201.47	11.4
T2	4	0	15	102.99	830.80	448.77	1382.57	378.39	21.5
NCC1	1	0	95	102.38	726.07	538.14	1366.59	394.37	22.4
NCC2	6	0	15	104.20	663.33	463.28	1230.81	530.15	30.1
Optimised	14	-10	20	121.39	351.39	448.84	921.62	839.34	47.7

3.2. Visual Performance

Results, in terms of selected daylight performance indicators for all the simulated cases are included in Table 5. It can be noticed that for traditional shading techniques (T1 and T2), annual average daylight availability remains almost similar to the one recorded in the base case, with an overall variability of −2%/+1%. The same happens for average, minimum and maximum values of UDI-a. On the contrary, the shading solution compliant with National Construction Code deemed-to-satisfy provisions (NCC1) shows a drop of about 12.2% of the average daylight availability. Interestingly average and maximum UDI-a values do not show a similar pattern (minimum UDI-a decreases of 7.3%, while maximum UDI-a increases of 2.9%). Moreover, for this solution, both UDI-f and UDI-s indicators increase significantly (respectively 6% and 6.3%). The improved solution (NCC2), on the contrary, experiences similar results to the ones obtained of T1 and T2, as it follows similar design principles.

The shading solution resulting from the optimization process determines an increase of useful daylight illuminance levels in the room. In fact, even though DA levels show a decrease in comparison to the base case ones, the percentage of time in which illuminance levels are in a comfortable range increases about 2%, if average values among all nine sensor points are considered. These are the best results among all scenarios.

Table 5. Daylight performance indicators for each tested scenario.

Scenario	Daylight Performance Indicators						
	DA (%)	UDI-f (%)	UDI-s (%)	UDI-a_{min} (%)	UDI-a_{ave} (%)	UDI-a_{max} (%)	UDI-e (%)
No shading	54.6	17.1	26.3	10.6	44.5	75.8	12.1
T1	55.5	17.4	25.2	8.2	45.5	77.9	11.9
T2	52.7	17.9	27.5	6.5	44.7	81.6	10.0
NCC1	42.4	23.1	32.6	0.7	37.2	78.7	7.1
NCC2	51.8	18.4	27.9	5.4	43.6	81.5	10.1
Optimised	51.4	19.3	27.4	5.3	46.4	81.1	6.8

Figure 8 shows the profile of UDI-a for each of the six tested scenarios. Each profile has been calculated considering the nine reference points evenly distributed across the entire depth of the typical office. Ideal values are used as targets for comparisons; they have been defined considering the hypothetical case of a façade system able to clear excessive indoor illuminances (*i.e.*, values over 3000 Lx), while maximising UDI-a percentage.

Base case (*i.e.*, "no shading"), T1, T2, and NCC2 scenarios show similar profiles, with UDI-a values very close to the ideal ones for reference points one to six. UDI-a percentages drop down rapidly for reference points seven to nine close to the window.

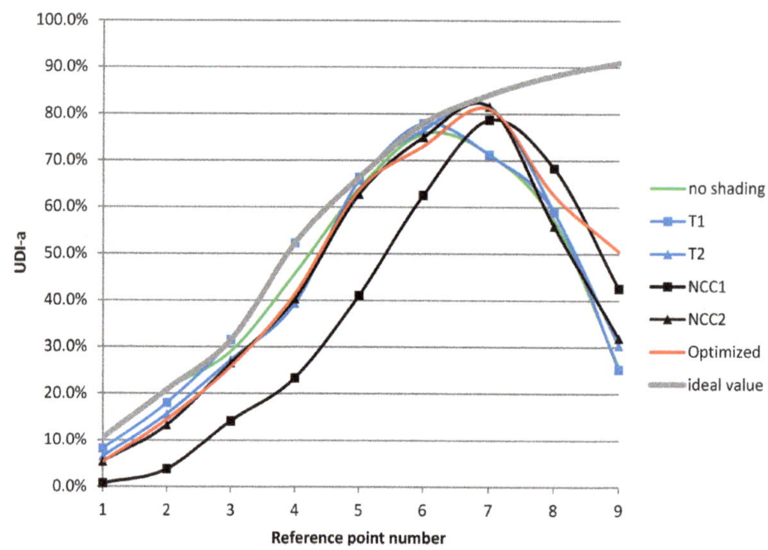

Figure 8. Profile of UDI-a for each tested scenario.

In order to explore the benefits of the optimized solutions under a visual viewpoint, horizontal illuminance levels of three selected points have been further analysed. Reference points two (located at 1.5 m from the back wall), five (located in the middle of the room), and eight (located at 1.5 m from the front wall) have been selected as significant for this analysis.

It should be noted that horizontal illuminance values constitute only one of the parameters for the assessment of indoor visual comfort, and more detailed studies should be carried out in the future in order to understand potential glare phenomena associated with the adoption of highly reflective external slats.

Figure 9 shows the annual distribution of indoor illuminances for the three selected reference points during daily occupancy hours (8 a.m. to 6 p.m.). The two red lines represent the lower and upper limit of UDI-a range. Focusing on the base-case scenario (Figure 9a), illuminance levels at reference point two (blue in the figure) are always lower than 500 Lx; moreover the probability of indoor illuminance levels lower than 300 Lx are high in early morning and late afternoon, especially in winter period. As a result, the annual overall percentage of UDI-a for this reference point is 20.9% and UDI-s contributes for an additional 52.5%. On the contrary, indoor illuminance levels at reference point eight (green in the figure) for about one third of the time exceed the upper threshold of 3000 Lx. On the contrary, in only 7.8% of time, this reference point experiences illuminance levels lower than 100 Lx.

The sensor located in the middle of the room (sensor point five, in black in the figure) experiences more constant illuminance values across all the year. For a wide majority of time (exceeding 85%) users located in this reference point would experience indoor illuminance levels on the workplane in the range between 100 Lx and 3000 Lx. Moreover illuminance levels never exceed the upper threshold of 3000 Lx.

Overall, comparing the distributions of indoor illuminance values of the base case and the optimized scenario, a significant improvement of comfortable horizontal daylight illuminance conditions can be noticed. Sensor points located in the middle and at the back of the room, already experiencing illuminance levels in the comfortable range for the majority of time, also continue to show the same pattern with the adoption of external shadings. Points located close to the window show a substantial increase of the percentage of hours with illuminance levels in the range between 100 and 3000 Lx.

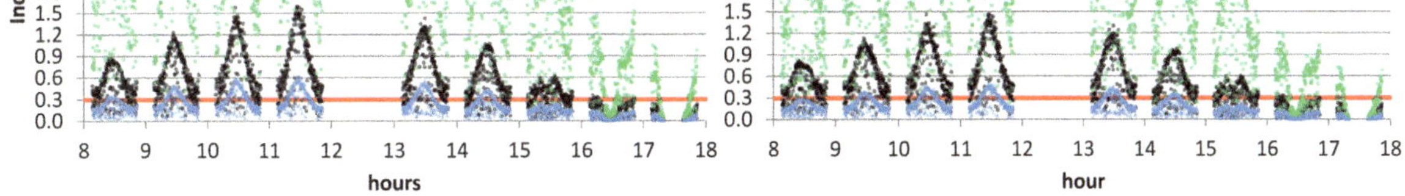

Figure 9. Distribution of indoor illuminance values for sensor points two (blue), five (black), and eight (green): (**a**) Scenario "no shading"; (**b**) Optimized scenario.

4. Discussion

The optimization problem, related to the selection of external shadings in an office building, was solved using an energy-related function (minimization of CO_2 emissions). However, it was found that the specific solution leads to an increase of daylight performances of the spaces. An overall increase of the minimum and average percentage of working hours experiencing comfortable daylight illuminances was found. Moreover, the increase of the uniformity of distribution of daylight illuminances across the entire space was another related advantage. It should be noted that predictions have been made with the assumption that no interior shading device will be needed. However, in some circumstances (for example, in the presence of low sun angles), interior shadings could be needed in order to avoid direct sunlight penetration. This could partially impact lighting energy savings, as well as cooling and heating loads.

Despite the fact that a low number of individuals per generation was used (10 out of 50 recommended), an optimized solution was found. The optimized solution showed a substantial increase of energy performance and environmental sustainability, even if compared to be best "traditional" solution. This, in the authors' viewpoint, is proof of the utility of the proposed methodology. A similar approach of using a reduced number of individuals per each generation has already been adopted in similar cases [14,15,22] and was found successful. This is a highly beneficial approach when time is a constraint due to the complexity of analyses to be performed. The obtained results are satisfactory, especially considering that Galapagos is an evolutionary solver adopting a generic Genetic Algorithms, which may not be the best optimiser for the specific proposed problem. Considering that each change in the settings affects each component of the energy loads differently, results in a multi-objective problem are difficult to predict. Thus, computer assisted optimization can help designers to solve complex problems in limited time, being that the main advantage of the studied workflow.

In DIVA, the combination of the advanced functionalities of daylight (Radiance/Daysim) and thermal (EnergyPlus) software, allows designer to obtain accurate results. However, each component has limitations that reduce its capabilities on dealing with complex geometries, large areas and/or multiple zones. DIVA thermal component allows the calculation of just one thermal zone per time. Thus, each thermal zone needs to be analysed independently, and the interaction of different zones within a building cannot be assessed. At the same time, this approach unavoidably offers limited possibilities for materials setting. Emissivity is not thermally considered for the shading setting. Thus, the radiation component of the energy emitted by external shadings is not accounted.

Moreover, daylight calculations based on Radiance raytracer algorithms [34] require long simulation time, even for a simple geometry like the studied one. Time required for performing daylight simulations has been demonstrated to be about 35 times longer than the one required for performing a full thermal dynamic analysis. Therefore, the feasibility of conducting an optimization process for large areas or complex geometries is limited, especially when time is a constraint.

Finally, optimized solution may not necessarily be the best solution for meeting other design criteria. In this specific case, parameters as external view factors and glare probability have not been considered. For the optimized scenario for example, negative shading angle and high number of louvres are likely to reduce views towards outside. Therefore, additional criteria need to be incorporated in the problem setting for real applications.

The proposed strategy explores just one component of the building envelope. The same approach or something similar can be used to optimise different design variables of building sub-systems, allowing informed decisions made at different stages of the design process.

It is expected that current limitations of the simulation software tools, in terms of the number of spaces able to be modelled and limited number of design parameters that can be considered, will be overcome in the future. In addition, simulation time is likely to be reduced by the constant evolution of computers processing capacity.

5. Conclusions and Recommendations

In conclusion, this work has demonstrated how:

1. The combination of parametric design with evolutionary optimization is a valid strategy for exploring solutions to building environmental performance problems. A comprehensive setting of the design problem and further improvement in the simulation tools could overcome the main difficulties found in this research.
2. Solutions optimised by the computational means can differ from traditional ones, providing new answers for design problems that can be more effective on solving specific tested issues. For the studied case, the optimised solution performs significantly better in terms of daylight availability, energy efficiency and CO_2 emissions than other studied alternatives, with different shading configurations.
3. Although the methodology was tested for one specific problem that could correspond to an advanced stage of the design process, a similar approach can be used in order to optimise problems such as—but not restricted to—building orientation and geometry, window-to-wall ratio, and thermal resistance of the external envelope.
4. Future studies can be built upon the current one and explore an expanded set of optimization criteria, combining energy-related indicator with visual comfort ones, such as glare probability, uniformity of daylight illuminances, and external view factor.

Acknowledgments

The authors would like to thank Richard de Dear of the University of Sydney and Michele Sauchelli of Politecnico di Milano for the beneficial discussions and for the advice received.

Author Contributions

Francesco Fiorito coordinated the research. Javier González built the model and performed combined energy and daylight simulations. Francesco Fiorito and Javier González defined research scope, objectives and jointly wrote the paper.

Conflicts of Interest

The authors declare no conflict of interest.

References

1. Attia, S.; Beltrán, L.; De Herde, A.; Hensen, J. "Architect Friendly": A Comparison of Ten Different Building Performance Simulation Tools. In Proceedings of the 11th International IBPSA Conference—Building Simulation 2009, Glasgow, Scotland, 27–30 July 2009; pp. 204–211.

2. Davidson, S. *Grasshopper: Algorithmic Modelling for Rhino*, NING/MODE Social: Brisbane, CA, USA, 2013. Avaliable Online: http://www.grasshopper3d.com/ (accessed on 23 September 2013).
3. Reinhart, C.F.; Lagios, K.; Niemasz, J.; Jakubiec, A. *Diva for Rhino*; Solemma LLC: Cambridge, MA, USA, 2012. Available Online: http://diva4rhino.com/ (accessed on 26 August 2013).
4. U.S. Department of Energy (DOE). *Energyplus 8.2.0*; U.S. DOE: Washington, DC, USA, 2014.
5. Renner, G.; Ekárt, A. Genetic algorithms in computer aided design. *Comput. Aided Des.* **2003**, *35*, 709–726.
6. Rutten, D. Evolutionary Principles Applied to Problem Solving Using Galapagos. Available online: http://www.grasshopper3d.com/profiles/blogs/evolutionary-principles (accessed on 23 September 2013).
7. Counsell, J.; Zaher, O.; Brindley, J.; Murphy, G. Robust nonlinear HVAC systems control with evolutionary optimization. *Eng. Comput.* **2013**, *30*, 1147–1169.
8. Wright, J.A.; Loosemore, H.A.; Farmani, R. Optimization of building thermal design and control by multi-criterion genetic algorithm. *Energy Build.* **2002**, *34*, 959–972.
9. Sahab, M.G.; Toropov, V.V.; Gandomi, A.H. A Review on Traditional and Modern Structural Optimization: Problems and Techniques. In *Metaheuristic Applications in Structures and Infrastructures*; Elsevier Inc.: Amsterdam, The Netherlands, 2013; pp. 25–47.
10. Turrin, M.; von Buelow, P.; Stouffs, R. Design explorations of performance driven geometry in architectural design using parametric modeling and genetic algorithms. *Adv. Eng. Inf.* **2011**, *25*, 656–675.
11. Jo, J.H.; Gero, J.S. Space layout planning using an evolutionary approach. *Artif. Intell. Eng.* **1998**, *12*, 149–162.
12. Rodrigues, E.; Gaspar, A.R.; Gomes, Á. An approach to the multi-level space allocation problem in architecture using a hybrid evolutionary technique. *Autom Constr.* **2013**, *35*, 482–498.
13. Grubeša, S.; Jambrošić, K.; Domitrović, H. Noise barriers with varying cross-section optimized by genetic algorithms. *Appl. Acoust.* **2012**, *73*, 1129–1137.
14. Caldas, L.G.; Norford, L.K. A design optimization tool based on a genetic algorithm. *Autom Constr.* **2002**, *11*, 173–184.
15. Trubiano, F.; Roudsari, M.S.; Ozkan, A. Building Simulation and Evolutionary Optimization in the Conceptual Design of a High-Performance Office Building. In Proceedings of the 13th Conference of the International Building Performance Simulation Association, Chambery, France, 25–28 August 2013; pp. 1306–1314.
16. Lobaccaro, G.; Fiorito, F.; Masera, G.; Prasad, D. Urban Solar District: A Case Study of Geometric Optimization of Solar Facades for a Residential Building in Milan. In Proceedings of the AuSES Solar 2012 Conference, Melbourne, Australia, 6–7 December 2012.
17. David, M.; Donn, M.; Garde, F.; Lenoir, A. Assessment of the thermal and visual efficiency of solar shades. *Build. Environ.* **2011**, *46*, 1489–1496.
18. Goia, F.; Haase, M.; Perino, M. Optimizing the configuration of a façade module for office buildings by means of integrated thermal and lighting simulations in a total energy perspective. *Appl. Energy* **2013**, *108*, 515–527.

19. Ochoa, C.E.; Aries, M.B.C.; van Loenen, E.J.; Hensen, J.L.M. Considerations on design optimization criteria for windows providing low energy consumption and high visual comfort. *Appl. Energy* **2012**, *95*, 238–245.
20. Favoino, F.; Jin, Q.; Overend, M. Towards an Ideal Adaptive Glazed Façade for Office Buildings. In Proceedings of the 6th International Conference on Sustainability in Energy and Buildings (SEB 2014), Cardiff, UK, 25–27 June 2014; pp. 289–298.
21. El Sheikh, M.; Gerber, D.J. Building Skin Intelligence: A Parametric and Algorithmic Tool for Daylighting Performance Design Integration. In Proceedings of the 31st Annual Conference of the Association for Computer Aided Design in Architecture: Integration Through Computation, ACADIA 2011, Calgary/Banff, Canada, 11–16 October 2011; pp. 170–177.
22. Gadelhak, M. Integrating Computational and Building Performance Simulation Techniques for Optimized Facade Design. In Proceedings of the 31st eCAADe Conference, Computation and Performance, Delft, The Netherlands, 18–20 September 2013.
23. Shi, X.; Yang, W. Performance-driven architectural design and optimization technique from a perspective of architects. *Autom Constr.* **2013**, *32*, 125–135.
24. Brotas, L.; Rusovan, D. Parametric daylight envelope. In Proceedings of the PLEA2013—29th Conference, Sustainable Architecture for a Renewable Future, Munich, Germany, 10–12 September 2013.
25. *NCC 2014. Building Code of Australia. Class 2 to Class 9 Buildings*. Australian Building Code Board: Canberra, Australia, 2014; Volume 1.
26. Roussac, A.C.; Steinfeld, J.; de Dear, R. A preliminary evaluation of two strategies for raising indoor air temperature setpoints in office buildings. *Arch. Sci. Rev.* **2011**, *54*, 148–156.
27. Standard Australia. AS/NZS 1680.1. Interior and Workplace Lighting. Part 1: General Principles and Recommendations; Standard Australia: Sydney, Australia, 2006; p. 129.
28. Mardaljevic, J. Validation of a lighting simulation program under real sky conditions. *Light. Res. Technol.* **1995**, *27*, 181–188.
29. Jakubiec, J.A.; Reinhart, C.F. Diva 2.0: Integrating Daylight and Thermal Simulations Using Rhinoceros 3D, Daysim and Energyplus. In Proceedings of the 12th Conference of International Building Performance Simulation Association Building Simulation 2011, Sydney, Australia, 14–16 November 2011; pp. 2202–2209.
30. Nabil, A.; Mardaljevic, J. Useful daylight illuminance: A new paradigm for assessing daylight in buildings. *Light. Res. Technol.* **2005**, *37*, 41–59.
31. Mardaljevic, J.; Andersen, M.; Roy, N.; Christoffersen, J. Daylighting Metrics: Is There a Relation between Useful Daylight Illuminance and Daylight Glare Probability? In Proceedings of the Building Simulation and Optimization Conference (BSO12), Loughborough, UK, 10–11 September 2012.
32. Australian Government. *National Greenhouse Accounts Factors. Australian National Greenhouse Accounts*; Department of the Environment: Canberra, Australia, 2014.
33. Akubiec, J.A.; Reinhart, C.F. The "adaptive zone"—A concept for assessing discomfort glare throughout daylit spaces. *Light. Res. Technol.* **2012**, *44*, 149–170.

34. Reinhart, C.F.; Wienold, J. The daylighting dashboard—A simulation-based design analysis for daylit spaces. *Build. Environ.* **2011**, *46*, 386–396.

10

Optimizing Whole House Deep Energy Retrofit Packages: A Case Study of Existing Chicago-Area Homes

Honnie Aguilar Leinartas and Brent Stephens *

Department of Civil, Architectural and Environmental Engineering, Illinois Institute of Technology, Chicago, IL 60616, USA; E-Mail: honnie.a.leinartas@gmail.com

* Author to whom correspondence should be addressed; E-Mail: brent@iit.edu

Academic Editor: David Dernie

Abstract: Improving the energy efficiency of the residential building stock plays a key role in mitigating global climate change. New guidelines are targeting widespread application of deep energy retrofits to existing homes that reduce their annual energy use by 50%, but questions remain as to how to identify and prioritize the most cost-effective retrofit measures. This work demonstrates the utility of whole building energy simulation and optimization software to construct a "tool-box" of prescriptive deep energy retrofits that can be applied to large portions of the existing housing stock. We consider 10 generally representative typology groups of existing single-family detached homes built prior to 1978 in the Chicago area for identifying cost-optimal deep energy retrofit packages. Simulations were conducted in BEopt and EnergyPlus operating on a cloud-computing platform to first identify cost-optimal enclosure retrofits and then identify cost-optimal upgrades to heating, ventilation, and air-conditioning (HVAC) systems. Results reveal that prescriptive retrofit packages achieving at least 50% site energy savings can be defined for most homes through a combination of envelope retrofits, lighting upgrades, and upgrades to existing HVAC system efficiency or conversion to mini-split heat pumps. The average upfront cost of retrofits is estimated to be ~\$14,400, resulting in average annual site energy savings of ~54% and an average simple payback period of ~25 years. Widespread application of these prescriptive retrofit packages across the existing Chicago-area residential building stock is predicted to reduce annual site energy use by 3.7×10^{16} J and yield approximately \$280 million USD in annual energy savings.

Keywords: BEopt; EnergyPlus; energy simulation; deep energy retrofit; residential

1. Introduction

Improving the energy efficiency of the building stock plays a key role in mitigating global climate change [1]. Importantly in the U.S., residential buildings are responsible for over 20% of primary energy consumption [2] and greenhouse gas emissions [3]. The vast majority of this share is associated with detached single-family dwellings, which make up approximately 70% of the existing residential building stock [4]. Correspondingly, several governmental and non-governmental organizations have proposed aggressive targets for applying deep energy retrofits to achieve a minimum of 50% reductions in annual energy use by existing residential buildings [5,6]. These goals create both a challenge and an opportunity for older homes in the U.S. and abroad.

There are a number of approaches commonly used to select and apply energy efficiency retrofits to buildings. In practice, it is common to apply a package, or "tool-box," of prescriptive deep energy retrofits [7–11]. This approach has the advantage of reducing the time and effort required for energy analysis on individual homes while allowing for efficient training of contractors to apply the same, or similar, retrofits to a large number of homes in a systematic way. However, prescriptive approaches can also miss important opportunities for energy savings. Conversely, other more advanced performance-based approaches, such as calibrated whole building energy models coupled with optimization algorithms [12–17], are also used to identify energy saving opportunities beyond those targeted by prescriptive measures, but they require more intensive analyses for individual buildings. Given the respective advantages of both prescriptive and simulation-based optimization approaches, this work attempts to unify both approaches by demonstrating the utility of simulation-based optimization methods to identify cost-effective prescriptive deep energy retrofit packages for a small number of generally representative homes that achieve at least 50% site energy savings, which can then be applied to a larger number of relatively similar homes in a particular area. A sample of older existing homes in Chicago, IL and surrounding areas is used as a case study for this demonstration using a combination of BEopt and EnergyPlus operating on a cloud computing services platform.

2. Methods

The next sections describe the identification of model homes that are generally representative of existing homes in the Chicago area (Section 2.1); translation of those characteristics to model inputs for the BEopt software package (Section 2.2); and methods for the energy simulation and retrofit optimization procedures (Section 2.3).

2.1. Identifying Representative Existing Homes in the Chicago Area

The Chicagoland region consists of seven counties and more than 3.3 million single-family homes, representing 63% of the population of single-family homes in Illinois [18]. There are over 900,000 older single-family homes in Cook County and Chicagoland areas that were built prior to the year 1978, when requirements for installing insulation in buildings were written into the local energy code.

Pre-1978 homes account for 82% of the single-family residential building population and approximately 59% of all natural gas consumption in the area [19]. These homes are notorious for being poorly insulated, having poor air sealing, and containing low-efficiency heating and air-conditioning equipment, and are thus found to be highly energy intensive relative to other home types.

A recent study by the Partnership for Advanced Residential Retrofit (PARR) surveyed the Chicagoland residential housing stock and identified 15 common single-family housing typology groups that accurately represent the vast majority of single-family homes in the area [18]. The groups were characterized based on data collected from the Cook County assessors, utility billing history, and prior energy efficiency programs. One of the goals of the PARR report was to identify energy efficiency solutions to yield 30% annual energy savings using a simulation-based optimization approach that relied on just one generally representative home model for each of the 15 groups. Here we advance this previous work by considering the 10 least efficient typology groups of single-family homes out of the original group of 15 home types (all built prior to 1978) for finding cost-optimal deep energy retrofit packages that can achieve 50% annual site energy reductions. The 10 typology groups are represented by individual home models with the following characteristics, listed according to exterior wall structure, time period of construction, number of stories, and average floor area:

- Group 1: Brick, 1942–1978, 1–1.5 stories (no split level), 110 m^2 (1180 ft^2)
- Group 2: Brick, Pre-1978, Split level (1.5 stories), 120 m^2 (1310 ft^2)
- Group 3: Brick, 1942–1978, 2 stories, 190 m^2 (2060 ft^2)
- Group 4: Brick, Pre-1942, 1–1.5 stories (no split level), 110 m^2 (1140 ft^2)
- Group 5: Brick, Pre-1942, 2 stories, 170 m^2 (1870 ft^2)
- Group 6: Frame, All years, Split level (1.5 stories), 120 m^2 (1340 ft^2)
- Group 7: Frame, 1942–1978, 1–1.5 stories (no split level), 110 m^2 (1180 ft^2)
- Group 8: Frame, 1942–1978, 2 stories, 150 m^2 (1580 ft^2)
- Group 9: Frame, Pre-1942, 1–1.5 stories, 120 m^2 (1240 ft^2)
- Group 10: Frame, Pre-1942, 2 stories, 190 m^2 (2060 ft^2)

Full details of these home characteristics are described in the original PARR report [18], which also demonstrates how the individual home models accurately represent the average home in each group. A few adjustments were made to the base model inputs to yield reasonably accurate predictions of annual source and site energy use when compared to average utility bills from each home group reported in [18]. We targeted pre-retrofit simulation accuracy of within 20% of the average annual source energy use, electricity use, and natural gas consumption from each group simulated in the PARR report and we were successful in replicating pre-retrofit results for each home group within this defined range. The only deviations from [18] included modeling Group 1 homes with slab construction, Group 4 with a floored attic, and Group 9 with R-3 attic insulation. Otherwise, the same model inputs for 10 of the representative home types were used for energy simulation and optimization in the BEopt software package, as described in Section 2.2. Screenshots of each home model are shown in Figure 1.

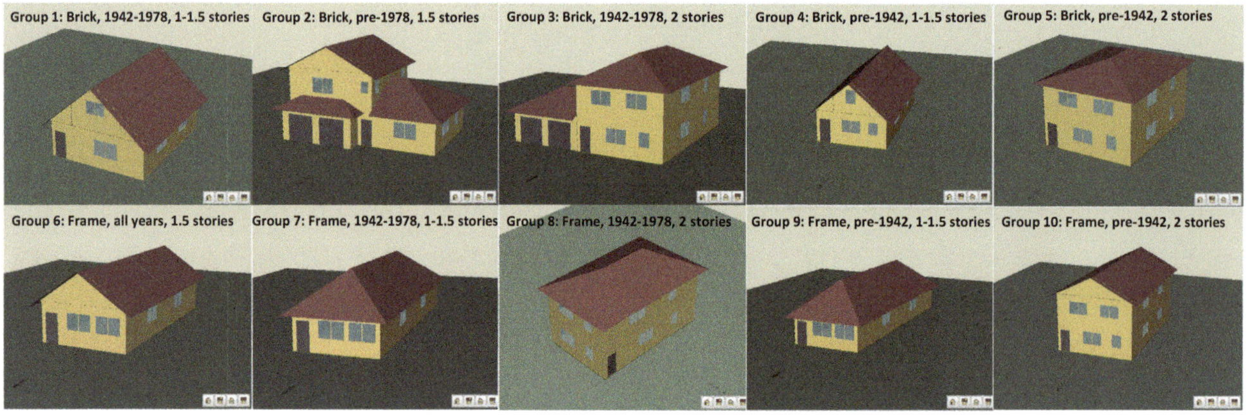

Figure 1. Screen capture of representative homes from the 10 typology groups as modeled in BEopt.

2.2. Model Inputs for BEopt

2.2.1. General Assumptions

BEopt Version 2.2.0.2 was used with EnergyPlus 8.1.0 as the simulation engine to model each home type. The following general assumptions were applied to all home models, in direct alignment with the PARR report [18]:

- Lot size, distance to neighbors, and orientation were based on homeowner input from the PARR research team and information of a standard city and suburban lot with all homes facing north.
- The heating set point was set at 21.1 °C (70 °F) and the cooling set point was 22.2 °C (72 °F).
- All windows were modeled as double-clear (*i.e.*, window glass plus storm window) in all cases. Note that there is not a "Double-Clear" window option in the BEopt version used for this work, but a custom option was created based on the same parameter from a previous version (BEopt version 1.0).
- Mechanical ventilation was modeled as "none" for pre-1942 homes and "spot ventilation" for the other vintages (*i.e.*, spot exhaust fans in kitchen and bathrooms). This parameter option was also not included in the newest version of BEopt and was created based on BEopt version 1.0 parameter with the same name.
- All homes were modeled with the assumption that they were fitted with relatively modern appliances to reflect common energy upgrades typically made to houses over time (*i.e.*, pre-1942 homes were not assumed to have appliances from 1942 but were assumed to have upgraded appliances more consistent with 10–20 year old vintages).
- Typical usage was assumed for miscellaneous electrical loads for all groups (*i.e.*, "Other Electrical Loads" were set at "1.00").

Next, assumptions for relevant characteristics of each home typology are described, including wall and roof construction type and insulation level, envelope airtightness, and HVAC system type and efficiency. They all assume the same "today" conditions from the PARR report [18] (*i.e.*, very old homes were assumed to have undergone at least minor energy efficiency retrofits within the last 50 years to bring their performance closer in line with what is observed in Chicago area homes).

2.2.2. Detailed Assumptions: Pre-1942 Construction (Groups 4–6, 9, and 10)

Detailed assumptions for existing homes built prior to 1942 are as follows. Brick walls (Groups 4 and 5) were double brick construction with enclosure layers from outside to inside as: 10.2 cm (4 inch) brick, 2.5 cm (1 inch) airspace (the weep space, largely mortar and bricks connecting the inside and outside layer), an inside brick layer, wood lath with no insulation in the spaces, and 1.6 cm (5/8 inch) drywall simulating plaster. A structural brick wall was not a default option in BEopt, so a custom option was made based on concrete masonry unit (CMU) structured walls (15.2 cm (6 inch) hollow CMU) with the same physical specifications as brick. Costs associated with these options were carried over from the CMU base. Wood frame walls (Groups 6, 9, and 10) were modeled from the outside to inside as: siding, sheathing, 38 × 89 mm (2 × 4) uninsulated walls, and 1.6 cm (5/8 inch) drywall. Attic insulation was assumed to be a minimal RSI-1.2 (R-7). Bungalows were modeled with finished attic space in the upper half story. Inter-zonal knee-walls were modeled with no insulation and the roof above the living space was also modeled with no insulation. Basements were modeled as uninsulated as well. Enclosure airtightness assumptions were "very leaky" for 1-story and "leaky" for 2-story homes. Boilers with 80% efficiency were used for heating systems (representing common efficiencies from the 1980s) and window air-conditioning units with a coefficient of performance (COP) of 2.93 (energy efficiency ratio, or EER, of 10) were assumed for cooling systems. Gas water heaters were assumed with an energy factor of 0.48, and a gas dryer was also assumed.

2.2.3. Detailed Assumptions: 1942–1978 Construction (Groups 1–3, 7 and 8)

Detailed assumptions for existing homes built between 1942 and 1978 were as follows. For brick walls (Groups 1–3), 2.5 cm (1 inch) of RSI-0.5 (R-3) fiberglass was included between furring strips. For wood frame (Groups 7 and 8) and inter-zonal walls, an RSI-1.2 (R-7) fiberglass batt was modeled as the typical insulation level for that era. Attic insulation was assumed to be RSI-3.4 (R-19). Envelope airtightness was upgraded to "leaky" for 1-story homes and "typical" for 2-story homes. Gas furnaces with AFUE of 78% were assumed to have replaced boilers in the average house in this time frame. Central cooling systems were modeled with a COP of 2.6 (EER = 8.9). Gas water heaters were assumed with energy factors (EF) of 0.48. Additionally, a special case assumption was applied to improve the accuracy of the existing model for Group 1. The majority of houses in this group had slab floor construction (no basement), which best fit the measured annual energy use intensity [18]; therefore, slab construction was modeled in this case only.

2.3. Retrofit Optimizations Using BEopt and EnergyPlus

BEopt was used in "optimization mode", which allows users to select pre-defined ranges of a number of model input parameters to consider for optimization (Figure 2 shows an example of ceiling assembly options from just one scenario). BEopt then performs energy simulations using a sequential search method to vary across all selected inputs and march towards a cost-optimal savings solution [20]. The sequential search method involves systematically applying permutations of all selected ranges of input parameters and searching for the most cost-effective option at each sequential point along the path towards zero net energy use. The method seeks the steepest downward slope along a least cost line to identify solutions that provide the most energy savings for the least amount of investment. Upfront material and construction costs for each retrofit measure were kept as default values in BEopt, which sources cost information from R.S. Means and other databases [20]. Based on the results, the most cost-effective option is selected as an optimal point on the path and included in an updated building configuration. The process is repeated until a specified desired outcome is reached (e.g., a user-defined source energy, site energy, or green house gas emissions reduction target). Each distinct simulation represents an "iteration point", or a unique building configuration.

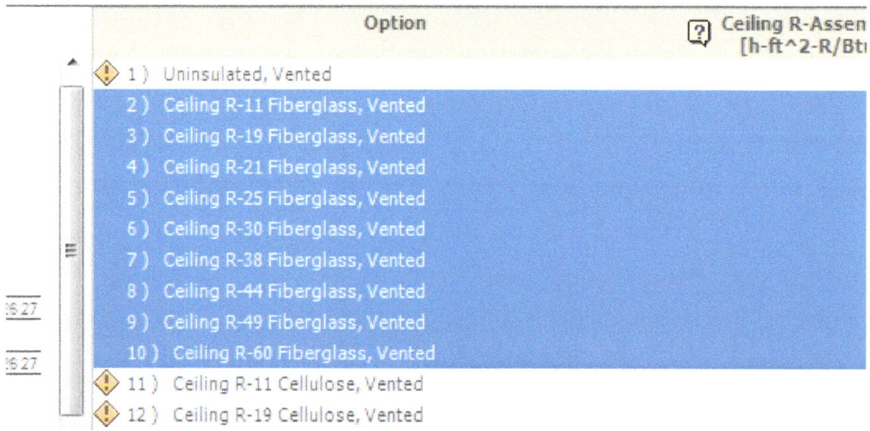

Figure 2. Screenshot of BEopt option list illustrating valid and invalid options for retrofit optimizations.

We began our optimizations for each typology group by selecting only building enclosure retrofit options in order to identify the least cost packages for reducing heating and cooling loads prior to seeking higher efficiency HVAC solutions. This is a common approach to identifying whole-house energy efficiency retrofit packages: reduce loads, and then improve equipment [21]. This approach also served to drastically reduce the computation time required to perform the simulations. Subsequently, optimizations based on the envelope-retrofitted models were then performed using a range of several HVAC system type and efficiency options. We then used the combined results to identify final cost-optimal packages for each home model representing its typology group as those that achieve at least 50% site energy savings while minimizing annualized energy related costs (AERC) and simple payback periods and providing the highest modified internal rate of return (MIRR). In each case, parameter options that BEopt deemed inappropriate or ineligible, including those for which costs could not be calculated, were denoted by an exclamation mark next to the option on the options selection screen (e.g., options 1, 11, and 12 in Figure 2). These options were not selected for use in the optimizations.

Possible building enclosure optimization options for each group included the following:

- **Exterior Wood Stud Walls (Groups 6–10):** For homes with frame exterior wall constructions, all higher thermal resistance options for exterior walls that were deemed valid/eligible by BEopt were chosen for optimization. These included most default 38 × 89-mm (2 × 4), 40.6 cm (16 inches) on center (o.c.) framing options insulated with fiberglass batt, blown-in cellulose, blown-in fiberglass, or spray foam insulation. Possible insulation levels ranged from RSI-1.2 $m^2 \cdot K/W$ (R-7 $h \cdot ft^2 \cdot {}^\circ F/Btu$) (fiberglass batts) to RSI-4.1 (R-23) (spray foam).

- **Exterior Brick Walls (Groups 1–5):** For homes with brick exterior wall constructions, new CMU options for the weep space plus 10.2-cm (4-inch) interior brick were created based off existing 15.2-cm (6-inch) hollow CMU options (BEopt does not have built-in double brick wall options). All specifications for the masonry (*i.e.*, block thickness, material conductivity, and density) were modified to reflect the specifications of brick, but all labor and material costs associated with the original default options were kept as defaults for the new brick options. These options include the double brick wall insulated with either of the following insulation types/levels within the interior side furring cavity: RSI-0.53 (R-3) fiberglass batt (5.1-cm or 2-inch furring cavity), RSI-1.76 (R-10) XPS (5.1-cm or 2-inch furring cavity), RSI-2.3 (R-13) closed cell spray foam (5.1-cm or 2-inch furring cavity), RSI-2.1 (R-12) polyiso (5.1-cm or 2-inch furring cavity), or RSI-3.3 (R-19) fiberglass batt within a 38 × 140 mm (2 × 6), 60.9 cm (24 inches) o.c. furring cavity.
- **Wall Sheathing (Groups 6–10):** For homes with frame exterior wall constructions, the following insulation options were chosen for optimization as being installed between the exterior finish and the frame wall: RSI-0.9 (R-5) XPS, RSI-1.8 (R-10) XPS, RSI-2.6 (R-15) XPS, RSI-1.1 (R-6) polyiso, and RSI-2.1 (R-12) polyiso.
- **Exterior Finish:** All colors for the appropriate finish were chosen (*i.e.*, light or medium/dark brick for brick constructed homes, and light or medium/dark wood siding for frame constructed homes).
- **Interzonal Walls:** In groups where interzonal walls are present, all eligible options for 38 × 89 mm (2 × 4) 40.6 cm (16 inches) o.c. interzonal walls were chosen. This included walls insulated with either fiberglass batt, blown-in cellulose, blown-in fiberglass, or spray foam insulation, with insulation levels ranging from RSI-1.2 (R-7) (fiberglass batts) to RSI-4.1 (R-23) (spray foam).
- **Unfinished Attic:** In groups where unfinished attic space was present, all eligible BEopt options were chosen for optimization. Specific options varied by group and were dependent mainly on vintage and existing insulation. Options were considered appropriate if they represented an improvement in thermal resistance. Selected options included but were not limited to options with insulation installed in either the attic floor space (ceiling of the finished space), within the cavity space between 38 × 140 mm (2 × 6) rafters (roof), or both ceiling and roof (in a few cases). Insulation in the form of vented, blown-in fiberglass; vented, blown-in cellulose; fiberglass batt; or closed cell spray foam in the ceiling space, and/or closed cell spray foam in the roof were chosen with R-values ranging from RSI-1.9 (R-11) to RSI 10.6 (R-60) for both types of insulation.
- **Finished Roof:** In groups where finished attic space was present, all eligible options with an R-value greater than or equal to the existing levels were considered, including the existing insulation, RSI 3.3 (R-19) blown-in fiberglass within 38 × 140 mm (2 × 6) rafters, both alone and in combination with XPS having thermal resistance of RSI-2.6, 3.5, or 4.4 (R-value of 15, 20, or 25).

- **Roof Material:** For all groups, the options chosen for roof material included "Asphalt shingles" in all default colors: dark, medium, light, or white or cool colors.
- **Radiant Barrier:** For all groups, both the available options for this parameter were chosen; "none" (the existing condition), or "double-sided, foil".
- **Unfinished Basement:** All default options for insulating and finishing of the basement interior perimeters were chosen for all groups.
- **Interzonal Floor:** Interzonal floors separate conditioned from unconditioned space (e.g., the floor separating the unfinished attic from the conditioned space below, or the ceiling area/floor area between the unconditioned garage and conditioned living space above). In models of groups for which interzonal floors were present, all default options were selected for optimization with this parameter.
- **Window Type:** For all groups, all eligible window options were chosen for optimization. This includes all double or triple pane; low, medium, or high gain low-e coated; insulated or non-metal frame; and air or argon filled configurations. An option for backside windows (*i.e.*, south facing) to have a high solar heat gain coefficient (SHGC) was also included among the options chosen for optimization.
- **Air Leakage:** Options selected for the air leakage parameter was dependent on group vintage and number of stories. Only the existing condition and the option that represented a single step up were chosen for each group (e.g., "very leaky" to "leaky", "leaky" to "typical", or "typical" to "tight"), as two-step upgrades were not assumed to be realistic for the pre-1978 vintage [18].
- **Mechanical Ventilation:** The selection of options for mechanical ventilation parameters was dependent on group vintage. For pre-1942 groups, all default options were selected, but for 1942–1978, all eligible options were selected for optimization. "Supply" and "supply, 50% of ASHRAE 62.2" options were not deemed eligible for simulation for the latter vintages.

Once optimal cost-effective enclosure retrofits were identified, another set of simulations was conducted using the following options as possible HVAC system types and characteristics:

- **Central A/C:** Options that represented improvements to the existing central air-conditioning (A/C) system efficiency were deemed eligible by BEopt. All default options with a COP equal to or higher than 2.6 (EER equal to or higher than 8.9) were chosen for optimizing central A/C systems.
- **Room A/C:** All BEopt default options for room A/C parameters were deemed eligible and were only utilized for pre-1942 groups with existing boiler and room A/C combinations (Groups 4–6, 9, and 10).
- **Furnace:** Only options that represented improvements to the existing furnace efficiency were deemed eligible and selected for optimization. Options include gas furnaces with AFUE ratings of 78% and above (up to 98%) and an electric furnace with an AFUE of 100%.
- **Boiler:** Similar to the A/C and furnace parameters, the options chosen for optimization with boilers included only those that represent an improvement to the existing boiler efficiency. This includes both condensing and non-condensing gas boilers with an AFUE of 80% and higher (up to 98%). Boiler parameters were only utilized for pre-1942 groups with existing boiler and room A/C combinations (Groups 4–6, 9, and 10).

- **Mini-Split Heat Pumps (MSHP):** All default options were selected for optimizing mini-split heat pump systems for all groups.
- **Electric Baseboards:** BEopt requires for electric baseboard heaters to be selected when simulating MSHP system options; thus, the default option of 100% efficiency was chosen for electric baseboard heater optimizations.
- **Ground-Source Heat Pumps (GSHP):** Chicago's close proximity to Lake Michigan has a large influence on the thermal conductivity of the area's soil and thus the potential to incorporate geothermal heating in the form of ground-source heat pumps (GSHPs) into homes. The soil of the entire Chicago area can be classified as illitic or silty/clayish [22]. The thermal conductivity of saturated silt/clay soil has a thermal conductivity within the range of 1–1.8 W/mK for saturated unfrozen soil and 2–2.5 W/mK for saturated frozen soil [23]. From this information, an average soil thermal conductivity of 1.8 W/mK was used, which corresponds to high conductivity soil in BEopt.
- **Ducts:** The options deemed eligible for optimization by BEopt depended on vintage and the presence of pre-existing ductwork. For groups without pre-existing ductwork (pre-1942 homes including Groups 4–6, 9, and 10), all default options were selected. For all remaining groups (all 1942–1978 groups), only options that represented an increase in thermal resistance or a decrease in duct leakage from the existing conditions were selected, including both insulated and uninsulated ducts with leakage fractions of 15% or less, as well as an option to relocate ducts to finished space. These parameters were only applied to ducted HVAC systems (e.g., furnace-central A/C combinations and GSHP scenarios).

An air-source heat pump (ASHP) case was also initially included among the HVAC systems simulated for optimization. However, conventional ASHPs have been recommended for use mostly in more moderate climates because the effectiveness of ASHPs begins to degrade substantially as outdoor temperatures fall below −8 °C (~17.6 °F) [24]. Moreover, it has been reported that most ASHP systems shut off when ambient temperatures reach freezing and switch to backup heating [25]. Since the average mean seasonal temperature for the Chicago area winters in the past decade is 26.1 °F (−3.2 °C) [26], it was determined that the implementation of ASHP systems would indeed necessitate a backup heating system. However, supplemental heat could not be accurately modeled in BEopt simultaneously with ASHPs, as the methods of heating are considered to be mutually exclusive, unfortunately. While advancements have been made to increase feasibility of their implementation in cold climates [27], the developers of BEopt have yet to make cold climate ASHP options available/appropriate for simulation [28]. Therefore, the results of the simulations involving this particular system were ultimately considered invalid and not used herein.

Finally, the following non-HVAC and non-enclosure options were selected for all cases in all groups:

- **Ceiling Fan:** All default options for ceiling fans were deemed eligible by BEopt; however, the options selected for optimization were limited to "none", "benchmark", "standard efficiency", "high efficiency", and "premium efficiency" ceiling fans.

- **Water Heater:** Options that represented an improvement in efficiency from the existing conditions were deemed eligible. Only those water heaters that operate using gas, electric, or heat pump water heater (HPWH) options were selected.
- **Solar Water Heating (SWH), SWH Azimuth, and SWH Tilt:** All default options for these parameters were considered appropriate and selected for optimization.
- **Lighting:** Lighting options were deemed eligible by BEopt if they consumed less electricity than the existing case and were thus selected for optimization. Included are options for 40% to 100% hardwired lighting to be converted to fluorescent lighting, 40%–100% hardwired and plug-in lighting to be converted to fluorescent lighting, conversion of 50% hardwired and plug-in lighting to fluorescent and 10% to Light-Emitting Diode (LED), and fixed annual lighting electricity consumption allowance of 1300 kWh/year (achieved by any combination of lighting technologies).

The simulations did not include analysis of any building user preference or behavior parameters such as natural ventilation, shading, or large appliances other than water heaters and HVAC equipment (e.g., refrigerators, washers, dryers, *etc.*). The simulations also did not include analysis of any building characteristics that are not practical to change, such as building orientation and distance from neighbors. The options for these parameters were selected based either on PARR assumptions, options that best fit measured or reported annual energy use in the PARR report (e.g., less efficient appliances were chosen to accurately reflect measured energy usage as needed), or BEopt defaults, and left unchanged throughout the simulation processes. For all simulations, the EnergyPlus weather location was set as "USA_IL_Chicago-OHare.Intl.AP.725300_TMY3.epw". Moreover, as the scope of this work includes the entire Chicagoland area, which includes several near suburbs, therefore, the terrain was set to "Suburban" for all simulations.

The most recent release of BEopt (version 2.2) was equipped with capabilities for user-specified utility rates. For the modeling purposes of this work, a Real-Time-Pricing (RTP) electricity cost profile was created based on actual RTP costs from the local electric utility, ComEd, for the year of 2012 [29]. RTP was used instead of average block pricing because utilities are changing to this method (albeit more slowly in some locations compared to others). For natural gas pricing, an average of the monthly residential cost of gas found on the website of the local natural gas utility (Nicor Gas) was used [30]. All other values for economic assumptions (e.g., inflation rate, discount rate, *etc.*) and payment assumptions (e.g., loan interest rate, loan period, marginal income tax rate, *etc.*) were left as the BEopt default values.

The goal of this work was to identify packages that achieve a 50% *site* energy savings (*i.e.*, only the energy used on-site without accounting for primary energy conversion efficiencies). However, the final BEopt optimization simulations were actually set to terminate when a 50% *source* energy savings was achieved, or otherwise continue the simulations until all possible input parameter combinations were exhausted. This was done primarily because a source energy reduction of 50% would ensure that at least a 50% site energy reduction was achieved, given that the source/site conversion ratios were kept at the BEopt default values of 3.15 for electricity and 1.09 for natural gas. This allowed for identifying packages with even higher site energy savings than 50% that were still cost-effective options.

Ultimately none of the simulations achieved 50% source energy savings, so each optimization case actually provided an exhaustive search over all selected input parameters.

Initial BEopt simulations were performed in a Virtual workstation running Windows 7 Professional on a MacBook Air equipped with a 1.8 GHz Intel Core i7 processor and 4 GB 1333 MHZ DDR3. The first simulation was allowed to run for about 15 h, during which only eight "iterations", or 1249 unique simulations (or "iteration points"), were completed before the test was manually terminated. Given this duration, we then decided to outsource simulations to a remote server on the Amazon Elastic Compute Cloud (EC2). For all subsequent simulations, a "C3 High-CPU Eight Extra Large" (c3.8xlarge) instance with 108 compute units, 32 cores, and 60 GB of memory was used to allow for more rapid remote simulations. The chosen instance was setup to use the latest Microsoft Windows Server operating system and BEopt was installed just as it would be on a typical PC computer. Simulations were managed though the Microsoft Remote Desktop application installed on the previously mentioned MacBook. The simulation run time was reduced down to about 1.5 h per optimization simulation case using Amazon EC2 (a case being defined as a set of simulations performed to optimize a particular home type and enclosure or type of HVAC retrofit combination). Ultimately, each group model involved the optimization of a minimum of four cases (e.g., one enclosure optimization case and at least three different HVAC system optimization cases), with each case having a minimum of nine iterations with any individual iteration points. Some cases required the simulation of upwards to 25 iterations with a mean number of about 15 iterations per case. Using the mean 15 iterations per case, it was calculated that it would have required a minimum of about 1125 h to complete simulations on the aforementioned MacBook personal computer. Using Amazon EC2, the required minimum run time was reduced to about 80 h, yielding a ~93% reduction in run time to complete simulations. Similar cloud computing approaches have been used in other recent studies as well [31,32].

3. Results and Discussion

Examples of typical inputs and outputs from the simulations are shown for just one home typology in Section 3.1 (Group 1). This serves to demonstrate the nature of the inputs required for this process and the resulting outputs for each group. Results from all home types are then explored and summarized in Sections 3.2–3.4.

3.1. Demonstration of Results from Group 1 (Brick, 1942–1978, 1–1.5 Stories, No Split Level)

Figure 3 shows full BEopt geometry inputs for the Group 1 representative home. After the base model and existing conditions were modeled for each home type and compared to results from those in the PARR report [18], building enclosure optimization simulations were then performed for each home. The simulations were set to terminate when the least cost package of all possible combinations of the selected parameter options was found. Figure 4 illustrates the results of the enclosure optimization simulation for the Group 1 home type based on annualized energy related costs (AERC) compared to site energy savings, along with the least cost fit line generated by the sequential search optimization method in BEopt. AERC combines the upfront costs of retrofits with the predicted resulting annual energy savings accumulated over the project analysis period (assumed to be the BEopt default of 30 years) and calculates an annualized cost metric for comparing scenarios. Each square symbol in Figure 4 represents a single combination of enclosure parameters or iteration point and the AERC for the existing home as-is (*i.e.*, savings = 0%) is shown with a black circle. Ideally, the most desirable packages would be present in the lower right portion of the graph, with large energy savings and low annualized costs.

In this case, the least cost option that was identified as cost-optimal represents an estimated site energy savings of about 32% with an AERC of about \$1740 per year (compared to an AERC in the existing home of about \$1915 per year). These results demonstrate that considerable energy savings may be achieved in this home type simply by upgrading the enclosure and keeping the existing HVAC systems (the selected enclosure retrofit details are described in full in Section 3.2). It can also be seen that the least cost option has an AERC that is almost \$200 per year less than the existing case for this group. In short, it pays in the long run to install the optimal enclosure retrofits.

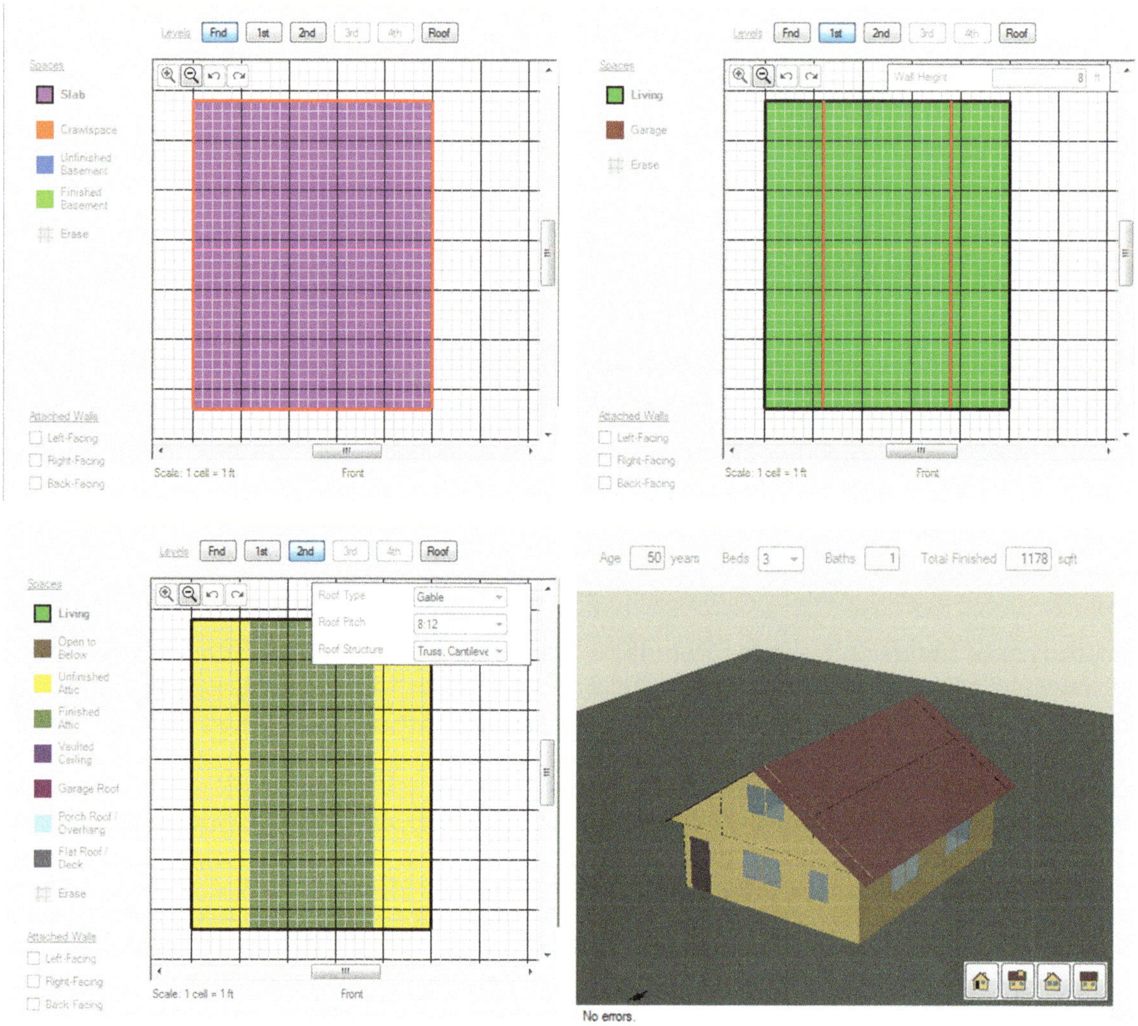

Figure 3. BEopt model for Group 1: Foundation, first, and second (attic) floor layouts and 3D view.

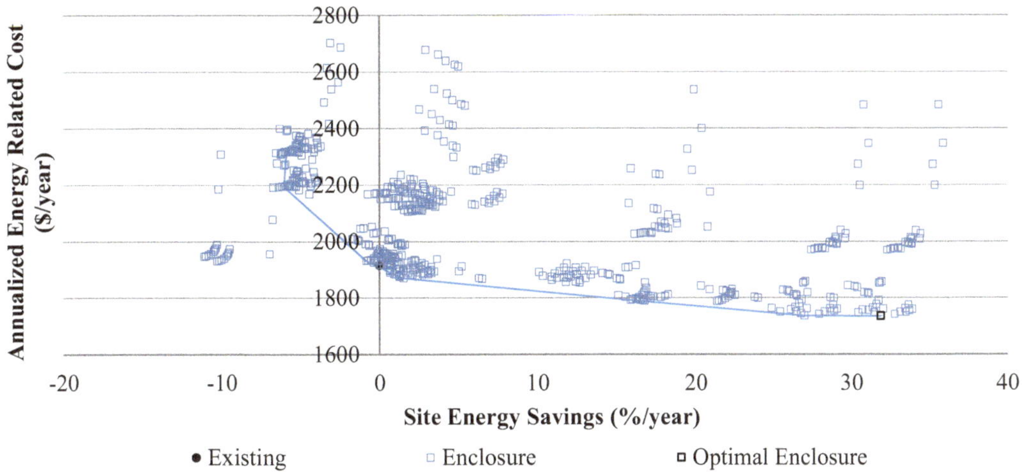

Figure 4. Building enclosure optimization results for Group 1 homes.

After the optimal enclosure retrofits were implemented, the model was then optimized for various HVAC systems in order to advance beyond 32% site energy savings. For Group 1 homes, this included the existing HVAC systems (combination of a furnace and central A/C), mini split heat pump (MSHP), and ground source heat pump (GSHP). Figure 5 shows the results for the HVAC optimization simulations for this group alongside the corresponding least cost fit line generated by the sequential search optimization method in BEopt. Similar to the enclosure results, optimal packages are those that maximize energy savings while minimizing AERC. For each HVAC case, the lowest cost iteration point along the least cost fit line that achieved an annual energy use reduction greater than 50% was chosen as the cost-optimal deep energy retrofit solution (*i.e.*, the MSHP case in Figure 5). This ensured that the package with the combined lowest AERC, highest energy savings, shortest payback period, and highest modified internal rate of return was chosen.

Note that in this example, the optimal package occurs at about 58% site energy savings for both the MSHP and the GSHP systems. Also note that in this example, there are several iteration points that represent options that have a lower AERC than the option chosen to be the optimal MSHP package but are not considered to be part of the least cost fit line. Further inspection revealed that BEopt does not consider these options to be least cost options because they were estimated to have higher payback periods than the cost-optimal package shown in Figure 5, despite the lower AERC. Moreover, the optimal package for the furnace-central A/C combination that represents the least cost at 50% savings occurs at an AERC that is higher than the absolute minimum of the least cost line for this HVAC system. Thus, Figure 5 demonstrates that the optimal retrofit package involves application of the enclosure retrofits from Figure 4 combined with conversion to a MSHP, which is predicted to yield an increase in AERC of only about $135 per year ($2050 per year compared to the original $1915 per year). In this particular home, enclosure retrofits are predicted to have net economic benefits (lower AERC than existing) but cannot achieve 50% energy savings, while additional upgrades to the HVAC system would yield greater than 50% energy savings albeit at a small additional annual cost (higher AERC).

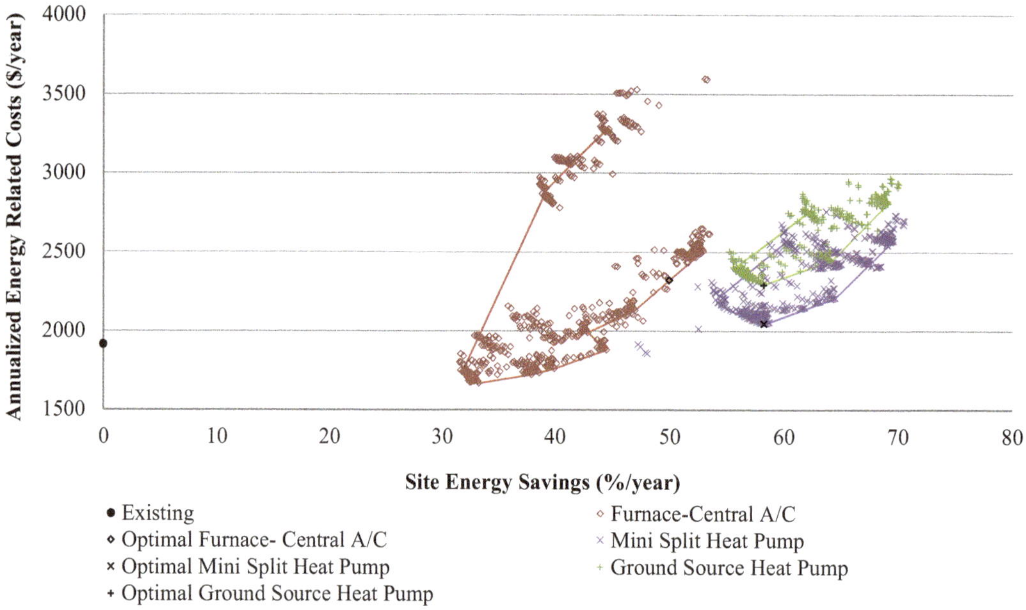

Figure 5. HVAC optimization results for Group 1 homes.

Next, Figure 6 illustrates the breakdown of estimated energy consumption by end use for the Group 1 model, including the predicted energy use at the existing/current condition, after the implementation of optimal enclosure, and for each of the optimal HVAC packages in combination with the optimal enclosure package. Again, it is clear that the conversion to a MSHP is estimated to result in the greatest energy savings out of all simulated options for this particular group. The majority of savings stem from a reduction in heating energy consumption.

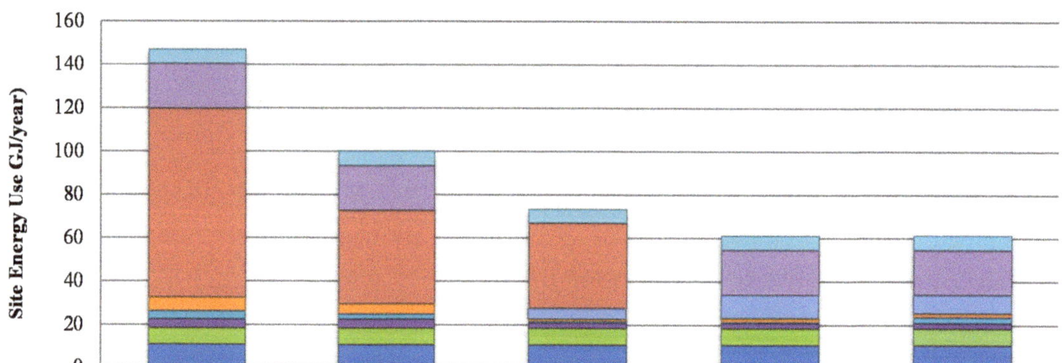

	Today	Enclosure: Iter 7, Pt 38	Furnace-Central A/C: Iter 14, Pt 15**	MSHP: Iter 6, Pt 7**	GSHP: Iter 9, Pt 3**
Lg. Appl. (G)	6.88	6.88	6.88	6.88	6.88
Hot Water (G)	20.86	20.84	0.00	20.84	20.78
Hot Water (E)	0.00	0.00	0.00	0.00	0.00
Heating (G)	87.08	43.08	39.25	0.00	0.00
Heating (E)	0.00	0.00	5.04	10.82	8.53
Cooling (E)	6.34	4.79	1.08	1.82	1.74
HVAC Fan/Pump (E)	3.79	2.38	0.58	0.23	2.71
Lights (E)	4.11	4.11	2.67	2.67	2.67
Lg. Appl. (E)	7.60	7.60	7.60	7.60	7.60
Vent Fan (E)	0.06	0.06	0.06	0.06	0.06
Misc. (E)	10.66	10.66	10.66	10.66	10.66

Figure 6. Site energy use of critical optimal iteration points for Group 1 homes.

Finally, there are two additional metrics to compare the overall costs and benefits of optimal retrofit packages for each home type. First, the "simple payback period" of a project represents the time required for a repayment of the original investment, but does not incorporate future costs (e.g., replacement costs or technology upgrades), future changes to cost savings (e.g., utility cost changes due to fuel price escalation), or the time value of money or loan financing. Figure 7 illustrates the simple payback periods for each of the optimal HVAC packages (combined with the optimal enclosure) for Group 1 homes. When combined with the optimal enclosure retrofit, a conversion to MSHP in this home is predicted to yield annual site energy savings of 58.2% (the same values shown in Figures 5 and 6) with a simple payback period of about 24 years. Further, this represents a modified internal rate of return (MIRR) of about 3.8%. These two metrics can help homeowners and decision makers better understand the costs and benefits of their investments. While a 24-year simple payback period seems like a long period of time, an MIRR of 3.8% is actually much higher than the current rate of return on, for example, U.S. savings bonds.

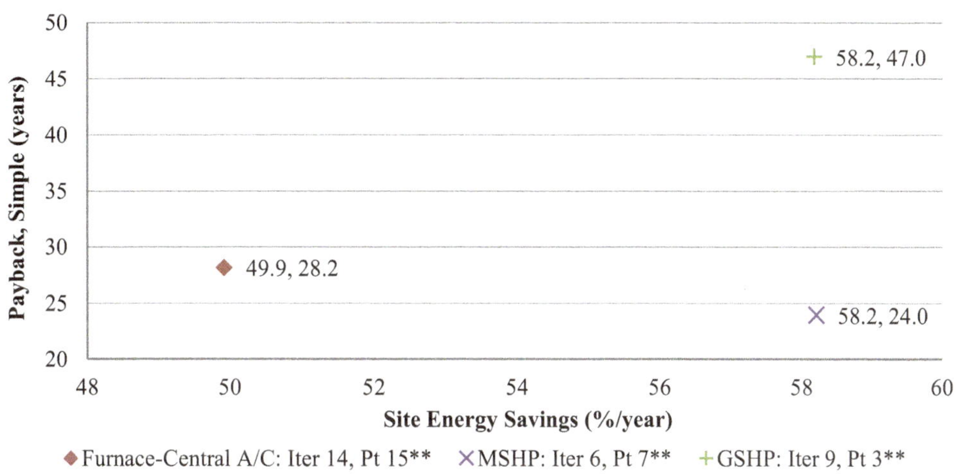

Figure 7. Simple payback period of critical optimal iteration points for Group 1 homes.

3.2. Optimal Cost-Effective Prescriptive Retrofit Packages for Each Pre-1978 Home Typology Group

A summary of the cost-optimal prescriptive deep energy retrofit packages resulting from the two-step optimization process applied to all 10 typology groups is shown in the following tables (Tables 1–10). Metrics for evaluating and comparing each retrofit package include: initial costs, annualized energy related costs (AERC), predicted site and source energy savings, simple payback periods, and the modified internal rate of return (MIRR).

Group 1: The cost-optimal retrofit package for Group 1 homes calls for an increase in exterior wall insulation (from RSI-0.53 to RSI-3.3 using fiberglass batts within a thicker wall construction) and unfinished attic insulation (improved to blown-in closed cell spray foam insulation), an increase in air tightness of the enclosure (from "leaky" to "typical"), and a conversion from the existing furnace and central A/C system to a high-efficiency MSHP system with electric baseboard heating (Table 1). It was also recommended that 100% of the lighting be converted to fluorescent bulbs for both hardwired and plug-in fixtures. A more energy efficient water heater was not found to be a cost-effective upgrade for this particular group. The total upfront costs of this optimal retrofit package is predicted to be about

$11,200, providing ~58% site energy savings with a simple payback period of 24 years and a MIRR of 3.8%. Ultimately, greater than 50% energy savings can likely be achieved in this home with additional annualized energy-related costs of only about $132 per year.

Table 1. Group 1: Brick, 1942–1978, 1–1.5 stories (no split level) Summary of retrofit package and cost and energy conservation implications.

Parameter name	Existing characteristics	Optimal enclosure + HVAC retrofit package
Exterior Wall (Masonry)	10.2-cm Hollow Brick, RSI-0.53 Fiberglass Batt, 2.5-cm furring, 61 cm o.c.	10.2-cm Hollow Brick, RSI-3.3 Fiberglass Batt, 38 × 140 mm, 61 cm o.c.
Unfinished Attic	Ceiling RSI-3.3 Fiberglass (Blown-in), Vented	Roof RSI-3.3 Closed Cell Spray Foam
Air Leakage	Leaky	Typical
Central A/C	COP 2.6	None
Furnace	Gas, 78% AFUE	None
Electric Baseboard	None	100% Efficiency
Mini Split Heat Pump	None	COP 6.1, Avg Seasonal COP 3.2
Ducts	Typical, Uninsulated (Unfinished Attic)	None
Water Heater	Gas 54% EF	No Change
Lighting	Benchmark	100% Fluorescent, Hardwired & Plug-in
Initial Cost ($)	$0	$11,242
AERC ($/year)	$1915	$2047
% Site (Source) Energy Savings	0	58.2% (39.9%)
Simple payback period (years)	N/A	24 years
Modified internal rate of return (MIRR, %)	N/A	3.8%

Group 2: Similar upgrades were recommended for the Group 2 home (see Table 2), including increasing thermal resistance in the exterior walls (to RSI-3.3 fiberglass batts in thicker wall), interzonal walls (to RSI-2.3 blown-in fiberglass), and the unfinished attic (to RSI-4.4 blown-in fiberglass), improving the air tightness of the enclosure (from "leaky" to "typical"), and converting to 100% fluorescent lighting. A conversion to a MSHP and electric baseboard heating was also recommended for Group 2. The total upfront costs of this retrofit package is predicted to be about $11,200, providing ~61% site energy savings with a simple payback period of 23 years and a MIRR of 3.9%. Thus, greater than 50% energy savings can likely be achieved in this home with additional annualized energy-related costs of only about $144 per year.

Group 3: For Group 3 homes, similar upgrades to the enclosure insulation and airtightness were suggested (RSI-3.3 and RSI-4.4 wall and attic insulation, respectively, and increasing airtightness from "typical" to "tight"), as well as a conversion to 100% fluorescent lighting (see Table 3). However, unlike the previous two groups, an update to the existing furnace-central A/C system (from a COP of

2.6 to 6.5 and from 78% AFUE to 98%) was found to be cost-optimal. It was also recommended that the existing ductwork be moved from the typical location in the unfinished basement to finished space. Additionally, an update from the existing water heater to a tankless water heater was found to be a cost effective upgrade. The total upfront costs of this retrofit package is predicted to be about $17,500, providing ~50% site energy savings with a simple payback period of 15 years and a MIRR of 5.9%. In this home type, we predict that these energy savings goals can be met at an annualized energy-related cost that is actually about $84 per year lower than the existing home.

Table 2. Group 2: Brick, Pre-1978, Split level (1.5 stories) Summary of retrofit package and cost and energy conservation implications.

Parameter name	Existing characteristics	Optimal enclosure + HVAC retrofit package
Exterior Wall (Masonry)	10.2-cm Hollow Brick, RSI-0.53 Fiberglass Batt, 2.5-cm furring, 60.9 cm o.c.	10.2-cm Hollow Brick, RSI-3.3 Fiberglass Batt, 38 × 140 mm, 60.9 cm o.c.
Interzonal Walls	RSI-1.2 Fiberglass Batt, GR-3, 39 × 89 mm, 40.6 cm o.c.	RSI-2.3 Fiberglass (Blown-in), Gr-1, 39 × 89 mm, 40.6 cm o.c.
Unfinished Attic	Ceiling RSI-1.9 Fiberglass (Blown-in), Vented	Ceiling RSI-4.4 Fiberglass (Blown-in), Vented
Air Leakage	Leaky	Typical
Central A/C	COP 2.6	None
Furnace	Gas 78% AFUE	None
Electric Baseboard	None	100% Efficiency
Mini Split Heat Pump	None	COP 6.6, Avg Seasonal COP 3.3
Ducts	Typical, Uninsulated (Unfinished Basement)	None
Water Heater	Gas, 54% EF	No Change
Lighting	Benchmark	100% Fluorescent, Hardwired & Plug-in
Initial Cost ($)	$0	$11,190
AERC ($/year)	$2081	$2225
% Site (Source) Energy Savings	0	61.6% (40.8%)
Simple payback period (years)	N/A	23 years
Modified internal rate of return (MIRR, %)	N/A	3.9%

Group 4: The optimal retrofit package for Group 4 again includes upgrades to the enclosure insulation and air tightness, a conversion to 100% fluorescent lighting, and conversion to a MSHP (see Table 4). Exterior wall, interzonal wall, and ceiling insulation should be increased to RSI-3.3, RSI-2.3, and RSI-3.3, respectively, using fiberglass batts or blown-in fiberglass. Airtightness should be increased from "very leaky" to "leaky", or higher. A water heater upgrade was not suggested for this group. The total upfront costs of this retrofit package is predicted to be about $10,500, providing ~61% site energy savings with a simple payback period of 31 years and a MIRR of 4.5%. For

approximately $250 more in AERC, this home type is predicted to achieve energy savings beyond the original 50% target.

Table 3. Group 3: Brick, 1942–1978, two stories Summary of retrofit package and cost and energy conservation implications.

Parameter name	Existing characteristics	Optimal enclosure + HVAC retrofit package
Exterior Wall (Masonry)	10.2-cm Hollow Brick, RSI-0.53 Fiberglass Batt, 2.5-cm furring, 60.9 cm o.c.	10.2-cm Hollow Brick, RSI-3.3 Fiberglass Batt, 38 × 140 mm, 60.9 cm o.c.
Unfinished Attic	Ceiling RSI-1.9 Fiberglass (Blown-in), Vented	Ceiling RSI-4.4 Fiberglass (Blown-in), Vented
Air Leakage	Typical	Tight
Central A/C	COP 2.6	COP 6.5
Furnace	Gas, 78% AFUE	Gas, 98% AFUE
Ducts	Typical, Uninsulated (Unfinished Basement)	In Finished Space
Water Heater	Gas, 54% EF	Gas Tankless
Lighting	Benchmark	100% Fluorescent Hardwired & Plug-in
Initial Cost ($)	$0	$17,506
AERC ($/year)	$2430	$2346
% Site (Source) Energy Savings	0	49.9% (45.5%)
Simple payback period (years)	N/A	15 years
Modified internal rate of return (MIRR, %)	N/A	5.9%

Table 4. Group 4: Brick, Pre-1942, 1–1.5 stories (no split level) Summary of retrofit package and cost and energy conservation implications.

Parameter name	Existing characteristics	Optimal enclosure + HVAC retrofit package
Exterior Wall (Masonry)	10.2-cm Hollow Brick, Uninsulated, 2.5-cm furring, 60.9 cm o.c.	10.2-cm Hollow Brick, RSI-3.3 Fiberglass Batt, 38 × 140 mm, 60.9 cm o.c.
Interzonal Walls	Uninsulated, 39 × 89 mm, 40.6 cm o.c.	RSI-2.3 Fiberglass (Blown-in), Gr-1, 39 × 89 mm, 40.6 cm o.c.
Unfinished Attic	Ceiling RSI-1.2 Fiberglass (Blown-in), Vented	Ceiling RSI-3.3 Fiberglass (Blown-in), Vented
Air Leakage	Very Leaky	Leaky
Room A/C	COP 2.9	None
Boiler	Gas, Hot Water, Forced, 80% AFUE	None
Electric Baseboard	None	100% Efficiency
Mini Split Heat Pump	None	COP 6.1, Avg Seasonal COP 3.25
Water Heater	Gas, 54% EF	No Change

Table 4. *Cont.*

Parameter name	Existing characteristics	Optimal enclosure + HVAC retrofit package
Lighting	Benchmark	100% Fluorescent, Hardwired & Plug-in
Initial Cost ($)	$0	$10,565
AERC ($/year)	$1840	$2090
% Site (Source) Energy Savings	0	60.6% (37.6%)
Simple payback period (years)	N/A	31 years
Modified internal rate of return (MIRR, %)	N/A	4.5%

Group 5: For Group 5 homes, similar increases in thermal resistance within the exterior walls and unfinished attic were suggested (see Table 5), including upgrading masonry walls to RSI-3.3 fiberglass batts and ceiling insulation to RSI-5.3 blown-in fiberglass. This particular group was found to benefit from an upgrade from the existing boiler system (80% AFUE) to a more energy efficient condensing unit (98% AFUE) and updated 3.1 COP window A/C units. A conversion to 100% fluorescent lighting was recommended but an updated water heater was not considered to be cost effective for this group. The total upfront costs of this retrofit package is predicted to be about $16,400, providing ~50% site energy savings with a simple payback period of 19 years and a MIRR of 6.4%. This retrofit package is predicted to achieve the target savings at a net benefit of $220 per year in AERC.

Table 5. Brick, Pre-1942, two stories—Summary of retrofit package and cost and energy conservation implications.

Parameter name	Existing characteristics	Optimal enclosure + HVAC retrofit package
Exterior Wall (Masonry)	10.2-cm Hollow Brick, Uninsulated, 2.5-cm furring, 60.9 cm O.C.	10.2-cm Hollow Brick, RSI-3.3 Fiberglass Batt, 38 × 140 mm, 60.9 cm o.c.
Unfinished Attic	Ceiling RSI-1.2 Fiberglass (Blown-in), Vented	Ceiling RSI-5.3 Fiberglass Batt, Vented
Air Leakage	Leaky	Typical
Room A/C	COP 2.9	COP 3.1, 20% Conditioned
Boiler	Gas, Hot Water, Forced, 80% AFUE	Gas, Hot Water, Condensing, 98% AFUE OAT Reset
Water Heater	Gas, 54% EF	No Change
Lighting	Benchmark	100% Fluorescent, Hardwired & Plug-in
Initial Cost ($)	$0	$16,387
AERC ($/year)	$2522	$2302
% Site (Source) Energy Savings	0	49.9% (45.3%)
Simple payback period (years)	N/A	19 years
Modified internal rate of return (MIRR, %)	N/A	6.4%

Group 6: For Group 6, the installation of continuous rigid insulation as exterior wall sheathing (RSI-2.1 polyiso) was found to be more cost effective than an increase in insulation within the interior of the wall cavity (see Table 6). Similar increases in thermal resistance within interzonal walls (RSI-2.3 blown-in fiberglass) and the unfinished attic space (RSI-3.3 blown-in fiberglass) were also recommended. Upgrades to the existing boiler (from 80% AFUE to 98%), room A/C units (from COP 2.9 to COP 3.1 units), and water heater (from 54% EF gas water heater to a tankless condensing gas water heater) were also found to be cost efficient retrofit measures, in addition to a conversion to 100% fluorescent lighting. The total upfront costs of this retrofit package is predicted to be about $15,400, providing ~49% site energy savings with a simple payback period of 27 years and a MIRR of 5.2%. For only an extra ~$63 per year in AERC, this home type is expected to achieve within 1% of the 50% energy savings goal.

Table 6. Group 6: Frame, All years, Split level (1.5 stories)—Summary of retrofit package and cost and energy conservation implications.

Parameter name	Existing characteristics	Optimal enclosure + HVAC retrofit package
Wall Sheathing	None	RSI-2.1 Polyiso
Interzonal Walls	Uninsulated, 39 × 89 mm, 40.6 cm O.C.	RSI-2.3 Fiberglass (Blown-in), Gr-1, 39 × 89 mm, 40.6 cm O.C.
Unfinished Attic	Ceiling RSI-1.2 Fiberglass (Blown-in), Vented	Ceiling RSI-3.3 Fiberglass (Blown-in), Vented
Air Leakage	Very Leaky	Leaky
Central A/C	None	None
Room A/C	COP 2.9	COP 3.1, 20% Conditioned
Furnace	None	None
Boiler	Gas, Hot Water, Forced, 80% AFUE	Gas, Hot Water, Condensing, 98% AFUE
Water Heater	Gas, 54% EF	Gas Tankless, Condensing
Lighting	Benchmark	100% Fluorescent, Hardwired & Plug-in
Initial Cost ($)	$0	$15,363
AERC ($/yer)	$2046	$2109
% Site (Source) Energy Savings	0	48.9% (42.5%)
Simple payback period (years)	N/A	27 years
Modified internal rate of return (MIRR, %)	N/A	5.2%

Group 7: The optimal retrofit package for Group 7 also includes the implementation of continuous rigid insulation as wall sheathing (RSI-2.1 polyiso), as well as an upgrade in the thermal resistance of the unfinished attic insulation (to RSI-5.3 blown-in fiberglass) and an increase in enclosure airtightness from "leaky" to "typical" (see Table 7). A conversion to a high-efficiency MSHP system and electric baseboard heaters is also predicted to be the more cost effective HVAC option. As with all other groups thus far, a conversion to 100% fluorescent lighting is also suggested. However, changes to the existing water heater were not found to be cost effective. The total upfront costs of this retrofit package

is predicted to be about $9000, providing ~53% site energy savings with a simple payback period of 39 years and an MIRR of 2.6%. This home group has the lowest expected MIRR and the second highest additional annualized energy-related costs to achieve the savings goal (increasing AERC by about $358 per year).

Table 7. Group 7: Frame, 1942–1978, 1–1.5 stories (no split level) Summary of retrofit package and cost and energy conservation implications.

Parameter name	Existing characteristics	Optimal enclosure + HVAC retrofit package
Wall Sheathing	None	RSI-2.1 Polyiso
Unfinished Attic	Ceiling RSI-3.3 Fiberglass (Blown-in), Vented	Ceiling RSI-5.3 Fiberglass (Blown-in), Vented
Air Leakage	Leaky	Typical
Central A/C	COP 2.6	None
Furnace	Gas, 78% AFUE	None
Electric Baseboard	None	100% Efficiency
Mini Split Heat Pump	None	COP 6.1, Avg Seasonal COP 3.25
Ducts	Typical, Uninsulated (Unfinished Basement)	None
Water Heater	Gas, 54% EF	No Change
Lighting	Benchmark	100% Fluorescent Hardwired & Plug-in
Initial Cost ($)	$0	$8977
AERC ($/year)	$1685	$2043
% Site (Source) Energy Savings	0	53.3% (30.7%)
Simple payback period (years)	N/A	39 years
Modified internal rate of return (MIRR, %)	N/A	2.6%

Group 8: Optimal Group 8 home upgrades include continuous rigid insulation (RSI-2.1 polyiso), improved unfinished attic insulation (RSI-5.3 blown-in fiberglass), and improved enclosure airtightness (from "leaky" to "typical") (see Table 8). An upgrade to the existing HVAC system (from 78% AFUE to 98% AFUE and from a COP of 2.6 to 6.5) was estimated to be the most cost effective option with the suggestion for ductwork to be moved from the uninsulated and unfinished basement to finished space. The implementation of a solar water heating system in combination with a tankless water heater was also predicted to be a cost effective upgrade for this group. As with all other groups, a conversion to 100% fluorescent lighting is also recommended. The total upfront costs of this retrofit package is predicted to be about $23,100, providing ~50% site energy savings with a simple payback period of 33 years and an MIRR of 3.5%. This group had the highest predicted AERC of an additional $523 per year required to achieve the savings goal.

Group 9: The optimal retrofit package for Group 9 calls for continuous insulation to be added to the exterior walls (RSI-2.1 polyiso), upgrades to the unfinished attic insulation (to RSI-4.4 blown-in fiberglass), and improved enclosure airtightness from "very leaky" to "leaky" (see Table 9). For this group, an upgrade to the existing HVAC system was found to be the most cost effective option (from an 80% AFUE boiler to a 98% AFUE boiler and from 2.9 COP room A/C units to 3.1 COP room A/C units). The cost to upgrade the water heater was not found to be offset by the energy savings for this group. As with all other groups, a conversion to 100% fluorescent lighting is also recommended. The total upfront costs of this retrofit package is predicted to be about $13,000, providing ~51% site energy savings with a simple payback period of 18 years and an MIRR of 6.6%. The 50% energy savings goal is predicted to be met in this home group with annual energy-related cost savings of about $112 per year.

Table 8. Group 8: Frame, 1942–1978, two stories Summary of retrofit package and cost and energy conservation implications.

Parameter name	Existing characteristics	Optimal enclosure + HVAC retrofit package
Wall Sheathing	None	RSI-2.1 Polyiso
Unfinished Attic	Ceiling RSI-1.9 Fiberglass (Blown-in), Vented	Ceiling RSI-5.3 Fiberglass (Blown-in), Vented
Air Leakage	Leaky	Typical
Central A/C	COP 2.6	COP 6.5
Furnace	Gas, 78% AFUE	Gas, 98% AFUE
Ducts	Typical, Uninsulated, (Unfinished Basement)	In Finished Space
Water Heater	Gas, 54% EF	Gas Tankless
Solar Water Heating	None	5.9 m^2 Closed Loop
SWH Azimuth	N/A	Back Roof
SWH Tilt	N/A	30 Degrees
Lighting	Benchmark	100% Fluorescent Hardwired & Plug-in
Initial Cost ($)	$0	$23,142
AERC ($/year)	$2102	$2625
% Site (Source) Energy Savings	0	50.1% (44.1%)
Simple payback period (years)	N/A	33 years
Modified internal rate of return (MIRR, %)	N/A	3.5%

Group 10: Finally, the optimal retrofit package for Group 10 involves the addition of continuous insulation on the exterior walls (RSI-2.1 polyiso), an increase in the thermal resistance of insulation in the unfinished attic (to RSI-5.3 blown-in fiberglass), and an improvement in airtightness from "leaky" to "typical" (see Table 10). Upgrades to the existing HVAC systems were suggested (from an 80% AFUE boiler to a 98% AFUE boiler and a small improvement in room A/C unit efficiency), as well as an upgrade to a tankless, condensing water heater. Consistent with upgrades to the other groups, a conversion to 100% fluorescent lighting was considered a cost effective upgrade as well. The total upfront costs of this retrofit package is predicted to be about $16,600, providing ~50% site energy savings with a simple payback period of 18 years and an MIRR of 6.6%. This retrofit package is expected to yield about $257 in annual energy-related cost savings.

Table 9. Group 9: Frame, Pre-1942, 1–1.5 stories Summary of retrofit package and cost and energy conservation implications.

Parameter name	Existing characteristics	Optimal enclosure + HVAC retrofit package
Wall Sheathing	None	RSI-2.1 Polyiso
Unfinished Attic	Ceiling RSI-0.53 Fiberglass (Blown-in)	Ceiling RSI-4.4 Fiberglass (Blown-in), Vented
Air Leakage	Very Leaky	Leaky
Central A/C	None	None
Room A/C	COP 2.9	COP 3.1, 20% Conditioned
Furnace	None	None
Boiler	Gas, Hot Water, Forced, 80% AFUE	Gas, Hot Water, Condensing, 98% AFUE
Water Heater	Gas, 54% EF	No Change
Lighting	Benchmark	100% Fluorescent, Hardwired & Plug-in
Initial Cost ($)	$0	$13,013
AERC ($/year)	$2047	$1935
% Site (Source) Energy Savings	0	51.4% (46.3%)
Simple payback period (years)	N/A	18 years
Modified internal rate of return (MIRR, %)	N/A	6.6%

Table 10. Group 10: Frame, Pre-1942, two stories Summary of retrofit package and cost and energy conservation implications.

Parameter name	Existing characteristics	Optimal enclosure + HVAC retrofit package
Wall Sheathing	None	RSI-2.1 Polyiso
Unfinished Attic	Ceiling RSI-1.2 Fiberglass (Blown-in), Vented	Ceiling RSI-5.3 Fiberglass Batt, Vented
Air Leakage	Leaky	Typical
Room A/C	COP 2.9	COP 3.1, 20% Conditioned
Boiler	Gas, Hot Water, Forced Draft, 80% AFUE	Gas, Hot Water, Condensing, 98% AFUE
Water Heater	Gas, 54% EF	Gas Tankless, Condensing
Lighting	Benchmark	100% Fluorescent, Hardwired & Plug-in
Initial Cost ($)	$0	$16,647
AERC ($/year)	$2664	$2407
% Site (Source) Energy Savings	0	50.2% (44.7%)
Simple payback period (years)	N/A	18 years
Modified internal rate of return (MIRR, %)	N/A	6.6%

3.3. Generalizing the Retrofit Packages

For all groups the optimal retrofit package involves either an upgrade to existing systems or the conversion to a MSHP and electric baseboard heating system when combined with several envelope retrofit measures. The envelope improvement measures typically included increasing exterior wall insulation with either exterior rigid insulation or additional cavity insulation; increasing attic insulation with blown-in fiberglass; and increasing building enclosure airtightness. Other common improvements include upgrades to the domestic hot water system (where appropriate) and a conversion to 100% compact fluorescent lighting (true for all homes). Other upgrades such as replacing windows or altering the finishing of basement spaces were modeled but were not found to be cost effective retrofits for any of the groups.

Although there appears to be no relation between the type of optimal retrofit package and vintage, other generalizations can be made based on exterior wall construction type and number of stories. The simulations demonstrate that once the corresponding optimal enclosure is implemented, frame construction homes are more likely to benefit from an upgrade to existing HVAC systems (80% probability), whereas homes of brick construction are slightly more likely to benefit from a conversion to a MSHP system (60% probability). Also, 1–1.5 story homes, including split level homes, are most likely to convert to MSHP systems (67% probability), whereas homes having two stories are most likely to upgrade existing HVAC systems (100% probability). These generalizations are interesting to consider in this approach to use cloud-based optimization simulations to identify prescriptive retrofit packages for these homes.

3.4. Predicted Energy and Monetary Savings Implications

The predicted energy savings for each group were estimated given that the optimum retrofit package was implemented. The estimated site energy consumption for the existing and the post-retrofit cases for each group is shown in Figure 8. An average estimated site energy consumption savings of about 105.2 GJ (99.7 MMBtu) is predicted, which equates to an average of 53.4% savings in site energy consumption across all 10 groups. The results here and in Section 3.2 clearly demonstrate that the 50% site energy savings targets can be met on average across all home groups in the Chicago area, often at a net annual cost savings.

Table 11 summarizes the same simulation results for both fuel types as well as total overall savings per home for each of the 10 groups. The monetary savings for per home were calculated using values for average electricity and gas prices ($0.12/kWh for electricity and ~$8/GJ or $8.5/MMBtu for gas) obtained from the U.S. Energy Information Administration [33,34]. If the appropriate optimal retrofit package is implemented in each home group, the average total energy savings achievable per household is estimated to be about $906 per year. Using these estimates of annual energy savings, we then estimated the potential annual energy savings for the entire Chicago area, assuming that the retrofits were applied to every existing home in each home group (Table 12). The predicted savings translate to an average of 105.2 GJ (99.7 MMBtu) in annual site energy savings and 116.3 GJ (110.2 MMBtu) in annual source energy savings for each group. Impressively, the total estimated savings across all homes would be about 37,000,000 GJ per year (Table 12), or 3.7×10^{16} J per year.

This equates to a 58% site energy savings and 39% source energy savings for the entire pre-1978 vintage single-family residential stock in the Chicagoland area.

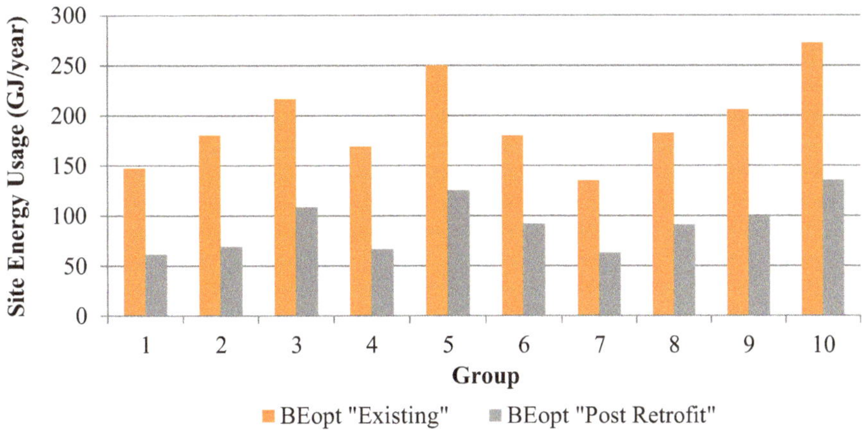

Figure 8. Summary of Site energy usage for all groups (per home).

Table 11. Summary of recommended optimal packages and estimated initial costs and savings.

Group	% Site (source) energy savings	Total site gas savings GJ/Year (MMBtu/year)	Total site gas savings ($/year)	Total site electricity savings (kWh/year)	Total site electricity savings ($/year)	Annual energy savings per home
1	58.2 (39.9)	87.0 (82.5)	$699	−359	−$42	$657
2	61.6 (40.8)	117.4 (111.3)	$943	−1659	−$195	$747
3	49.9 (45.5)	94.3 (89.4)	$757	3845	$453	$1210
4	60.6 (37.6)	111.9 (106.1)	$899	−2604	−$306	$592
5	49.9 (45.3)	114.5 (108.5)	$919	2919	$344	$1263
6	48.9 (42.5)	80.9 (76.7)	$650	1978	$233	$882
7	53.3 (30.7)	80.1 (75.9)	$643	−2261	−$266	$377
8	50.1 (44.1)	82.1 (77.8)	$659	2608	$307	$966
9	51.7 (46.3)	98.4 (93.3)	$790	2119	$249	$1040
10	50.2 (44.7)	127.3 (120.7)	$1022	2580	$304	$1326

Finally, we estimated the potential monetary savings for each group and across the entire pre-1978 existing single-family housing stock in the Chicagoland area, as shown in Table 13. Although the estimated average site electricity usage savings for each group is actually a net additional cost of $1.2 million USD per year (mostly due to switching to electric heating systems in many of the optimal retrofit packages), the average estimated savings for on-site natural gas usage by each group is about $28.9 million USD per year, providing an overall average group energy cost savings of about $27.7 million USD per year per group. When summed across the entire pre-1978 single-family residential housing stock in the Chicago area, the total potential monetary savings as a result of application of the optimal retrofit packages identified herein is about $280 million USD per year. The average upfront cost of retrofits is estimated to be about $14,400 per home, providing a total investment opportunity across all of these existing homes of about $4.6 billion USD. If these total estimated upfront costs of retrofits are compared to the total estimated energy cost savings across all of the homes in the 10 home typology groups, the simple payback period on the initial $4.6 billion investment would be less than 17 years.

Table 12. Summary of total energy savings predicted for each group and the entire Chicago area.

Group	**% of Population** †	**Number of homes**	**"Today" total site consumption GJ/year (MMBtu/year)**	**"Today" total source consumption GJ/year (MMBtu/year)**	**"Post-retrofit" total site savings GJ/year (MMBtu/year)**	**"Post-retrofit" total source savings GJ/year (MMBtu/year)**
1	17.9	77,435	11,414,275 (10,819,218)	17,641,803 (16,722,088)	6,641,716 (6,295,466)	7,066,525 (6,698,128)
2	6.1	26,445	4,782,249 (4,532,938)	7,080,608 (6,711,476)	2,946,185 (2,792,592)	2,884,806 (2,734,413)
3	4.8	20,755	4,498,860 (4,264,322)	6,679,316 (6,331,105)	2,244,394 (2,127,388)	3,039,238 (2,880,794)
4	11.6	50,239	8,509,495 (8,065,872)	12,345,260 (11,701,668)	5,151,808 (4,883,231)	4,642,988 (4,400,936)
5	4.1	17,629	4,412,145 (4,182,128)	6,148,138 (5,827,619)	2,202,074 (2,087,274)	2,782,350 (2,637,298)
6	2.1	9225	1,662,484 (1,575,815)	2,400,198 (2,275,069)	812,653 (770,288)	1,020,926 (967,703)
7	23.6	101,957	13,771,500 (13,053,555)	20,445,886 (19,379,987)	7,335,908 (6,953,467)	6,281,775 (5,954,289)
8	3.8	16,411	3,000,101 (2,843,698)	4,427,608 (4,196,785)	1,502,821 (1,424,475)	1,952,975 (1,851,161)
9	11.2	48,365	8,706,919 (8,253,004)	12,425,116 (11,777,361)	6,347,519 (6,016,606)	5,128,020 (4,860,683)
10	2.9	12,479	3,396,791 (3,219,707)	4,688,574 (4,444,146)	1,704,912 (1,616,031)	2,097,239 (1,987,905)
Chicago land Total	88.1	380,940	64,154,819 (60,810,255)	94, 282,506 (89,367,305)	36,889,990 (34,966,816)	36,896,841 (34,973,309)

† Relative to a dataset of 432,605 homes.

Table 13. Summary of potential monetary savings for each group and the entire Chicago area.

Group	% of population †	Number of homes	Total site gas savings GJ/year (MMBtu/year)	Total site gas savings ($/year)	Total site electricity savings (kWh/year)	Total site electricity savings ($/year)	Total site energy savings ($/year)
4	17.9	77,435	6,739,749 (6,388,388)	$54,109,642	−27,799,165	−$3,271,962	$50,837,680
5	6.1	26,445	3,105,212 (2,943,329)	$24,929,992	−43,872,255	−$5,163,764	$19,766,228
6	4.8	20,755	1,957,549 (1,855,497)	$15,716,060	79,802,975	$9,392,810	$25,108,870
7	11.6	50,239	5,623,528 (5,330,358)	$45,148,131	−130,822,356	−$15,397,791	$29,750,340
8	4.1	17,629	2,017,948 (1,912,747)	$16,200,963	51,459,051	$6,056,730	$22,257,693
10	2.1	9225	746,473 (707,558)	$5,993,012	18,247,050	$2,147,678	$8,140,690
12	23.6	101,957	8,164,156 (7,738,536)	$65,545,402	−230,524,777	−$27,132,766	$38,412,636
13	3.8	16,411	1,346,998 (1,276,776)	$10,814,291	42,799,888	$5,037,547	$15,851,838
14	11.2	48,365	4,760,639 (4,512,455)	$38,220,490	102,485,435	$12,062,536	$50,283,025
15	2.9	12,479	1,589,057 (1,506,215)	$12,757,644	32,195,820	$3,789,448	$16,547,092
Chicagoland Total	88.1	380,940	36,051,309 (34,171,857)	$289,435,627	−106,028,334	−$12,479,535	$276,956,092

† Relative to a dataset of 432,605 homes.

3.5. Limitations and Uncertainty

There are several limitations and uncertainties that have the potential to affect the resulting optimal packages herein that must be mentioned. First, this work investigates just 10 different individual home models selected to represent more than 400,000 homes in the Chicago-area single-family building stock. Having only one model per group may not yield an accurate representation of the statistical distribution of homes within a typology group. Therefore, the optimal packages detailed herein may not result in the same level of energy savings or initial costs of retrofits for all homes in each group. Second, the results herein are limited to the assumptions for upfront costs of individual retrofits and retrofit packages. Although this could present a limitation to our findings, recent work has shown that while cost discrepancies often exist between actual retrofit packages and the same retrofit packages modeled in BEopt, BEopt cost assumptions actually tend to overestimate market costs and thus can be considered conservative pricing estimates [35]. Moreover, while any discrepancies in assumptions for upfront costs will affect output variables such as AERC, payback periods, and MIRR, they do not affect the predicted potential energy savings for any of the groups. Third, this work does not take into account changes in future energy costs and it is limited to a 30-year project analysis period. Different assumptions for energy costs or the analysis period would yield different values of AERC, payback periods, and MIRR. Last, there may be other secondary retrofit costs that have not been captured herein. For example, in groups where the prescriptive retrofit package involves a conversion to a MSHP system, our cost assumptions do not account for the potential upgrade to many existing electric service panels that may be required to be able to accommodate the added electrical load. Given these limitations, further field investigation may be required to improve the accuracy of the cost estimates associated with the prescriptive deep energy retrofit packages.

4. Conclusions

Results from this work reveal that prescriptive deep energy retrofit solutions that achieve at least 50% annual site energy reductions can indeed be defined for each of the 10 typology groups investigated, largely through common envelope retrofit measures for all groups, upgrades to lighting, and either upgrades to existing HVAC system efficiency or a conversion to mini-split heat pump (MSHP) systems. The building envelope retrofit measures typically included increasing exterior wall insulation with either exterior rigid insulation or additional cavity insulation; increasing attic insulation with blown-in fiberglass; and increasing building enclosure airtightness. Other common improvements include upgrades to the domestic hot water system and a conversion to 100% compact fluorescent lighting. Other upgrades such as replacing windows or altering the finishing of basement spaces were not found to be cost effective retrofits for any of the groups. Once the cost-optimal enclosure retrofits were implemented, frame construction homes were more likely to benefit from an upgrade to existing HVAC systems and brick construction homes were slightly more likely to benefit from a conversion to a MSHP system. Overall, the prescriptive deep energy retrofit packages described herein are expected to save between $400 and $1300 on energy costs per year per home, depending on typology. The average upfront cost of retrofits is estimated to be about $14,400, but would yield average annual site energy savings of ~54% and an average simple payback period of about 25 years per home. A scaling analysis suggests that widespread application of these deep energy retrofit packages to the entire existing pre-1978 Chicago-area residential building stock would provide a total investment opportunity of about $4.6 billion USD with approximately $280 million in annual energy cost savings and a simple payback period of less than 17 years.

Acknowledgments

We are grateful to Ralph Muehleisen for his initial guidance in the development of this project and for his helpful review of this work.

Author Contributions

Honnie Aguilar Leinartas conceived, designed, and performed the simulations; Honnie Aguilar Leinartas and Brent Stephens analyzed the data and wrote the paper.

Conflicts of Interest

The authors declare no conflict of interest.

References

1. Ürge-Vorsatz, D.; Danny Harvey, L.D.; Mirasgedis, S.; Levine, M.D. Mitigating CO_2 emissions from energy use in the world's buildings. *Build. Res. Inf.* **2007**, *35*, 379–398.
2. Part Data of DOE Buildings Energy Data Book. Available online: http://buildingsdatabook.eren.doe.gov/TableView.aspx?table=1.1.3 (accessed on 23 March 2015).
3. Part Data of DOE Buildings Energy Data Book. Available online: http://buildingsdatabook.eren.doe.gov/TableView.aspx?table=2.4.1 (accessed on 23 March 2015).
4. Part Data of DOE Buildings Energy Data Book. Available online: http://buildingsdatabook.eren.doe.gov/TableView.aspx?table=2.2.2 (accessed on 8 September 2010).
5. Cluett, R.; Amann, J. *Residential Deep Energy Retrofits*; ACEEE: Washington, DC, USA, 2014.
6. Architecture 2030. The 2030 Challenge. Available online: http://www.architecture2030.org/2030_challenge/the_2030_challenge (accessed on 23 March 2015).
7. McIlvaine, J.; Sutherland, K.; Martin, E. *Energy Retrofit Field Study and Best Practices in a Hot-Humid Climate*; BA-PIRC/Florida Solar Energy Center: Cocoa, FL, USA, 2013.
8. Lyons, J. *Short-Term Test Results: Multifamily Home Deep Energy Efficiency Retrofit*; BA-PIRC: Cocoa, FL, USA, 2013.
9. Wigington, L. Staged Approaches for Deep Energy Reductions in Existing Homes. In Proceedings of the 2010 ACEEE Summer Study on Energy Efficiency in Buildings, Pacific Grove, CA, USA, 15–20 August 2010; pp. 339–354.
10. Less, B.; Fisher, J.; Walker, I. *Deep Energy Retrofits: Eleven California Case Studies*; Lawrence Berkeley National Laboratory: Berkeley, CA, USA, 2012.
11. Liaukus, C. *Reducing Energy Use in Existing Homes by 30%: Learning from Home Performance with Energy Star*; Building America Research Alliance: Kent, WA, USA, 2014.
12. Asadi, E.; da Silva, M.G.; Antunes, C.H.; Dias, L. A multi-objective optimization model for building retrofit strategies using TRNSYS simulations, GenOpt and MATLAB. *Build. Environ.* **2012**, *56*, 370–378.
13. Asadi, E.; da Silva, M.G.; Antunes, C.H.; Dias, L. Multi-objective optimization for building retrofit strategies: A model and an application. *Energy Build.* **2012**, *44*, 81–87.
14. Junghans, L. Sequential equi-marginal optimization method for ranking strategies for thermal building renovation. *Energy Build.* **2013**, *65*, 10–18.
15. Chantrelle, F.P.; Lahmidi, H.; Keilholz, W.; Mankibi, M.E.; Michel, P. Development of a multicriteria tool for optimizing the renovation of buildings. *Appl. Energy* **2011**, *88*, 1386–1394.
16. Ma, Z.; Cooper, P.; Daly, D.; Ledo, L. Existing building retrofits: Methodology and state-of-the-art. *Energy Build.* **2012**, *55*, 889–902.
17. Heo, Y.; Choudhary, R.; Augenbroe, G.A. Calibration of building energy models for retrofit analysis under uncertainty. *Energy Build.* **2012**, *47*, 550–560.

18. Spanier, J.; Scheu, R.; Brand, L.; Yang, J. *Chicagoland Single-Family Housing Characterization*; Department of Energy. Energy Efficiency & Renewable Energy: Washington, DC, USA, 2012.
19. CNT Energy. *Chicago Regional Energy Snapshot: Profile and Strategy Analysis*; Elevate Energy: Chicago, IL, USA, 2009.
20. Christensen, C.; Anderson, R.; Horowitz, S.; Courtney, A.; Spencer, J. *BEopt™ Software for Building Energy Optimization: Features and Capabilities*; National Renewable Energy Lab (NREL): Golden, CO, USA, 2006.
21. Bichiou, Y.; Krarti, M. Optimization of envelope and HVAC systems selection for residential buildings. *Energy Build.* **2011**, *43*, 3373–3382.
22. USDA. *Soil Survey of Cook County, Illinois*; U.S. Department of Agriculture: Washington, DC, USA, 2012.
23. Farouki, O. *Thermal Properties of Soils*; United States Army Corp of Engineers, Cold Regions Research and Engineering Laboratory: Hanover, NH, USA, 1981.
24. Sami, S.M.; Tulej, P.J. A new design for an air-source heat pump using a ternary mixture for cold climates. *Heat Recovery Syst. CHP* **1995**, *15*, 521–529.
25. Bertsch, S.S.; Groll, E.A. Two-stage air-source heat pump for residential heating and cooling applications in northern U.S. climates. *Int. J. Refrig.* **2008**, *31*, 1282–1292.
26. *National Weather Service Chicago Seasonal Temperature, Precipitation, and Snow Rankings*; National Oceanic Atmospheric Administration: 2014. Available online: http://www.crh.noaa.gov/lot/?n=CHI_winter_temps (accessed on 23 March 2015).
27. Guoyuan, M.; Qinhu, C.; Yi, J. Experimental investigation of air-source heat pump for cold regions. *Int. J. Refrig.* **2003**, *26*, 12–18.
28. NREL Cold Climate ASHPs? Available online: http://beopt.nrel.gov/node/186 (accessed on 1 April 2014).
29. ComEd Commonwealth Edison Company: Live Prices 2014. Available online: https://rrtp.comed.com/live-prices/ (accessed on 1 April 2014).
30. Nicor Gas Nicor Gas Price, 2014. Available online: http://nicorgas.aglr.com/Repository/Files/Nicor_Rider_6_History.pdf (accessed on 1 April 2014).
31. Richman, R.; Zirnhelt, H.; Fix, S. Large-scale building simulation using cloud computing for estimating lifecycle energy consumption. *Can. J. Civ. Eng.* **2014**, *41*, 252–262.
32. Coffey, B. Approximating model predictive control with existing building simulation tools and offline optimization. *J. Build. Perform. Simul.* **2013**, *6*, 220–235.
33. EIA Natural Gas Sumary: Annual Residential Price. Available online: http://www.eia.gov/dnav/ng/ng_sum_lsum_a_EPG0_PRS_DMcf_a.htm (accessed on 1 April 2014).
34. EIA Average Retail Price of Electricity. Pre-Generated report 5.6: Average retail price of electricity to ultimate customers by end-use sector, by state. Available online: http://www.eia.gov/electricity/data/browser/ (accessed on 1 April 2014).

35. Yee, S.; Milby, M.; Baker, J. *Evaluation of Missed Energy Saving Opportunity Based on Illinois Home Performance Program Field Data: Homeowner Selected Upgrades Versus Cost-Optimized Solutions*; Partnership for Advanced Residential Retrofit: 2014. Available online: http://apps1.eere.energy.gov/buildings/publications/pdfs/building_america/illinois_home_performance_upgrades.pdf (accessed on 1 March 2015).

11

Comparison of NZ's Energy Efficiency Regulation and Verification Assumptions to Real Building Loads and Operation

Shaan Cory [1,*], Michael Donn [1] and Andrew Pollard [2]

[1] Centre for Building Performance Research, Victoria University of Wellington, P.O. Box 600, Wellington 6140, New Zealand; E-Mail: michael.donn@vuw.ac.nz

[2] Building Research Association of New Zealand (BRANZ) Ltd, Private Bag 50908, Porirua 5240, New Zealand; E-Mail: andrew.pollard@branz.co.nz

* Author to whom correspondence should be addressed; E-Mail: coryshaa@gmail.com

Academic Editor: George Baird

Abstract: The New Zealand building design industry assumes various building model inputs for the consumption of energy through lighting and appliances. It also makes assumptions regarding when these energy consumers are considered to be "turned on". This paper aims to better inform industry energy modellers about the real load and operation of real commercial buildings in New Zealand when compared to New Zealand Standard energy efficiency requirements and assumptions. The paper presents a set of New Zealand relevant commercial building operation information. Typical operation information is provided for three commercial building types: (1) Office; (2) Retail; and (3) Mixed/Other. The information provides low, typical, and high installed building load and operation pattern scenarios for the three building types. The typical data presented in this paper is significantly different to the load requirement and operation modelling assumptions presented in the New Zealand Building code. The results established in this paper are informed by data gathered in the Building Research Association of New Zealand (BRANZ) Building Energy End-Use Study (BEES). The purpose of BEES is to increase knowledge on energy use patterns for the entire New Zealand building stock. The intention of this paper is to disseminate the established knowledge that will eventually update the assumptions used in New Zealand commercial energy models.

Keywords: building energy modelling; building load; operation patterns; Building Energy End-Use Study (BEES); monitored data

1. Introduction

This paper presents typical, high and low building load densities and patterns of use for commercial buildings in New Zealand. Patterns of use refers to the daily use profiles of the load presented as the percentage of load "on" at different time intervals throughout the day. The goal is to better inform building energy modellers about typical lighting and equipment end uses. In the process, this work highlights the differences between the assumptions made when using the modelling verification method to meet the New Zealand Building Code (NZBC) and what is occurring in real buildings [1].

The research is a part of the Building Research Association of New Zealand (BRANZ) Building Energy End-Use Study (BEES). The purpose of BEES is to monitor and analyse the energy and water consumed by non-residential buildings around New Zealand. The project ran for 6 years [2]. BEES aims to provide more insightful knowledge of energy use patterns for the entire New Zealand commercial building stock [3]. Real data is collected within selected premises through the BEES programme from the monitoring of temperature, humidity, light levels, CO_2 levels, occupant and equipment schedules, internal loads and fuel consumption.

This paper is one of two reports which documents better informed building energy models for the New Zealand design industry. The building load and patterns of use conclusions determined in this study informed the creation of prototypical commercial building energy models. The prototypical models are documented on the BRANZ BEES website. The models provide a base case for the building design industry to assess the impact of building load and operation on their buildings energy performance.

Lack of Measured Building Operation Information

Currently, when analysing building energy performance, the New Zealand building design industry must make assumptions about various building model inputs based on New Zealand Standard (NZS) verification method for displaying compliance with NZBC. Inputs to their performance calculation models rely on assumptions about the energy consumed by lights and appliances. Energy modellers assessing energy may use the NZS 4243 Lighting Power Density (LPD) value as a base scenario for what is currently installed in commercial buildings. NZS values are used because there is no other reliable information on the typical building load power densities found in real commercial buildings.

Half of the commercial buildings in New Zealand were built before 2000. This pre-dates the NZBC and NZS 4243 clause regulating the LPD. This can cause modellers to enter into their performance calculation of an existing building retrofit analysis assumed values that may not be typical. Additionally, there is no such NZBC regulated density value for other building loads, such as office equipment and hot water. This results in modellers inputting a value for an Equipment Power Density (EPD) and Hot Water Power Density (HWPD) that is a best guess or an assumption based on what is included in the NZS 4243 modelling method to prove a building design meets the NZBC. Table 1 displays the power density "assumptions" found in the NZS 4243.

Table 1. New Zealand Standard (NZS) 4243 Power densities. Table created using [4,5].

Building Type	Power Densities (W/m^2)		
	Regulated Lighting Power Density (LPD)	Assumed Equipment Power Density (EPD)	Assumed Hot Water Power Density (HWPD)
Office	12 W/m^2	8.1 W/m^2	Not required
Retail	8–16 W/m^2 depending on retail type	1.1–2.7 W/m^2 depending on retail type	Not required
Mixed/Other	8–18 W/m^2 depending on use type	1.1–8.1 W/m^2 depending on use type	Not required

Note: The Retail and Mixed/other power densities have a range due to the different building types.

The same problem occurs for the operation patterns of building loads. No measured information for the patterns of use of building loads in real buildings is currently published. As a result, an energy modeller is left with little option but to assume when equipment is turned "on" and when it is turned "off" based on the modelling assumptions set in the NZS 4243 energy modelling method. Table 2 presents the operation pattern assumptions found in the NZS 4243 energy modelling method for different building types. There is no evidence that these assumptions will be relevant to how real commercial buildings in New Zealand consume energy.

Table 2. NZS 4243 Lighting and Plug Load patterns of use. Table adapted from [4].

Building Type	Day Type	Patterns of Use (% of Load "on")				
		12 p.m.–8 a.m.	8 a.m.–11 a.m.	11 a.m.–6 p.m.	6 p.m.–10 p.m.	10 p.m.–12 p.m.
Office	Week	5%	90%	90%	30%	5%
	Saturday	5%	30%	15%	5%	5%
	Sunday	5%	5%	5%	5%	5%
Retail (Restaurant)	Week	5% (15%)	90% (40%)	90% (90%)	50% (90%)	5% (50%)
	Saturday	5% (15%)	90% (30%)	90% (80%)	30% (90%)	5% (50%)
	Sunday	5% (15%)	40% (30%)	40% (70%)	5% 60%)	5% (50%)
Mixed/Other	Week	5%	90%	90%	5%	5%
	Saturday	5%	24%	5%	5%	5%
	Sunday	5%	5%	5%	5%	5%

Notes: Retail has two patterns of use presented as restaurants a different to general retail. Also, Mixed/Other are the assumptions for a warehouse, but it could be a mixture of all three use types.

New Zealand is not alone with regard to these assumptions and lack of information. The current state of the art in prototypical building models can be found in a US set of prototypical building models [6]. These are based on informed engineering judgements about "typical" or "design" values for building loads and their operation. As part of the BEES modelling research, a set of EnergyPlus Template models [7] were developed which followed the same format as the U.S. prototypical models. The aim of the template models was to reduce the difficulties associated with energy modelling, particularly relating to the ease and speed of creating a model to produce reliable results [8]. The BEES models have the same short comings in that they are built on assumptions about the building loads and their operation. If the BEES models were updated using the measured data presented in this report, they would be more advanced than any prototypical models currently found. The biggest single advantage of updating the models with measured building load and operation patterns is that they offer the potential for building design teams to examine the risk that a predicted building performance is dependent on the assumptions made about people, lighting and equipment used in the design modelling. There have been criticisms of high performance designs due to the fact that they only perform well in the particularly narrow focused situation "assumed" during the performance modelling [9]. The low and high building loads data from BEES allows designers to quickly test realistic design scenarios and establish how robust their design concept is.

2. Methodology

This study uses the measured data for lighting, equipment and hot water energy use from the BEES survey of commercial premises. The BEES team collected data over a four year period from 2008 to 2012. Energy was measured at one minute intervals for a 2–3 weeks' period in a representative sample of New Zealand commercial building premises. The weekly periods were spread throughout the year for different buildings. This means that the energy results could be from a two week period in summer or a two week period in the winter. Because this report only deals with lighting, equipment and hot water the outdoor conditions will have a minimal impact on energy use. This was the case for lighting also as there were no daylight induced electric lighting dimming in any of the premises. For site measurement technique and protocol used for monitoring refer to BEES Years 1 and 2 report [3].

This report presents the power density and patterns of use for lighting, equipment, and hot water. The lighting used is strictly indoor lighting to provide task lighting to the occupants. Equipment is made up of a number of appliances that are used by occupants to undertake day to day tasks and enable businesses to provide a service. Appliances include computers, printers, servers, refrigerators, chillers, water coolers, water boilers, phone systems, security systems, ovens, stove tops, deep fryers, and other appliances used in the day to day use of commercial buildings. Hot water energy use covers the energy used to provide hot water for domestic use, such as hand washing and showers, and commercial use, such as dishwashing.

All results were calculated using the measured average weekday and weekend 10 min interval load. Each 10 min interval measured was averaged against other weekday and weekend 10 min intervals. The power densities were calculated by dividing the maximum measured load (for lighting, equipment, and hot water), by the monitored floor area. The operation patterns were calculated by dividing the measured load by the maximum measured load to establish the percentage of load "on" during that 10 min interval.

A typical, a low and a high energy load scenario is presented in this report. The typical scenario is the Median (50th percentile) load and pattern of use found across the sample of building premises. Because the sample has outliers which can differ greatly from other values, the median provides a good indicator of the most typical value in the sample [10]. The Low scenario is the 10th percentile of measured load and pattern of use across the sample of premises. The High scenario is the 90th percentile of measured load and pattern of use across the sample of premises. The sample sizes vary between the lighting, equipment and hot water assessments and are displayed in Table 3.

The time intervals used in the patterns of use analysis were decided by assessing natural breaks across the whole data set in the amount of load "on" during each hour. For example, 12 a.m. to 5 a.m. is used as there was less than 10 percent difference in load across these hours. The hours were then averaged to get a percentage value of load "on" across these 5 h. These data are presented for 3 different building types. The load was split by building use type because NZS 4243 regulates load for different building types, and the use of different commercial buildings results in very different building load attributes. Table 4 displays the three building types assessed (Office, Retail, and Mixed/Other) and the detailed uses found within each of the three building types.

A further breakdown was made when assessing the lighting, equipment and hot water power densities. The breakdown assessed the impact of building floor area size. Building floor area size was assessed because NZS 4243 regulates building loads for only "large buildings" which it deems are 300 m^2 or greater. The assessment of size indicates whether building size has a significant impact on the building loads. Table 5 displays the breakdown of the different building floor area sizes assessed (small, medium and large), and the percentage of the total commercial building floor area each building size makes up.

Table 3. Sample sizes used to calculate typical, low and high values.

End-Use	All Buildings	Office	Retail	Mixed/Other
Lighting	101	35	29	37
Equipment	83	28	22	33
Hot Water	30	9	7	14

Table 4. Building type categories.

Building Type	Office	Retail	Mixed/Other
Specific building use types found in each building type	Office-type use	Retailing use, Motor vehicle sales and services, Liquor outlets including taverns, Service stations, Tourist-type attractions	A mixture of office and retail use types. As well as, warehouses, and service buildings.

Table 5. Building size groupings.

Building Size	Small	Medium	Large
Building floor area range	5 to 649 m^2	650 to 3499 m^2	Over 3500 m^2
Percentage of all commercial floor area	20%	40%	40%

3. Results and Discussion

All results displayed in the graphs below are documented on the BEES website [7]. The results presented on the website enable energy modellers to enter the values into energy models more precisely.

3.1. Lighting Power Density (LPD)

Figure 1 displays the typical, high and low LPDs' for an average commercial building (white), Offices (light grey), Retail (dark grey), and Mixed/Other commercial use types (black). As can be seen, Retail use types have the highest installed LPD's. This is followed by Offices, and Mixed/Other commercial use types. This trend mimics the trend of the maximum allowable LPD set by the NZS

4243 verification method [5]. It is worth noting that the typical LPDs found in Offices are slightly lower (11 W/m^2) than the value set by NZS 4243 of 12 W/m^2 [5]. By comparison, the LPDs found in Retail premises are similar to the minimum set out in NZS 4243. For Retail NZS 4243 requirement is 8–16 W/m^2 depending on retail use type, and the typical measured LPD across all retail use types is 14 W/m^2 [5]. The typical LPD measured in Mixed/Other commercial use type premises is much lower than the NZS 4243 requirement. NZS 4243 allows for a maximum of 8–18 W/m^2 depending on use type, and the typical measured LPD was 6 W/m^2. Mixed/Other commercial LPDs are below the sample average. Also, the high scenario (20 W/m^2) for Mixed/Other commercial LPD's is not significantly different to the Retail (26 W/m^2) and Office (21 W/m^2) high values as the low and typical scenarios.

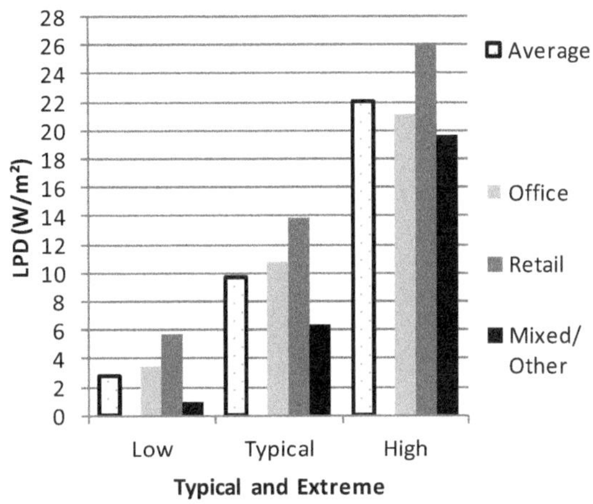

Figure 1. Typical, High and Low LPDs' for different commercial building types.

Figure 2 presents the typical, high and low LPDs' for Small buildings (light grey), Medium sized buildings (dark grey), and Large buildings (black) to assess if size has an impact on the installed LPD. As can be seen, Medium sized buildings have the highest measured LPDs with a typical value of approximately 13 W/m^2. The LPDs found in Small and Large buildings were the same with 8 W/m^2. The high scenario results suggest that lighting was most dense in Medium and Small sized buildings due to the large building high LPD value being half the Small and Medium sized buildings LPDs. This was likely attributed to smaller buildings being more associated with Retail uses which have higher lighting demands.

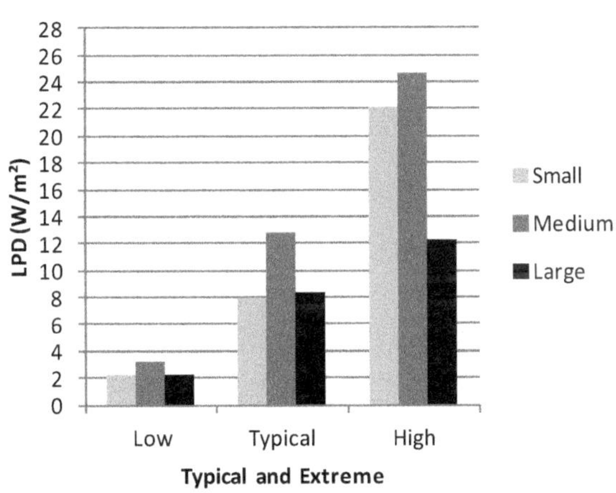

Figure 2. Typical, High and Low LPDs' for different sized commercial buildings.

3.2. Lighting Operation Patterns

Figure 3 displays the Typical (green), High (orange), Low (blue) and NZS 4243 (black) patterns of lighting use for Offices, Retail and Mixed/Other commercial use types.

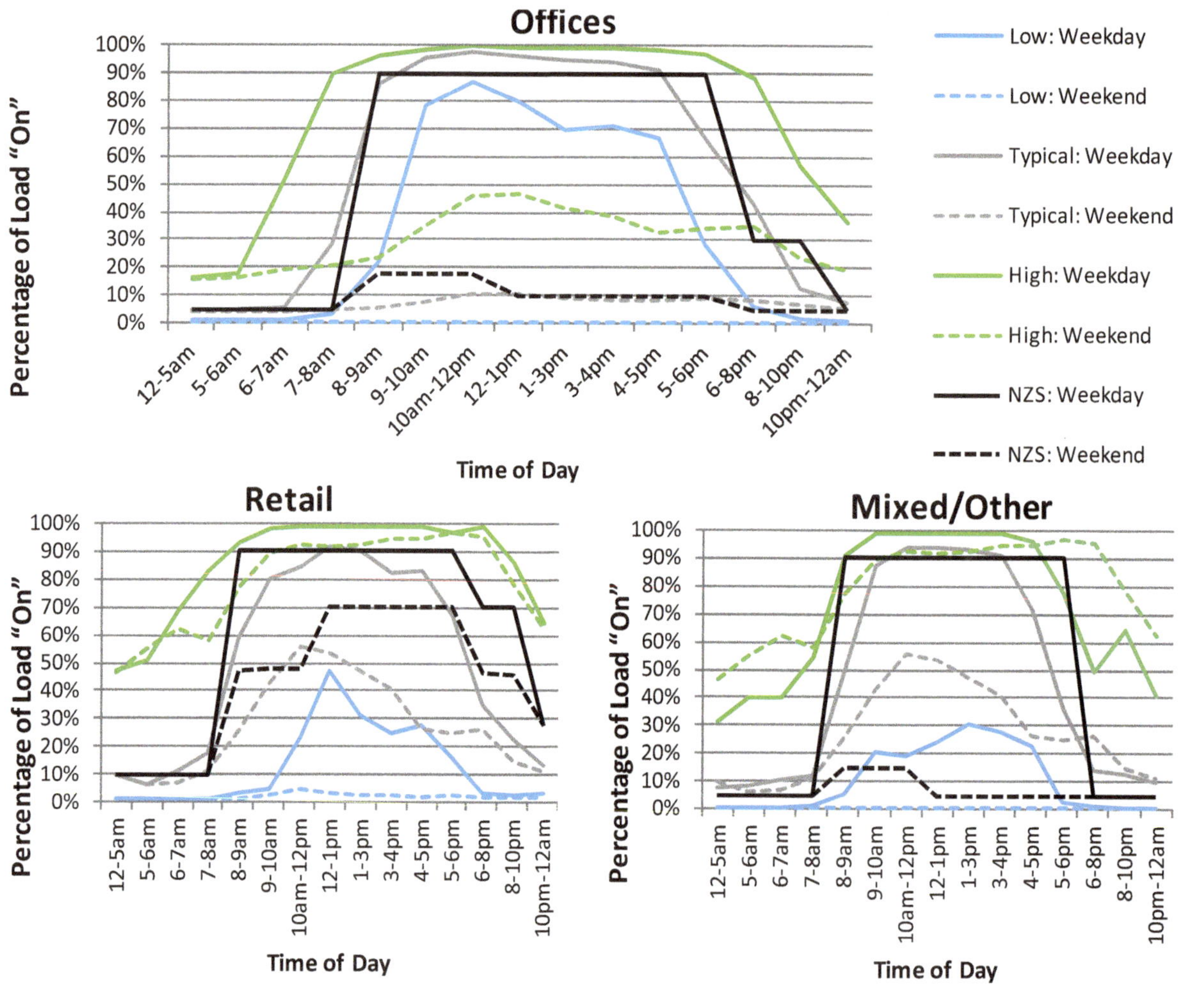

Figure 3. Typical and Extreme Lighting Patterns' of use for different commercial building types. Notes: The Retail pattern of use was an average between the NZS 4243 retail and restaurant patterns. The weekend patterns were an average of the NZS 4243 Saturday and Sunday patterns of use.

As can be seen, Offices have the largest difference in usage patterns between weekdays and weekends. The typical lighting usage of an Office was higher (80%–90% turned "on") than Retail and Mixed/Other commercial uses during the weekdays (70%–80% turned "on"). Retail and Mixed/Other commercial uses have less difference in usage patterns between weekdays and weekends (20%–40% difference) when compared to Offices (60%–70% difference). This fits with expectations that Retail and Mixed/Other commercial use buildings have more intense and longer weekend hours compared to Office buildings. Interestingly, the high use Retail and Mixed/Other use types have almost as intense weekends as weekdays compared to the lighting in high use Offices.

The typical lighting use pattern for Offices was similar to the NZS 4243 assumed pattern of use. However, the Retail and Mixed/Other patterns sit between the typical and high scenario patterns of use. If the NZS 4243 assumptions were used in an energy model, then the model would slightly overestimate the amount of lighting turned "on" in a typical building.

3.3. Equipment Power Density (EPD)

Figure 4 displays the Typical, High and Low EPDs' for the average commercial building (white), Offices (light grey), Retail (dark grey), and Mixed/Other commercial use types (black). Mixed/Other commercial use types have the lowest typical EPD with 5 W/m^2. Offices were the second lowest equipment focused use type with a typical EPD of 8 W/m^2. Retail has the highest typical EPD of 15 W/m^2. This suggests that Retail buildings are the highest equipment energy intensive building type. This finding makes sense as Retail buildings contain premises which are often used for food sales that include refrigeration and cooking. Retail also has a significantly larger high scenario EPD than the other building types, with an EPD of 58 W/m^2. The typical EPD of Offices was similar to the NZS 4243 assumption of 8 W/m^2. However, the other two building types typical EPD's were well above the NZS 4243 assumed EPDs (refer to Table 1 for NZS 4243 assumptions). This highlights the difference between the theoretical EPD and the real buildings EPD. The impact this would have on energy models would be large considering the amount of internal heat gains that would not be modeled using the NZS 4243 assumptions.

Figure 5 presents the typical, high and low EPDs' for Small buildings (light grey), Medium sized buildings (dark grey), and Large buildings (black) to assess the impact of building size on the installed EPD. Small sized buildings were the most equipment focused with a typical EPD of 13 W/m^2. Medium and Large buildings have a substantially lower density of equipment installed with a typical EPD of 6 and 7 W/m^2 respectively. This trend follows in the high EPD scenario. However, small buildings have a much larger EPD (59 W/m^2) when compared to both Medium and Large sized buildings, which have high EPDs of 38, and 18 W/m^2 respectively. This suggests that Small and Medium sized buildings are dominated by Retail loads while larger buildings are dominated by Office loads.

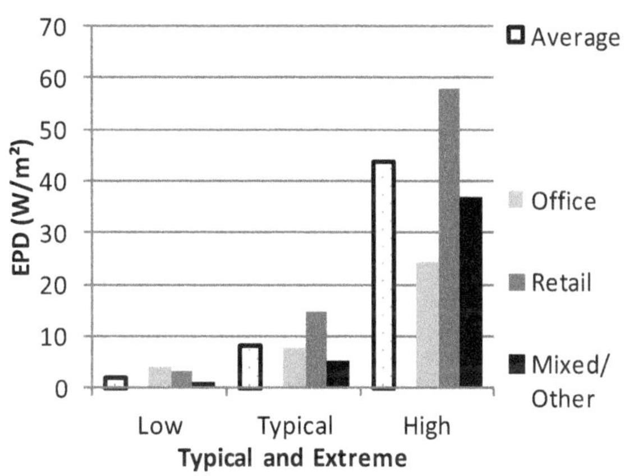

Figure 4. Typical, High and Low EPDs' for different commercial building types.

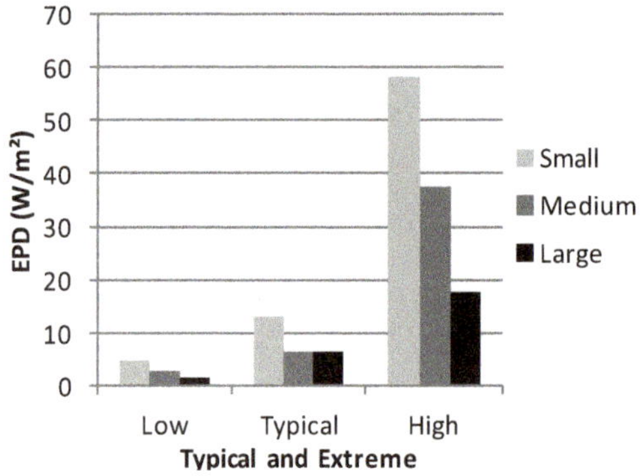

Figure 5. Typical, High and Low EPDs' for different sized commercial buildings.

3.4. Equipment Operation Patterns

Figure 6 displays the Typical (green), High (orange), Low (blue) and NZS 4243 (black) patterns of equipment use for Offices, Retail and Mixed/Other commercial use types.

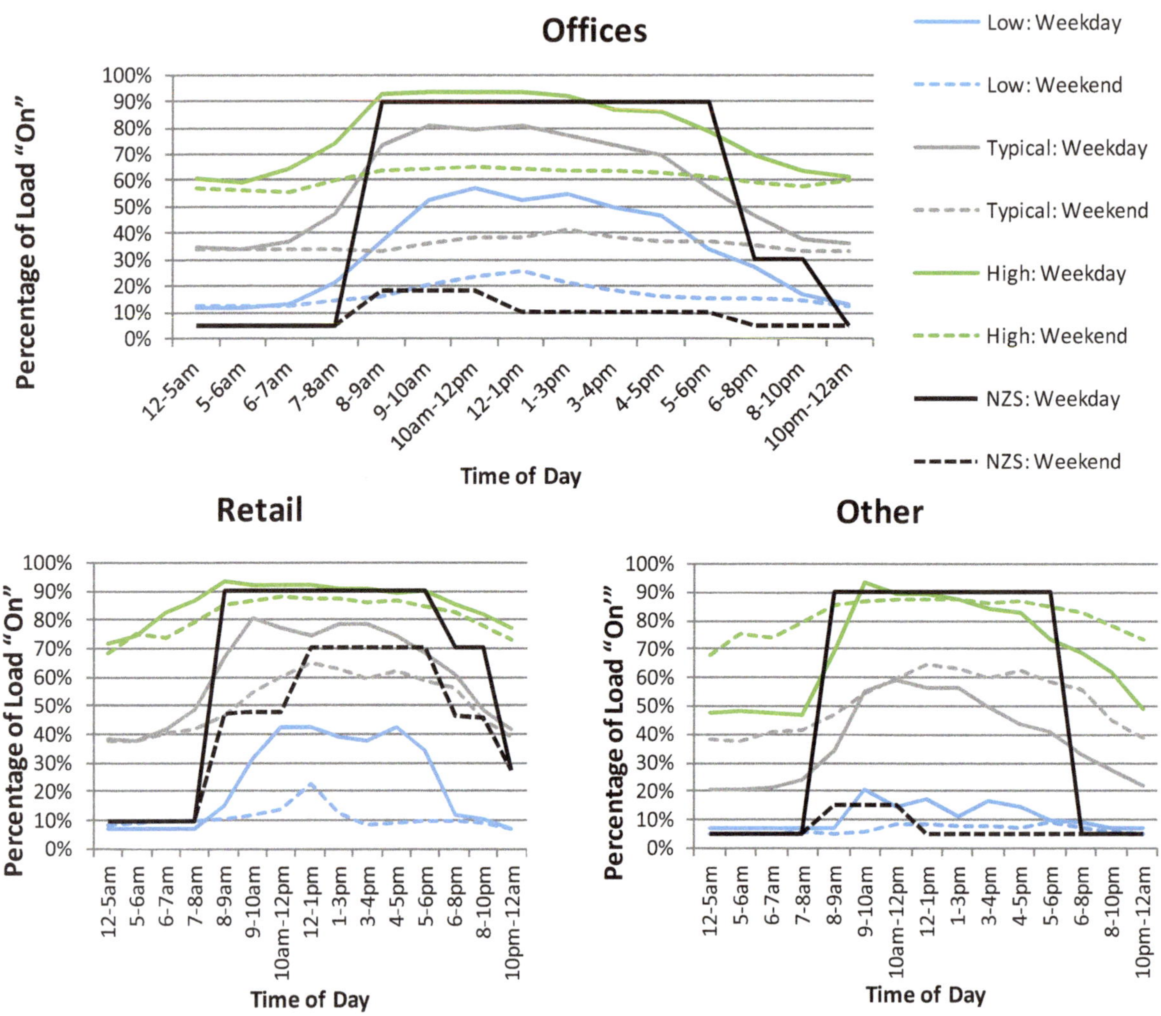

Figure 6. Typical and extreme equipment patterns' of use for different commercial building types. Note: The Retail pattern of use was an average between the NZS 4243 retail and restaurant patterns. The weekend patterns were an average of the NZS 4243 Saturday and Sunday patterns of use.

Offices have the least energy intensive patterns of use as seen by the amount of load that was left "on" during unoccupied periods. This was shown by the weekend and night load percentages. Also, Offices have the biggest difference between night and daytime load patterns. This was highlighted by the larger peaks in equipment energy use of approximately 30%–50% during daytime hours. Office weekday patterns of use have a consistent peak of use whereas their weekend schedules were more or less consistently flat throughout the weekend. This was shown by the weekend day loads not being significantly different to the overnight use. Retail buildings have the most energy intensive patterns of use. This was highlighted by the larger percentage of equipment "on" and for longer periods of time. This was further reinforced by the fact that Retail buildings have higher EPD's (refer to Figure 4). Furthermore, the greater the energy intensity of a Retail premise, the closer the weekend pattern of use was to the weekday pattern. The daytime pattern of use peak in Retail premises (20%–30% more equipment "on") was not as large when compared to Offices (30%–50% more equipment "on"). This could be due to the large refrigeration loads running consistently throughout the whole day, while only a small number of appliances were turned "on" during occupied hours. Unlike Office and Retail, the weekend patterns of use were higher than the weekday patterns of use in Mixed/Other commercial use buildings. The weekday and weekend on Mixed/Other use buildings also have a definite daytime peak of equipment use. Consistent with Retail, the daytime peaks were not as large as for Offices; however, there was a bigger daytime peak during weekdays when compared to the weekends. Also, the weekend loads were more consistently "on" and the weekend night loads were greater than the weekday night loads. This could be due to premises with restaurant and other food type having longer weekend hours.

The results indicate that the common perception among energy auditors, but not in building codes or standards, of equipment being left "on" during unoccupied hours is true. Half of the installed equipment load in Offices was left on overnight. Over a yearly period this equates to a large sum of energy that is essentially wasted. Coupled with this being typical across the commercial building stock, it would seem there is large potential for energy savings if equipment is turned off over night.

The NZS 4243 patterns of use assume that 10%–20% more equipment is turned "on" during daytime hours in Office and Retail buildings, and 10%–30% more in Mixed/Other buildings. The NZS 4243 patterns of use sit between the Typical and High scenarios for all three building types during daytime hours and around the Low scenario during night hours. Therefore, if the NZS 4243 assumptions are used in the model, they would overestimate the amount of equipment turned "on" during daytime and underestimate equipment turned "on" during night hours in a typical building.

3.5. Hot Water Power Density (HWPD)

Figure 7 displays the typical, high and low HWPDs' for average commercial buildings (white), Offices (light grey), Retail (dark grey), and Mixed/Other commercial use types (black). Retail buildings have the largest HWPD with 7 W/m^2 for a typical building and 24 W/m^2 for high use buildings. This was most likely due to the food use types associated with Retail using larger amounts of hot water compared to Offices and Mixed/Other commercial use types. Office and Mixed/Other use type HWPD's were below the average for commercial buildings with 3 and 2 W/m^2 respectively. The high scenario HWPDs for Office and Mixed/Other building types were significantly lower than the Retail HWPD.

Figure 8 presents the typical, high and low HWPDs' for Small buildings (light grey), Medium sized buildings (dark grey), and Large buildings (black) to assess if size has an impact on the installed HW-PD. The typical HWPD was relatively similar for each of the three building size groups. However, the "high" scenario indicates that Medium-to-Large sized buildings have higher demand for hot water when compared to Small buildings. This may occur from larger buildings needing to service a greater number of occupants when compared to smaller buildings.

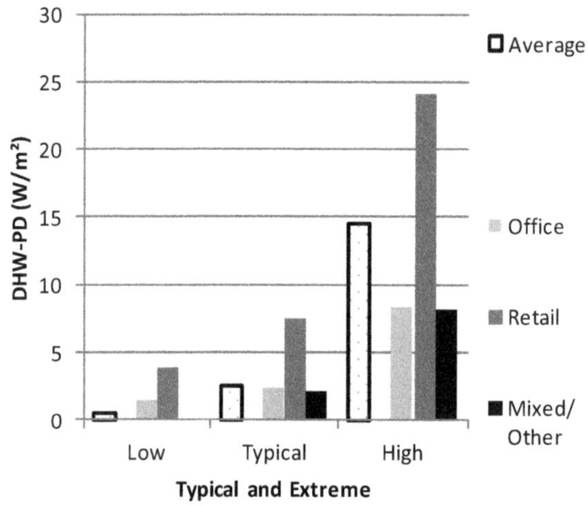

Figure 7. Typical, High and Low HWPDs' for different commercial building types.

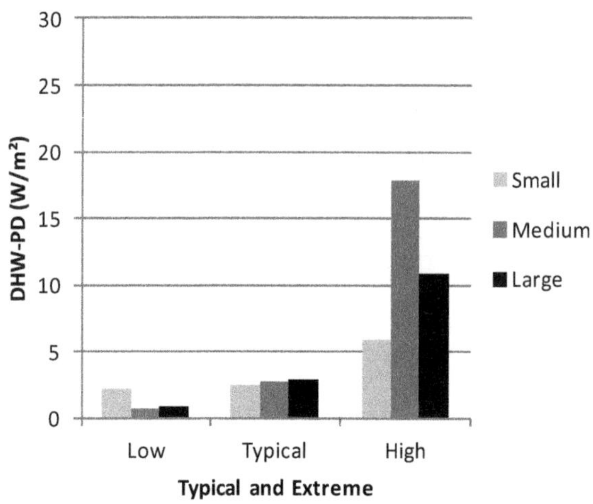

Figure 8. Typical, High and Low HWPDs' for different sized commercial buildings.

3.6. Hot Water Operation Patterns

Figure 9 displays the typical (green), high (orange) and low (blue) patterns of hot water use for Offices, Retail and Mixed/Other commercial use types. Office hot water energy use was very different between weekday and weekend, but also from nighttime to daytime use. Office hot water use was much lower overnight, by approximately 30%–40%, as well as during the weekend with a 20%–40% smaller daytime peak of hot water energy usage. Retail hot water energy use patterns have consistent trends across both weekdays and weekends. The intensity of use between weekdays and weekends also does not change considerably (less than 10%–20%). Mixed/Other commercial use types consume more hot water energy in the weekends than during weekdays. This was consistent with the equipment patterns of use for Mixed/Other use building types.

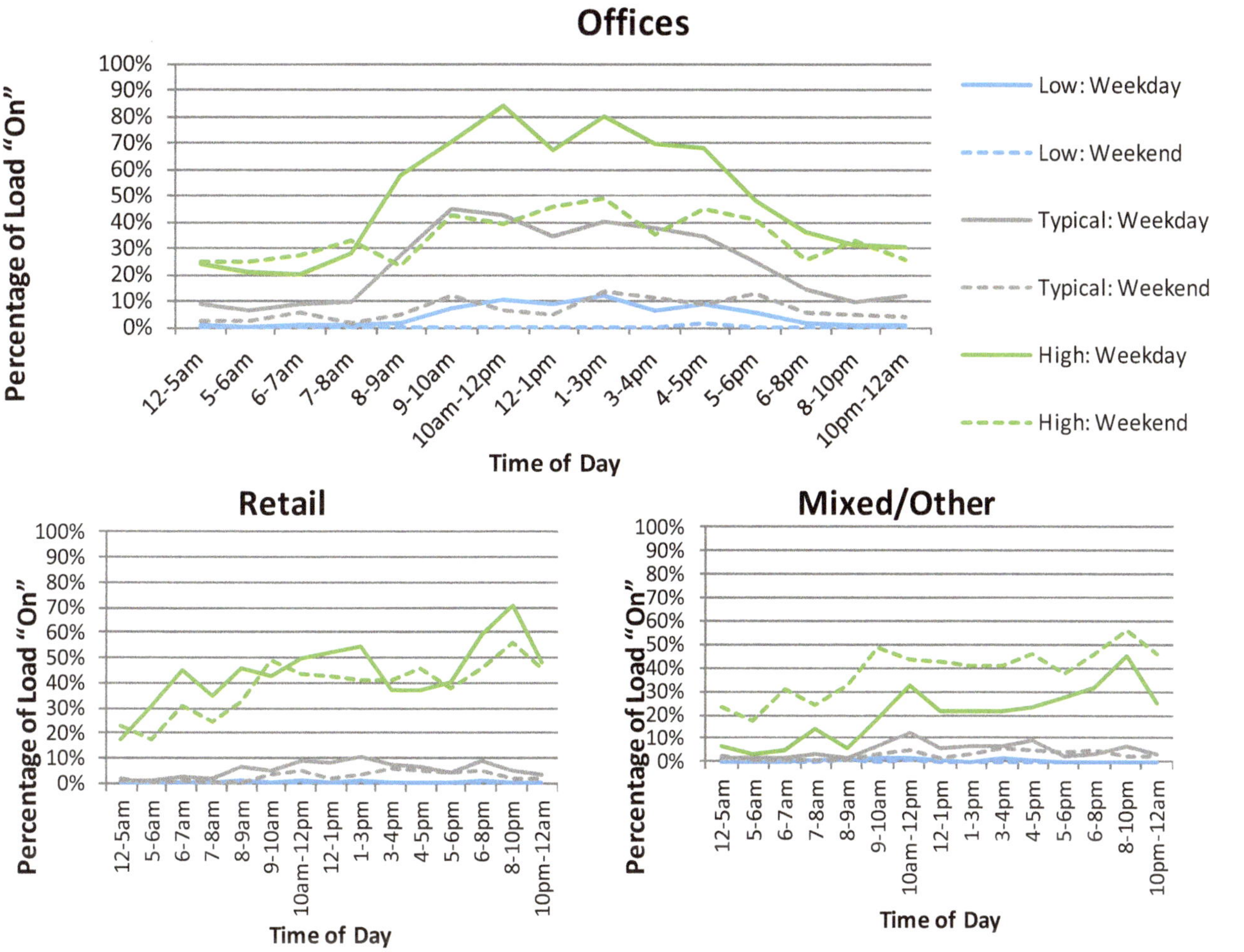

Figure 9. Typical and extreme DHW patterns' of use for different commercial building types.

4. Conclusions

This paper presents the typical, high and low building load densities and their operation patterns for commercial buildings in New Zealand. It fills the gap for the shortfall in information available for commercial building designers and environmental engineers. The shortfall relates to the lack of information regarding building loads found in real commercial buildings. Additionally, this paper compares the building loads found in real commercial buildings to the loads regulated in the NZS 4243 and the assumptions used in the modelling method for comparing energy use to prove the building code has been met.

The results established in this paper indicate that the current NZS 4243 values for power densities and associated schedules are not representative of existing building design in New Zealand. It has been found that the NZS 4243 values for the LPD are indicative of the typical LPDs found in existing buildings. It suggests that typical commercial buildings are designed to just meet the code and not to be any more energy efficient than they need to be. The Retail and Mixed/Other lighting patterns of use overestimate the amount of weekday lighting turned "on". The EPDs found in real buildings were significantly higher for Retail and Mixed/Other commercial use type buildings than the assumptions made in the NZS 4243 modelling method. This results in an underestimation of equipment energy use and internal heat gains if the NZS 4243 values were used. If concerned about assumptions affecting the performance of the building design, an energy modeller could double the power densities or reduce them by a third. By doing so, it roughly estimates the extreme lighting and equipment scenarios which occur across the commercial building stock.

Additionally, the common perception that equipment is left "on" during unoccupied hours is true. Half of the installed equipment load in Offices was left "on" overnight. Over a yearly period this equates to a large sum of energy that is essentially wasted. This combined with the typical values across the commercial building stock indicates that there is large potential for energy savings if equipment is turned off overnight in Offices.

The conclusions are relevant internationally. Building codes and standards suffer the same limitations as the New Zealand building standards which incorrectly assume no equipment or lighting use overnight. The U.S. prototypical models have same assumption limitations and it is likely these assumptions are used elsewhere internationally. These operational patterns were input as "good engineering judgement". It shows that good judgement needs to be informed and updated using the results presented in this thesis in the absence of a countries own experimental data (which is advised to be gathered). Furthermore, the High and Low scenario can be used as a way of altering prototypical models internationally. They can be used as a ratio to be used as a basis for overseas design as well.

These conclusions provided the basis for the creation of prototypical models, as well as the update of existing template models. The models are founded on the real building performance established in this study and the BEES studies. Each model contains New Zealand relevant materials, constructions, building loads, patterns of use, and HVAC system type and properties.

Acknowledgments

The authors would like to acknowledge the BEES team members and BEES project funders who enabled the analysis of the load and patterns of use data.

Author Contributions

Shaan Cory and Michael Donn conceived and designed the experiment; Shaan Cory performed the experiment; Shaan Cory analysed the data; Andrew Pollard contributed materials; Shaan Cory and Michael Donn wrote the paper.

Conflicts of Interest

The authors declare no conflict of interest.

References

1. Department of Building and Housing. *Compliance Document for New Zealand Building Code Clause H1: Energy Efficiency*, 3rd ed.; Department of Building and Housing: Wellington, New Zealand, 2011.

2. BRANZ Ltd. Building Energy End-use Study (BEES), 31 March 2013. Available Online: http://www.branz.co.nz/BEES (accessed on 26 August 2013).
3. Isaacs, N.; Saville-Smith, K.; Bishop, R.; Camilleri, M.; Jowett, J.; Hills, A.; Moore, D.; Babylon, M.; Donn, M.; Heinrich, M.; *et al. Building Energy End-Use Study (BEES) Years 1 & 2*; BRANZ Study Report SR 224 (2009); BRANZ: Judgeford, New Zealand, 2009.
4. Standards New Zealand. *NZS 4243: Part 1: Energy Efficiency—Large Buildings: Building Thermal Envelope*; Standards New Zealand: Wellington, New Zealand, 2007.
5. Standards New Zealand. *NZS 4243: Part 2: Energy Efficeincy—Large Buildings: Lighting*; Standards New Zealand, Wellington, New Zealand, 2007.
6. Torcellini, P.; Deru, M.; Griffith, B.; Benne, K.; Halverson, M.; Winiarski, D.; Crawley, D. *DOE Commercial Building Benchmark Models*; NREL/CP-550-43291; National Renewable Energy Laboratory (NREL): Golden, CO, USA, 2008.
7. BRANZ Ltd. BEES for Energy Modelling, 1 April 2013. Available Online: http://www.branz.co.nz/cms_display.php?sn=169&st=1&pg=9690 (accessed on 13 November 2013).
8. Cory, S.; Hsu, C.; Donn, M. *Template Files for Commercial Building Stock Energy Simulations*; Centre for Building Performance Research: Wellington, New Zealand, 2009.
9. Bordass, B.; Cohen, R.; Field, J. Energy Performance of Non-Domestic Buildings: Closing the Credibility Gap. In Proceedings of the Buildings Performance Congress, Frankfurt, Germany, 26–27 April 2006.
10. Urdan, T.C. *Statistics in Plain English*, 3rd ed.; Routledge: New York, NY, USA, 2010.

Assessment of the Potential to Achieve very Low Energy Use in Public Buildings in China with Advanced Window and Shading Systems

Eleanor S. Lee [1,*], Xiufeng Pang [1], Andrew McNeil [1], Sabine Hoffmann [1], Anothai Thanachareonkit [1], Zhengrong Li [2] and Yong Ding [3]

1 Building Technology and Urban Systems Division, Energy Technologies Area, Lawrence Berkeley National Laboratory, Mailstop 90-3111, 1 Cyclotron Road, Berkeley, CA 94720, USA; E-Mails: xpang@lbl.gov (X.P.); amcneil@lbl.gov (A.N.); sabine.hoffmann@bauing.uni-kl.de (S.H.); athanachareonkit@lbl.gov (A.T.)

2 College of Mechanical Engineering, Tongji University, 1239 Siping Road, Shanghai 200092, China; E-Mail: lizhengrongtongji@126.com

3 Urban Construction and Environment Engineering, Chongqing University, 83 Shabei Street, Shapingba, Chongqing 400045, China; E-Mail: dingyongqq@163.com

* Author to whom correspondence should be addressed; E-Mail: eslee@lbl.gov

Academic Editor: Richard Mistrick

Abstract: As rapid growth in the construction industry continues to occur in China, the increased demand for a higher standard living is driving significant growth in energy use and demand across the country. Building codes and standards have been implemented to head off this trend, tightening prescriptive requirements for fenestration component measures using methods similar to the U.S. model energy code American Society of Heating, Refrigerating and Air-Conditioning Engineers (ASHRAE) 90.1. The objective of this study is to (a) provide an overview of applicable code requirements and current efforts within China to enable characterization and comparison of window and shading products, and (b) quantify the load reduction and energy savings potential of several key advanced window and shading systems, given the divergent views on how space conditioning requirements will be met in the future. System-level heating and cooling loads and energy use performance were evaluated for a code-compliant large office building using the EnergyPlus building energy simulation program. Commercially-available, highly-insulating, low-emittance windows were found to produce 24%–66% lower perimeter zone HVAC electricity use compared to the mandated energy-efficiency standard in force (GB 50189-2005) in cold climates like Beijing. Low-e windows with operable exterior shading produced up to 30%–80% reductions in perimeter zone HVAC electricity use in Beijing and 18%–38% reductions in Shanghai compared to the standard. The economic context of China is unique since the cost of labor and materials for the building industry is so low. Broad deployment of these commercially available technologies with the proper supporting infrastructure for design, specification, and verification in the field would enable significant reductions in energy use and greenhouse gas emissions in the near term.

Keywords: energy efficiency; building; windows; shading; China

---

1. Introduction

The building construction industry in China has no parallel in terms of growth. Between 1985 and 2004, 10 Bm^2 in new construction was added resulting in 40 Bm^2 of total floor area in China by around 2005 [1]. About 30% of China's total energy use can be attributed to buildings. In addition, although per capita carbon emissions is low, energy use in commercial buildings can be high. Xiao *et al.* [2] conducted a statistical analysis of annual energy use data from 4600 office buildings across China and observed a unique dual sector characteristic in energy use intensity reflecting a large proportion of buildings with low energy use intensity (EUI) and a small proportion of buildings with a significantly higher EUI. An example is given where two large office buildings in the same city, one with and one without operable windows, attributed 0.2 kWh/m^2-year and 5.8 kWh/m^2-year, respectively, to heating, ventilation, and air-conditioning (HVAC) energy use, the former having opened windows during the transitional seasons to meet comfort requirements. Growth in the proportion of buildings in the latter category is anticipated and can be attributed to an increased demand for a higher service standard. As the demand for a higher standard of living increases, demand for amenities such as large area windows, tighter ranges on indoor setpoint temperatures, 24-h HVAC operation, *etc.*, will increase. These drivers point to significant growth in the energy use intensity per building and energy demand overall across the country.

To counter this increased demand, the Chinese government has and continues to implement aggressive energy policies to improve the energy efficiency of buildings. Growth in commercial building primary energy use is expected to increase from 5 Quads (10^{12} Btu) in 2005 to more than 13 Quads by 2020, where about 50%–60% of energy use is due to heating, ventilation, and air-conditioning (HVAC) and 20%–30% is due to lighting [3,4]. The "Design Standard for Energy Efficiency of Public Buildings" (GB 50189-2005, [5]) was implemented in July 2005 with the goal of reducing public sector building energy use levels by 50% compared to consumption levels in 1983 (public buildings include office buildings, schools, hotels, hospitals, retail, *etc.*). Seven years later, major cities (e.g., Beijing, Shanghai, Tianjin, Chongqing, and Shenzhen) have or are implementing more stringent energy efficiency standards that target a 65% reduction in energy use. There are also voluntary certification programs such as the Evaluation Standard for Green Buildings (GB/T 50378-2006) initiated in 2006 and the Three-Star Green Buildings program. These programs are similar to the US Green Building

Council's Leadership in Energy and Environmental Design (LEED) program and have been developed to award sustainable building design and operations.

In the face of rapid growth in the buildings sector over the past two decades, such policies are a powerful and practical means of enabling China to meet greenhouse gas emission reduction goals. China's 12th Five Year Plan (2011–2015) seeks to reduce all sector energy intensity (primary energy use per unit of gross domestic product (GDP)) by 16% and commits to reduce CO_2 intensity (CO_2 per unit GDP) by 40%–45% below 2005 levels by 2020 [6]. Implementation of energy efficiency policies will also enable China to address critical situations brought on by rapid urbanization, reducing energy demand given limited supply, improving environmental quality through the reduction of air pollution, and enhancing market competitiveness by reducing operating costs.

Like the US, China building energy efficiency codes are developed in collaboration with The Ministry of Construction, The Ministry of Housing and Urban Development (MOHURD), industry or manufacturer associations, and the design and engineering community. The GB 50189-2005 Standard was developed and is supported by the China Academy of Building Research. The Standard is mandatory nationwide and is enforced by local governments. Like the development of the US model energy code American Society of Heating, Refrigerating and Air-Conditioning Engineers (ASHRAE) 90.1, the GB 50189-2005 Standard was developed by modeling a typical early 1980s public building using the US Department of Energy's DOE-2 building energy simulation program, then applying energy efficient component measures to achieve 50% savings [7]. Selection of component measures is driven by a balance between efficiency goals and practical constraints, identifying which energy-efficiency strategies and measures will deliver adequate, sustained savings, which are commercially available and supported by a relatively mature infrastructure for efficient dissemination across the country, which are likely to succeed given the preferences of the building industry, and which will meet the cost-effective criteria defined by a competitive market.

Fenestration systems—windows, solar control shading systems, and daylighting systems—affect two of the largest energy end uses in Chinese public buildings: HVAC and lighting. The objective of this study is to: (a) provide an overview of applicable code requirements and current efforts within China to enable characterization and comparison of window and shading products, and (b) quantify the load reduction and energy savings potential of several key advanced window and shading systems, given the divergent views on how space conditioning requirements will be met in the future. We investigate the 50% and 65% energy efficiency code requirements as related to windows, then survey which basic window glazing products meet the prescriptive requirements mandated by the codes. We then investigate the energy savings potential of several alternate fenestration systems to gauge performance relative to the GB 50189-2005 Standard and discuss the potential success of such measures within the pragmatic context of the commercial building construction market in China.

2. Energy-Efficiency Requirements for Fenestration

2.1. Energy-Efficiency Standards

As a component measure, windows are subject to efficiency requirements dictated by the code. Maximum whole window U-values and center-of-glass shading coefficient (SC) requirements are specified in the mandatory section of the GB 50189-2005 code. Similar to the ASHRAE 90.1 building energy efficiency standard [8], the requirements are defined by climate zone, window orientation, and window-to-wall ratio (WWR, defined without the window frame). See Table 1. To enable reduction of lighting energy use, small area windows with a WWR less than 0.4 must have a center-of-glass visible transmittance (Tvis) greater than or equal to 0.4, even though there are no mandatory requirements for daylighting controls. Exterior shading is recommended in some climates. A minimum of 30% of the windows must be operable.

Table 1. Prescriptive building envelope requirements defined by the GB 50189-2005 Standard.

Element		Max. Uw, Actual Uw, Actual Ug ($W/m^2 \cdot K$)	Max SC Value, Actual SHGC
Climate Zone Severe Cold A (Harbin)			
Window	$WWR \leq 0.2$	3.0, 2.51, 1.8	-
	$0.2 < WWR \leq 0.3$	2.8, 2.37, 1.8	-
	$0.3 < WWR \leq 0.4$	2.5, 2.30, 1.8	-
	$0.4 < WWR \leq 0.5$	2.0, 1.84, 1.3	-
	$0.5 < WWR \leq 0.7$	1.7, 1.59, 1.1	-
Climate Zone Cold (Beijing)			
Element		**Max. Uw, Actual Uw, Actual Ug ($W/m^2 \cdot K$)**	**Max SC Value, Actual SHGC**
Window	$WWR \leq 0.2$	3.5, 3.09, 2.7	-
	$0.2 < WWR \leq 0.3$	3.0, 3.00, 2.7	-
	$0.3 < WWR \leq 0.4$	2.7, 2.14, 1.6	$SC \leq 0.7$, 0.44
	$0.4 < WWR \leq 0.5$	2.3, 2.08 1.6	$SC \leq 0.6$, 0.44
	$0.5 < WWR \leq 0.7$	2.0, 1.99, 1.6	$SC \leq 0.5$, 0.44
Climate Zone Hot Summer/Cold Winter (Shanghai)			
Element		**Max. Uw, Actual Uw, Actual Ug ($W/m^2 \cdot K$)**	**Max SC Value, Actual SHGC**
Window	$WWR \leq 0.2$	4.7, 3.09, 2.7	-
	$0.2 < WWR \leq 0.3$	3.5, 2.95, 2.5	$SC \leq 0.55$, 0.41
	$0.3 < WWR \leq 0.4$	3.0, 2.77, 2.4	$SC \leq 0.50/0.60$ *, 0.37
	$0.4 < WWR \leq 0.5$	2.8, 2.10, 1.6	$SC \leq 0.45/0.55$ *, 0.32
	$0.5 < WWR \leq 0.7$	2.5, 1.99, 1.6	$SC \leq 0.40/0.50$ *, 0.30
Climate Zone Hot Summer/Warm Winter (Guangzhou)			
Element		**Max. Uw, Actual Uw, Actual Ug ($W/m^2 \cdot K$)**	**Max SC value, Actual SHGC**
Window	$WWR \leq 0.2$	6.5, 5.79, 5.8	-
	$0.2 < WWR \leq 0.3$	4.7, 3.12, 2.4	$SC \leq 0.5/0.6$ *, 0.41
	$0.3 < WWR \leq 0.4$	3.5, 3.03, 2.4	$SC \leq 0.45/0.55$ *, 0.41
	$0.4 < WWR \leq 0.5$	3.0, 2.96, 2.4	$SC \leq 0.4/0.5$ *, 0.37
	$0.5 < WWR \leq 0.7$	3.0, 2.20, 1.6	$SC \leq 0.35/0.45$ *, 0.32

Max Uw and SHGC are prescribed values; Actual Uwindow, Ucenter-of-glass, and SHGC are the properties of windows modeled using full spectral data from actual commercially available glazings in the International Glazing Database in Window 6; * South, east, west / north; WWR and SC does not include the window frame. *U*-value includes the frame.

The stricter 65% Standards were developed and piloted for select cities as mentioned above. As an illustration of these more stringent standards, we list the prescriptive window requirements defined by the Shanghai Design Standard for Energy Efficiency in Public Buildings (DGJ08-107-20) in Table 2.

Table 2. Prescriptive window requirements defined by the DGJ08-107-20 code for Climate Zone Hot Summer/Cold Winter (Shanghai).

Element		Max. U values ($W/m^2 \cdot K$)	Window Shading Coefficient
Roof		0.5	-
Walls		0.8	-
Floor (contacted with ambient air)		0.8	-
Window	WWR ≤ 0.2	3.5	-
	0.2 < WWR ≤ 0.3	3.2	$SC_W \leq 0.45$
	0.3 < WWR ≤ 0.4	2.8	$SC_W \leq 0.40/0.50$ *
	0.4 < WWR ≤ 0.5	2.5	$SC_W \leq 0.35/0.45$ *
	0.5 < WWR ≤ 0.7	2.2	$SC_W \leq 0.30/0.40$ *

*: South, east, west/north.

For example, the 65% Shanghai Standard requires that public buildings greater than 5000 m^2 in floor area with central air-conditioning have a total window shading coefficient (SC_W) value of 0.30 or less and a whole window *U*-value of 2.2 W/m^2-K or less for large-area, vertical windows with a WWR (without frame) between 0.50 and 0.70 facing south, east, or west. Here, the SC_W is the product of the Shading Coefficient of Windows (SW) and the External Shading Coefficient (SD). SW is the product of the center-of-glass shading coefficient (SC) and the window frame factor, f (e.g., $f = 0.70$ for wood frame, $f = 0.75$ for aluminum frame). SD is determined by exterior (outdoor) shading type, geometry, orientation of window, and degree of opacity using a look-up table. The 65% Shanghai Standard enables the designer to reduce the solar control requirements for the glazed window if exterior or between-pane shading systems are used. For the same example above, if a horizontal overhang that is as deep as the height of the window is used, SD would be 0.7 for a south-facing window and the maximum center-of-glass shading coefficient SC of an aluminum-framed window could be raised from 0.40 to 0.57 (*i.e.*, $SCw = SC \times f \times SD$ so with no overhang, $0.3 = SC \times 0.75 \times 1$ and with the overhang, $0.3 = SC \times 0.75 \times 0.70$).

If an operable exterior or between-pane shading system is used and covers the whole window, the Shanghai Standard requires no calculation of the SC_W, and the SC_W is simply considered as code-compliant (User Instructions for DGJ08-107-20). The U-value requirements can also be reduced with a correction factor, g (e.g., $g = 0.9$ for a roller shade, $g = 0.95$ for a blind). For example, if a roller shade is used, the SCw requirement is considered as met regardless of the window properties, and the *U*-value requirement is increased to 2.4 W/m^2-K compared to 2.2 W/m^2-K if no exterior shade was used. The operation of the shade is not mandated—the shade can be manually operated or automated with motorized controls. In the case of ASHRAE 90.1, no credit is given for exterior shading unless the shading is proven to be permanently installed and if operable, automated at all times to control for the worst case sunny condition (even if it is overcast).

2.2. Prescriptive Glazing and Shading Measures

The primary means of meeting the GB 50189-2005 code is to specify unshaded glazed windows with the requisite properties. There are ten Chinese glass manufacturers, several of which are foreign firms manufacturing glass in partnership with Chinese manufacturers (*i.e.*, Pilkington, Taiwan Glass, Asahi). China Southern Glass is by far the largest Chinese manufacturer. Domestic manufacturing is critical for compliance materials and products in large, rapidly growing construction markets like China. Not only is there a 40% tariff imposed on foreign products making imported products very expensive, demanding construction schedules result in low tolerance for delays in product delivery. Because of these constraints, the code development process accommodates cost-effective products that are commercially available from Chinese manufacturers.

The process of characterizing window properties using standardized methods was initiated in China to enable consumers to select, compare, and verify products for code compliance. Since the early 1990s, glass manufacturers from around the world have measured their products using protocols defined by international standards and submitted these data to the International Glazing Database (IGDB) [9]. The IGDB, which was developed and is maintained by the Lawrence Berkeley National Laboratory (LBNL), now contains over 4100 glazing products. MOHURD and industry organizations initiated discussions with LBNL in early 2010 on how to measure window properties to enable entry of Chinese products to the IGDB, with a first training session occurring in August 2010 at LBNL. In October 2010, MOHURD agreed to begin to adopt the International Standards Organization ISO 15099, which specifies the detailed calculations procedures for determining the thermal and optical transmission properties of window systems. Data submissions were first received in November 2010 and a second international training workshop was held at LBNL in February 2011. Manufacturers were then qualified to submit data after successfully participating in an interlaboratory comparison.

All ten manufacturers are now contributing measured spectral, conductance, and emittance data for their glazing products to the IGDB, coordinated through the China National Safety Glass and Quartz Glass Test Center. As of the latest release of the IGDB (version 23), there were a total of 246 submittals from Chinese manufacturers (Table 3). In parallel, a Chinese version of LBNL's Window 6 software [10] was rewritten in collaboration with LBNL and is now being distributed by the Guangdong Provincial Academy of Building Research (PABR). This software tool computes whole window solar heat gain coefficient (SHGC) and *U*-values for any arbitrary combination of glazings from the IGDB, spacers and framing systems, gas fills, and some common shading devices in accordance with ISO 15099. Framing and spacer products are offered by numerous vendors in China. Some foreign companies with manufacturing plants in China import thermally-broken spacers for use in the fabrication of insulating glass units. LBNL's Therm software for modeling framing and spacer systems is publicly available and free for use (in English).

To understand what range of window products are commercially available from Chinese manufacturers, glazing layers from the IGDB were combined into an insulating glass unit (IGU), assuming a clear 3 mm inboard lite, thermally broken spacer, and a 12.7 mm air gap. Center-of-glass SHGC, U-value, and visible transmittance (Tvis) data for these IGUs are given for all Chinese manufacturers and for all commercial products available worldwide in Figure 1 (SC values can be converted to SHGC by multiplying the value by 0.87 (e.g., SC = 0.30 is SHGC = 0.26). Products with a daylight-to-solar ratio (Tvis/SHGC) between 2.0 to 2.5 are amongst the high-performance glazings on the market today for commercial building applications since they simultaneously reject solar gain and admit daylight. As shown in Figure 1, there are more than a few glazing products that meet the more stringent requirements (assuming the center-of-glass *U*-value is nearly the same as the whole window *U*-value if the window is large).

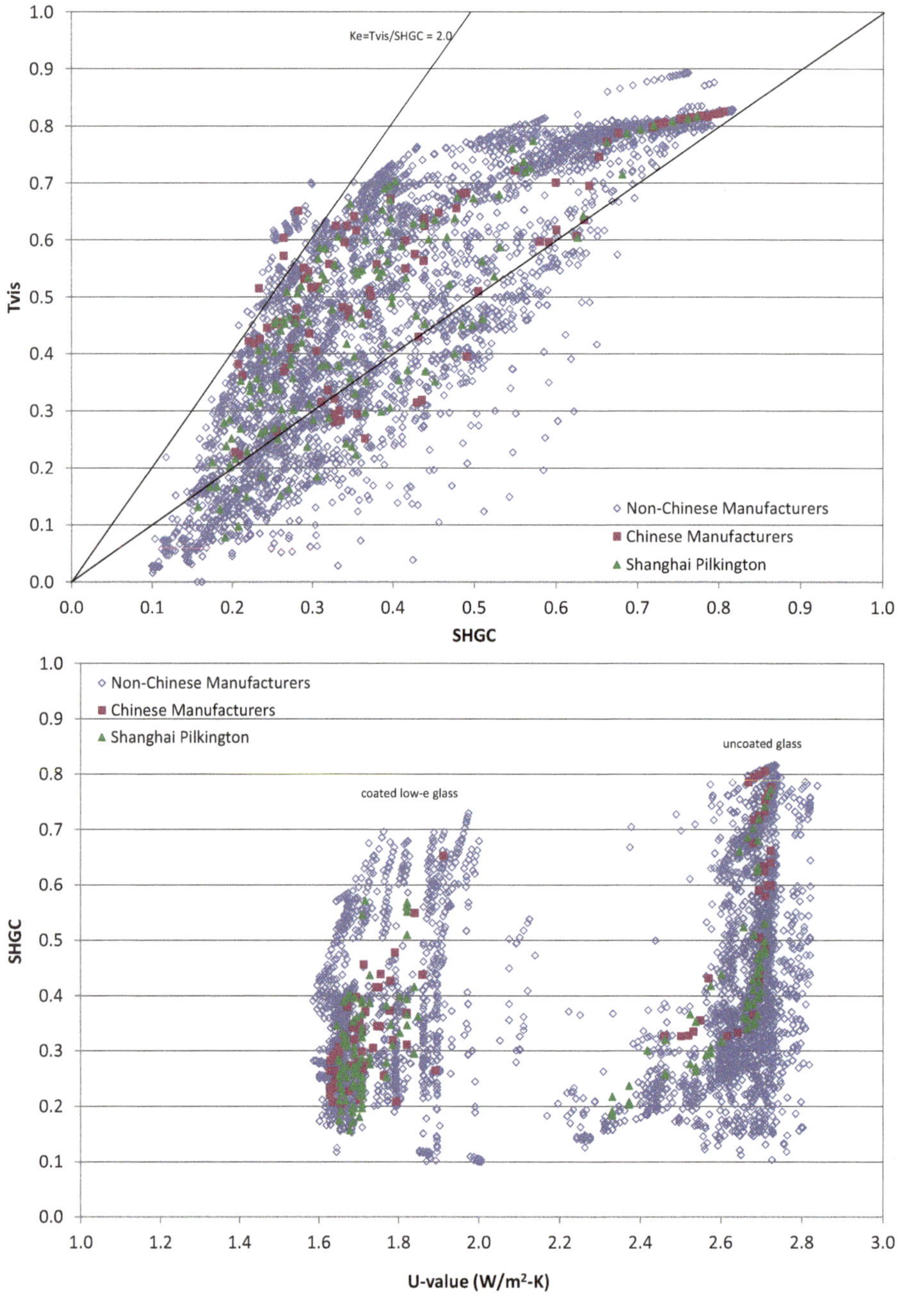

Figure 1. Center-of-glass solar heat gain coefficient, *U*-value, and visible transmittance (Tvis) data for all glass layers in the IGDB (version 23) combined with a clear 3 mm inboard lite and a 12.7 mm air gap. Lines depict a Ke ratio (Tvis/SHGC) of 2.0 and 2.5.

Table 3. Chinese glass manufacturers with submittals in the International Glazing Database (IGDB).

Manufacturer	No. of products in IGDB ver.23
AGC Flat Glass (Dalian) Co., Ltd (company of Asahi) (AFD)	1
China Southern Glass (CSG)	61
Grand Engineering Glass (GRA)	8
Guangdong Avic Special Glass Technology Co., Ltd (SGT)	0
Intex Glass (Xaimen) Co., Ltd (INX)	0
Luoyang NewJingrun Architectural Glass Co., Ltd (NJR)	1
Shanghai Yaohua Pilkington Glass Co., Ltd (SYP)	155
Taiwan Glass Ind. Corp. (TGI)	14
Yantai Jialong Nano Industry Co., Ltd (YJN)	3
ZNG Glass Co., Ltd. (ZNG)	3
Total submittals	246

There are no comparable single value performance indices for comparing shading and daylighting systems. For glazings, a normal incidence value for SC and Tvis is sufficient to compare different products. For shading systems, SC and Tvis can vary significantly, depending on solar angle of incidence and position of the shading system (tilt angle of louvers, height of shade, *etc.*). There is significant research activity underway to characterize and model the solar heat gain and daylight performance of light scattering or optically complex fenestration systems using bidirectional transmittance and reflectance or scattering distribution function (BSDF) data [11,12]. At present, the energy performance of common systems like roller shades or Venetian blinds can be modeled in software like EnergyPlus assuming hemispherical diffusing or Lambertian surface reflectance properties.

3. Energy Savings Potential of Fenestration Measures

3.1. Definition of Office Building Prototype

To understand how advanced window and shading technologies will perform relative to the GB 50189-2005 Standard, we use building energy simulations to evaluate the energy savings potential for a prototypical office building. In a prior study by Lam *et al.* [13], building energy simulations were conducted for the five major climate zones in China to investigate the thermal and energy performance of office buildings and identify measures that could increase building energy efficiency. Findings indicated that passive solar designs would yield significant energy savings in severe cold and cold climates and control of solar heat gains through windows in hot climates were required. In a related study by Yang *et al.* [14], heat gains and losses of three generic office building envelope designs were assessed for the five major climates in China to determine what building envelope measures could be used to improve building energy efficiency during major retrofits of existing buildings in China. This second study provided more specific guidance for the different climates, such as use of solar control films to reduce solar gains or using reduced *U*-value windows for heating-dominated climates. Unlike these prior studies, this study provides a more detailed evaluation of the various window and shading parameters affecting perimeter zone energy performance.

In the US, building energy-efficiency codes were initially developed by conducting parametric building energy simulations of prototypical buildings [15]. Building prototypes are often abstract, synthetic buildings, not real buildings, that have been developed for the purpose of being representative of a population of buildings of a given type; e.g., office, hospital, *etc.*, Data are collected on real buildings and these data are used to formulate a statistical representation of building construction, systems, and operations [16]. The US Department of Energy (DOE) has invested in the development of such prototypes over the past few decades, and has made such prototypes publicly available for use by the buildings industry in order to standardize methods for evaluating technical measures and policies [17]. The models were developed for use with the EnergyPlus building energy simulation program [18]. We made inquiries as to whether the Chinese had similar models and were answered with the affirmative, but the models are proprietary and not available for public use. Therefore, in order to estimate energy savings of alternate energy efficiency measures, we modified the DOE large office prototype so that we had one EnergyPlus model that complied with the 1983 baseline building model ("baseline") and a second model that complied with the GB 50189-2005 standard ("code"). Assumptions for these models are given in Appendix A (Tables A1–A5).

The large office prototype is a 12-story, 48 m high building with a rectangular floor plate that is 73 m (north-south) by 49 m (east-west). Perimeter zones are 4.57 m deep, with a 2.7 m ceiling height and a floor-to-floor height of 4 m. Exterior walls are of mass wall construction. No exterior obstructions were modeled. Internal loads were 18 W/m^2 for lighting, 13 W/m^2 for equipment loads, and 8 m^2 per person for occupant loads. Occupancy was primarily between the hours of 9:00 to 17:00 on weekdays. The HVAC schedule assumed setpoint temperatures of 20 °C for heating and 26 °C for cooling during occupied hours, and 12 °C for heating and 37 °C for cooling during unoccupied hours. Detailed schedules are given in Appendix Table A5.

In China, HVAC systems for typical existing construction tend to be distributed, not centralized [19,20] and for new, low-energy building design, innovative heating and cooling strategies will tend to rely more on distributed hydraulic, perimeter zone solutions. For example, two- or four-pipe fan coil unit (FCU) systems with centralized outdoor air unit are more popular than variable-air-volume (VAV) systems. These FCUs are typically mounted in the ceiling at the entrance of the room, then used by the occupants on demand. Two-pipe systems cost less than four-pipe systems, but the two-pipe system is either in the cooling or heating mode which results in thermal discomfort if the system is meeting cooling demands in the south zone but heating is required in the north zone. Four-pipe systems enable more flexibility in maintaining individual zone comfort and are therefore more appropriate for large, high quality office buildings.

For heating, the solution depends on the climate. In Northern China, most buildings utilize district heating. In central China, the heating equipment is diverse, ranging from district heating, hot water boilers, to heat pumps. In Southern China, heating equipment is not allowed. Therefore, district heating is defined for the office prototype for the Harbin and Beijing climates, and no heating system is defined for Guangzhou. According to colleagues in China, the air-source heat pump is a better choice for providing heating in Shanghai because natural gas is expensive. However, modeling an air-source heat pump as central plant equipment is not supported in the current version of EnergyPlus, so we decided to use the electric hot water boiler as the heating equipment for the Shanghai climate.

Cooling is typically provided by water-cooled chillers for commercial buildings in China, so we used two equal-sized, water-cooled chillers in the prototype. The core of the office building as well as the facades with large-area windows and significant sun exposure require cooling even in winter. To provide chilled water to the building in winter with the proper handling of the condenser water for anti-freezing, we can either run the chiller or use the water-side economizer to generate the chilled water. In the prototype, we use the chiller to produce the chilled water in the winter for cooling.

3.2. Parametric Analysis of Window Systems

Parametric simulations were run on the large office prototype, varying the climate, window area, and window type (Table 4). The windows for the 1983 baseline building were modeled using the Simple Glazing System method in EnergyPlus, and the GB 50189-2005 code and alternate windows were modeled using detailed glazing system data obtained from using full spectral data in the Window 6 software tool, which implements the ISO 15099 Standard for the calculation of SHGC and *U*-value.

Table 4. EnergyPlus parametric simulations.

Parameter	**Values**	
Climate zones	Harbin, Beijing, Shanghai, Guangzhou	
WWR	0%, 15%, 30%, 45%, 60%, 70%	
Window type	Reference	1983 construction
	Code	GB 50189-2005 compliant window
	E1	Low-e window
	E2	Low-e window with interior roller shade
	E3	Low-e window with exterior roller shade
	F1	Highly insulated window
	F2	Highly insulated window with interior roller shade
	F3	Highly insulated window with exterior roller shade

Because we used data from commercially available, not hypothetical windows, the maximum SC and *U*-value requirements for the 2005 code were not matched exactly. Actual values are given in Table 4 for the windows that best matched the code requirements. As noted above, *U*-value and SC requirements defined by Chinese codes are not in complete agreement with the definitions used by the ASHRAE 90.1 standard. The Chinese code defines WWR and SC without the frame but includes the frame in the definition of *U*-value. The ASHRAE standard was used to define the WWR. No adjustments were made to the SC value, assuming that the frame area had a small effect.

Two alternate window types were modeled using Window 6 *idf* format descriptions as input to EnergyPlus: a spectrally selective dual pane window (E) and a triple pane insulating window (F) (Table 5). Each window was modeled with no shade (case 1), with an interior shade (case 2), or with an exterior shade (case 3); e.g., E2 is window E with an interior shade.

Table 5. Advanced window properties.

Type	Window E	Window F
	Spectrally-Selective, Dual-Pane, Low-e	**Highly Insulating, Triple-Layer, Low-e**
Layer 1	6 mm low-e (IGDB ID# 5439)	6 mm low-e (IGDB ID# 5439)
Gap	1.27 cm Argon	1.27 cm Argon
Layer 2	6 mm Clear (IGDB ID# 103)	Suspended film (IGDB ID# 350)
Gap	-	1.27 cm Argon
Layer 3	-	6 mm low-e (IGDB ID# 5439 flipped)
U-value *	2.17(15%), 2.01(30%), 1.92(45%), 1.85(60%), 1.76(70%)	1.50(15%), 1.35(30%), 1.26(45%), 1.20(60%), 1.11(70%)
SHGC	0.272	0.226
SC	0.313	0.260
Tvis	0.639	0.465
Tvis/SHGC	2.35	2.06

* *U*-value is given as whole window value, and the value in the parenthesis is the window to wall ratio. SHGC, SC, and Tvis are given as center of glass values. The emittance of IGDB ID#5439 and #350 is 0.018 and 0.711, respectively.

Both the interior and exterior shades were operated at each time step to reduce HVAC perimeter zone loads. A light gray fabric roller shade with an openness factor of 3% was modeled using the standard models in EnergyPlus. Spectral measurements for solar and visible absorption, reflectance and transmittance were used as input to the model. Operation of the shade was modeled using the "On night if heating and on day if cooling" mode (based on the setpoint and zone temperatures) to control window heat gains and losses and minimize HVAC energy use. Other measures (e.g., furniture placement, location of task and view direction) were assumed to be used to address other constraints such as thermal and visual discomfort. The algorithm was implemented as follows: (a) if it is nighttime (sun is down), then if the zone heating rate of the previous time step is non-zero (e.g., 10 W), then the shade is fully lowered over the window; and (b) if it is daytime (sun is up) and the zone cooling rate of the previous time step is non-zero, then the shade is lowered over the window.

3.3. Definition of Energy Use Metrics

The various window systems were evaluated using two metrics: (1) perimeter zone sensible heating and cooling load; and (2) perimeter zone HVAC energy use. These metrics were based on the loads and energy use for all twelve floors of the office prototype.

The perimeter zone sensible heating and cooling loads represent the sensible energy needed to maintain the room air temperature at the desired level as the zone thermal load resulting from heat flow through the building envelope and internal loads from equipment, people, and lighting varies. Inefficient mechanical system design and operations, such as how the HVAC system is sized, the operation and air flow rate from the dampers, and how outside air is conditioned, can obscure the performance impact of fenestration efficiency measures. These reported loads are independent of the HVAC system in use and are therefore useful as an indicator of how well the building envelope passively mitigates the thermal effects of the outdoor environment as would occur in very low energy buildings. Because there is significant uncertainty as to whether the HVAC models closely represent the systems used in China, these data provide fundamental information on the relative effectiveness of each window system in managing loads. The loads were computed assuming the same cooling and heating setpoint schedule as the HVAC schedule, where the load was zero if the zone temperature was within the setpoint deadband.

Annual HVAC energy use was also determined to provide a complete assessment on energy performance impacts. For this prototype, the ratio of the perimeter area to the total floor area was low (Ap/Atotal = 0.30) causing core energy use to dominate the whole building energy use profile. Therefore, perimeter zone district heating energy use or electric boiler energy use was determined by allocating energy use in proportion to the thermal demands within the zones. Perimeter zone HVAC electricity consumption was also determined in proportion to the thermal demands within the zones and included the total energy use required by all electric-based equipment (central plant chiller, condenser, pumps, fan coil units) needed to provide heating, cooling, and ventilation to these zones.

3.4. Results for a Cold Climate

Beijing was chosen as the representative city in a cold climate. Results are given in Tables 6–9, and Figures 2 and 3. In Figures 2 and 3, the *x-y* plot depicts window-to-wall ratio (WWR) on the *x*-axis and the sensible zone loads or annual energy use (MJ/m^2-floor-year) on the *y*-axis. For reference, conversion factors are: 11.35 MJ/m^2-year = 1 kBtu/ft^2-year; 38.74 MJ/m^2-year = 1 kWh/ft^2-year. Performance of the two types of windows is benchmarked against the GB 50189-2005 code. Both unshaded windows E1 and F1 meet the SC and *U*-value requirements of the 2005 code across the full range of WWR.

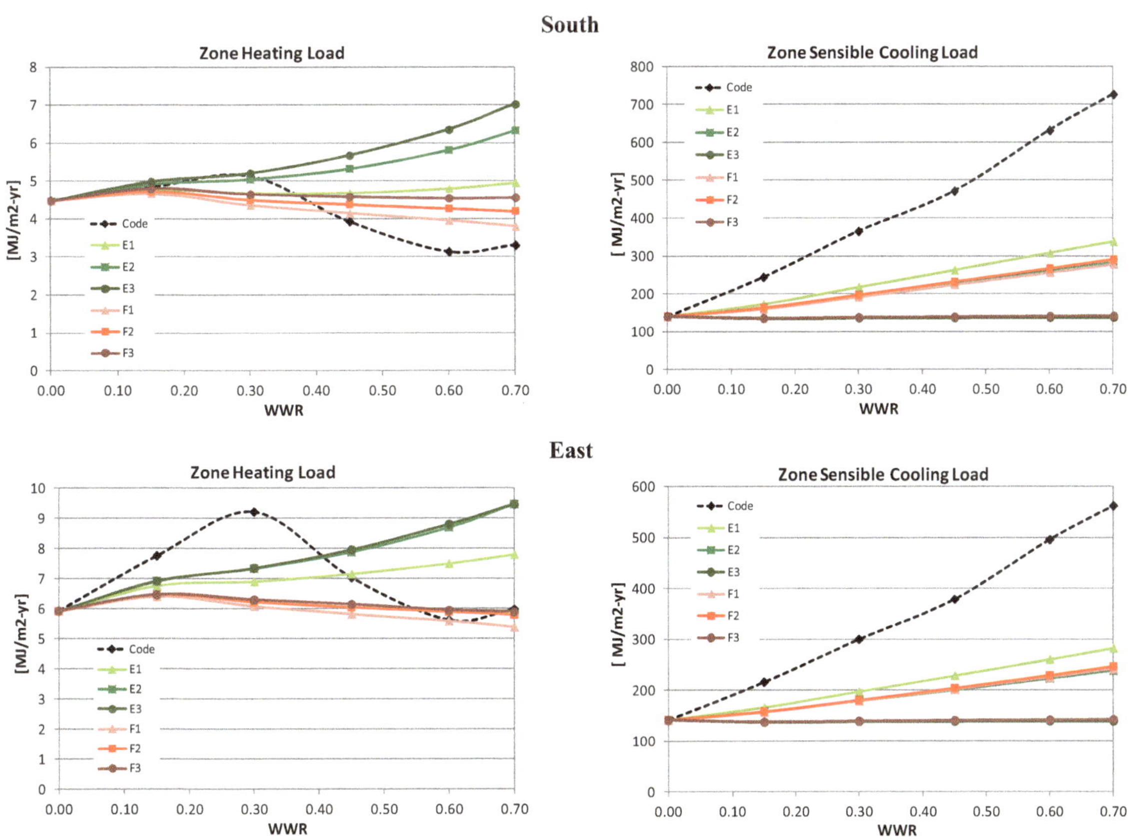

Figure 2. *Cont.*

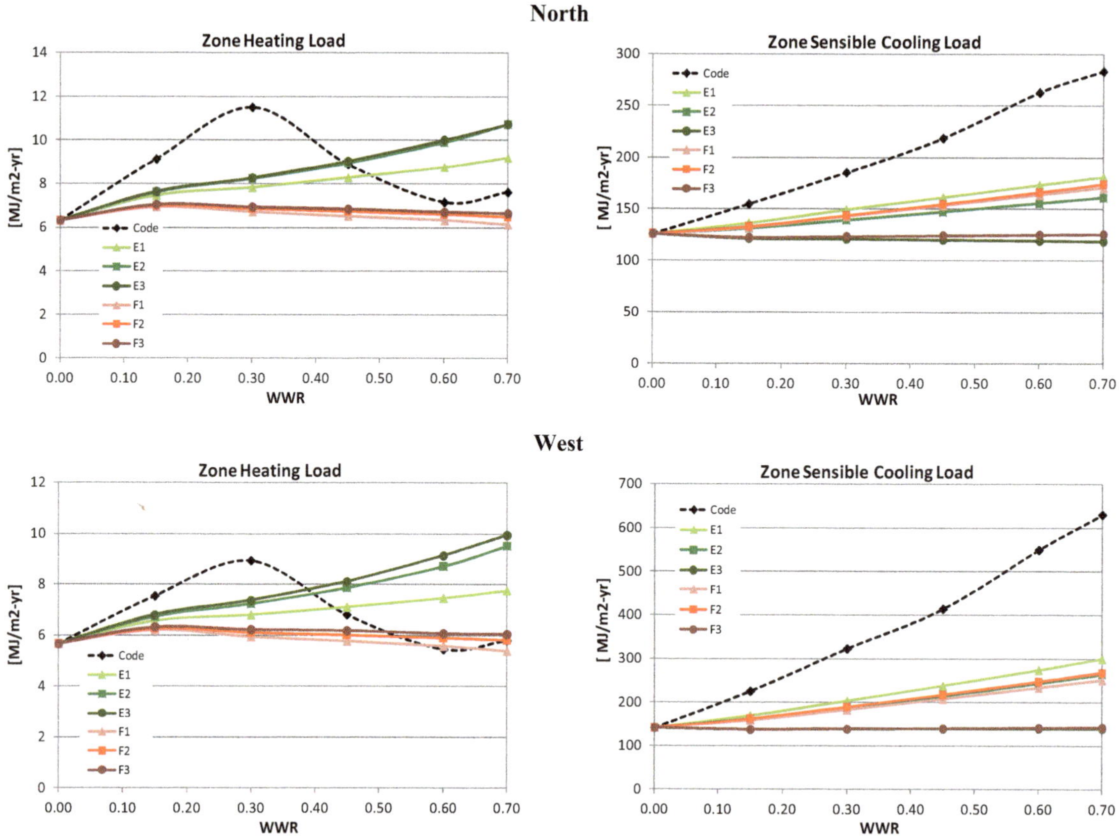

Figure 2. Heating and sensible cooling loads (MJ/m^2-year) for perimeter zones in Beijing.

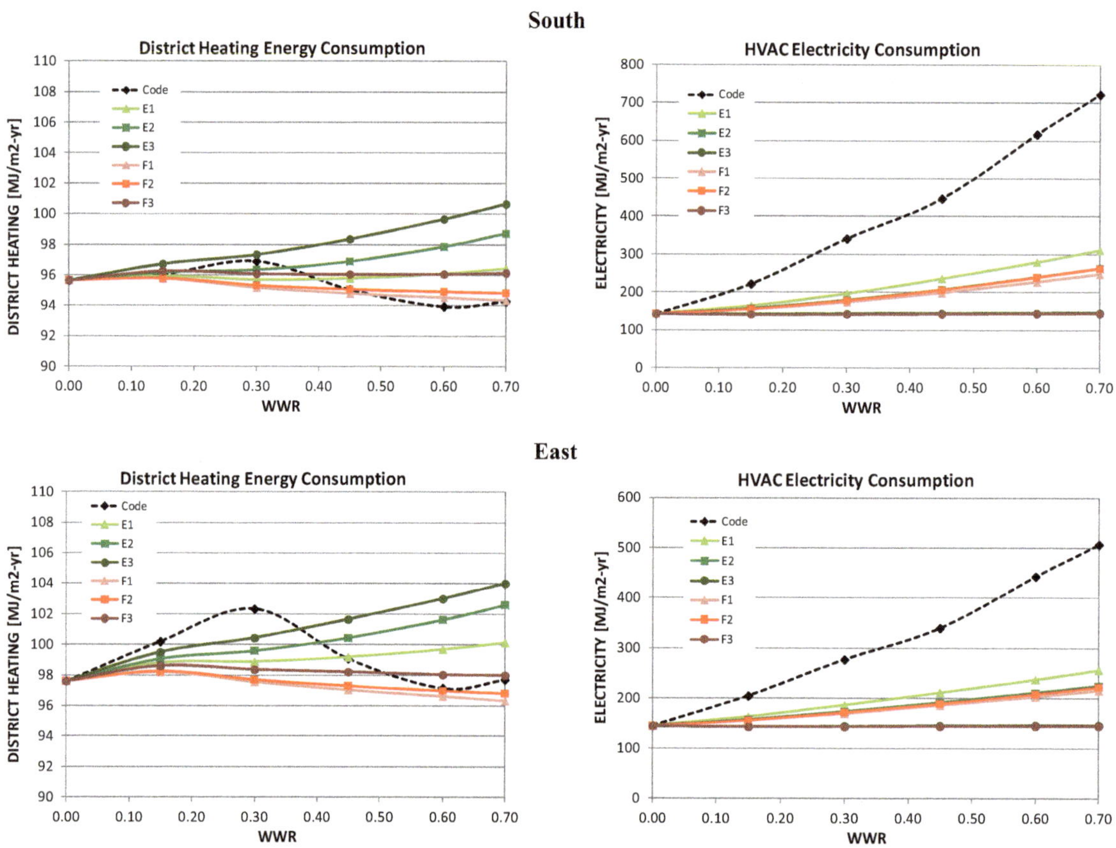

Figure 3. *Cont.*

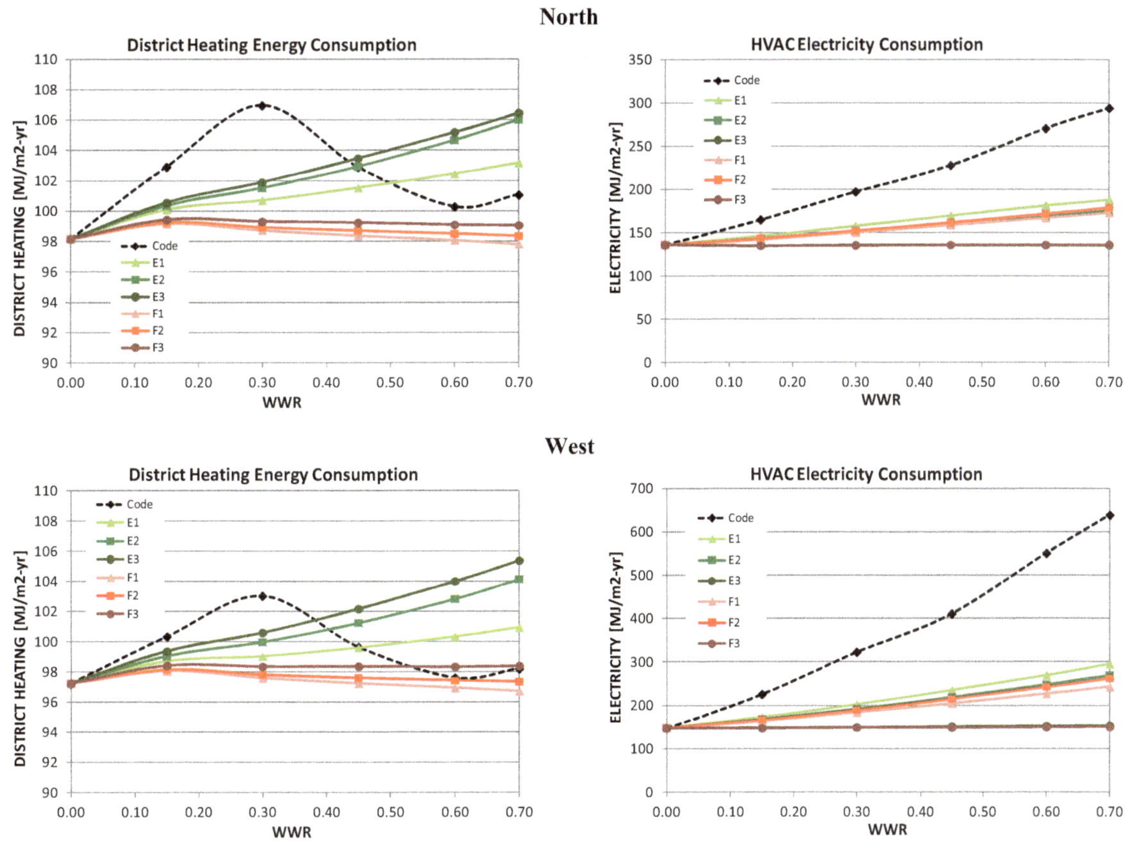

Figure 3. Disaggregated energy use (MJ/m^2-year) for perimeter zones in Beijing.

Table 6. Zone heating load (MJ/m^2-year) and percent savings compared to the GB 50189-2005 Standard for Beijing.

WWR	Code	E1		F1		E2		F2		E3		F3	
	EUI	EUI	%	EUI	%	EUI	%	EUI	%	EUI	%	EUI	%
South zone													
0.00	4.48	4.48	0%	4.48	0%	4.48	0%	4.48	0%	4.48	0%	4.48	0%
0.15	4.83	4.77	1%	4.67	3%	4.92	−2%	4.73	2%	4.99	−3%	4.80	1%
0.30	5.13	4.68	9%	4.36	15%	5.05	2%	4.50	12%	5.21	−1%	4.65	9%
0.45	3.93	4.69	−19%	4.15	−6%	5.33	−35%	4.38	−11%	5.67	−44%	4.58	−17%
0.60	3.14	4.80	−53%	3.95	−26%	5.82	−85%	4.27	−36%	6.35	−103%	4.54	−45%
0.70	3.31	4.95	−50%	3.80	−15%	6.33	−91%	4.20	−27%	7.02	−112%	4.56	−38%
East zone													
0.00	5.93	5.93	0%	5.93	0%	5.93	0%	5.93	0%	5.93	0%	5.93	0%
0.15	7.77	6.78	13%	6.42	17%	6.93	11%	6.47	17%	6.93	11%	6.48	17%
0.30	9.22	6.90	25%	6.08	34%	7.34	20%	6.22	33%	7.35	20%	6.30	32%
0.45	7.04	7.16	−2%	5.81	17%	7.89	−12%	6.04	14%	7.96	−13%	6.15	13%
0.60	5.61	7.50	−34%	5.59	0%	8.70	−55%	5.89	−5%	8.80	−57%	5.95	−6%
0.70	5.98	7.80	−30%	5.38	10%	9.47	−58%	5.79	3%	9.47	−58%	5.89	2%

Table 6. *Cont.*

	Window Type												
	Code	E1		F1		E2		F2		E3		F3	
WWR	EUI	EUI	%	EUI	%	EUI	%	EUI	%	EUI	%	EUI	%
North zone													
0.00	6.31	6.31	0%	6.31	0%	6.31	0%	6.31	0%	6.31	0%	6.31	0%
0.15	9.13	7.45	18%	6.94	24%	7.59	17%	6.98	23%	7.64	16%	7.04	23%
0.30	11.51	7.82	32%	6.70	42%	8.23	29%	6.84	41%	8.27	28%	6.93	40%
0.45	8.89	8.27	7%	6.51	27%	8.95	−1%	6.72	24%	9.04	−2%	6.84	23%
0.60	7.15	8.76	−22%	6.32	12%	9.89	−38%	6.59	8%	10.01	−40%	6.71	6%
0.70	7.62	9.19	−21%	6.12	20%	10.73	−41%	6.47	15%	10.74	−41%	6.64	13%
West zone													
0.00	5.66	5.66	0%	5.66	0%	5.66	0%	5.66	0%	5.66	0%	5.66	0%
0.15	7.54	6.54	13%	6.19	18%	6.72	11%	6.25	17%	6.79	10%	6.33	16%
0.30	8.95	6.79	24%	5.93	34%	7.23	19%	6.09	32%	7.37	18%	6.22	30%
0.45	6.81	7.11	−4%	5.77	15%	7.86	−15%	6.00	12%	8.12	−19%	6.19	9%
0.60	5.45	7.46	−37%	5.58	−2%	8.72	−60%	5.88	−8%	9.15	−68%	6.07	−11%
0.70	5.80	7.75	−34%	5.39	7%	9.52	−64%	5.80	0%	9.95	−71%	6.05	−4%

Table 7. Zone sensible cooling load (MJ/m^2-year) and percent savings compared to the GB 50189-2005 Standard for Beijing.

	Window Type												
WWR	Code	E1		F1		E2		F2		E3		F3	
	EUI	EUI	%	EUI	%	EUI	%	EUI	%	EUI	%	EUI	%
South zone													
0.00	140.44	140.44	0%	140.44	0%	140.44	0%	140.44	0%	140.44	0%	140.44	0%
0.15	243.84	173.28	29%	160.87	34%	162.88	33%	163.26	33%	135.57	44%	136.02	44%
0.30	364.42	217.49	40%	191.83	47%	193.99	47%	196.50	46%	136.60	63%	137.96	62%
0.45	470.15	262.45	44%	224.36	52%	226.45	52%	231.15	51%	137.16	71%	139.68	70%
0.60	630.80	307.71	51%	256.89	59%	259.65	59%	266.05	58%	137.53	78%	141.20	78%
0.70	725.44	337.87	53%	278.68	62%	282.22	61%	289.76	60%	137.63	81%	142.14	80%
East zone													
0.00	141.45	141.45	0%	141.45	0%	141.45	0%	141.45	0%	141.45	0%	141.45	0%
0.15	216.02	166.62	23%	157.11	27%	158.02	27%	158.07	27%	137.58	36%	137.79	36%
0.30	299.75	197.51	34%	178.96	40%	178.79	40%	180.33	40%	138.54	54%	139.46	53%
0.45	378.48	228.57	40%	201.73	47%	200.24	47%	203.96	46%	139.01	63%	140.89	63%
0.60	495.64	260.55	47%	225.20	55%	222.70	55%	229.14	54%	139.30	72%	142.13	71%
0.70	561.97	282.35	50%	241.36	57%	238.17	58%	246.72	56%	139.36	75%	142.86	75%

Table 7. *Cont.*

	Window Type												
	Code	E1		F1		E2		F2		E3		F3	
WWR	EUI	EUI	%	EUI	%	EUI	%	EUI	%	EUI	%	EUI	%
North zone													
0.00	125.83	125.83	0%	125.83	0%	125.83	0%	125.83	0%	125.83	0%	125.83	0%
0.15	154.65	136.08	12%	132.10	15%	131.05	15%	132.64	14%	121.39	22%	122.36	21%
0.30	185.26	149.29	19%	142.80	23%	139.15	25%	143.51	23%	120.90	35%	123.41	33%
0.45	218.60	161.40	26%	153.27	30%	147.08	33%	154.54	29%	120.13	45%	124.34	43%
0.60	262.87	173.33	34%	163.48	38%	155.32	41%	166.00	37%	119.31	55%	125.11	52%
0.70	283.57	181.24	36%	170.31	40%	160.79	43%	173.81	39%	118.67	58%	125.57	56%
West zone													
0.00	140.91	140.91	0%	140.91	0%	140.91	0%	140.91	0%	140.91	0%	140.91	0%
0.15	224.95	168.97	25%	158.72	29%	162.18	28%	161.75	28%	136.95	39%	137.02	39%
0.30	322.51	203.40	37%	183.06	43%	188.39	42%	188.91	41%	138.17	57%	138.69	57%
0.45	413.71	238.02	42%	208.18	50%	215.41	48%	217.29	47%	138.95	66%	140.14	66%
0.60	549.67	273.82	50%	233.54	58%	244.17	56%	246.61	55%	139.46	75%	141.40	74%
0.70	630.12	299.17	53%	250.58	60%	264.60	58%	266.72	58%	139.54	78%	142.16	77%

Table 8. HVAC electricity energy use intensity (EUI, MJ/m^2-year) and percent savings compared to the GB 50189-2005 Standard for Beijing.

	Window Type												
WW	Code	E1		F1		E2		F2		E3		F3	
R	EUI	EUI	%	EUI	%	EUI	%	EUI	%	EUI	%	EUI	%
South zone													
0.00	142.58	142.58	0%	142.58	0%	142.58	0%	142.58	0%	142.58	0%	142.58	0%
0.15	221.88	163.95	26%	154.99	30%	156.86	29%	155.94	30%	141.89	36%	141.32	36%
0.30	341.34	196.59	42%	174.22	49%	179.18	48%	178.12	48%	143.60	58%	142.22	58%
0.45	446.93	236.37	47%	199.30	55%	207.20	54%	206.36	54%	144.62	68%	143.43	68%
0.60	617.05	279.94	55%	227.54	63%	239.74	61%	239.29	61%	145.72	76%	144.19	77%
0.70	722.40	311.30	57%	247.81	66%	262.95	64%	262.57	64%	146.30	80%	144.69	80%
East zone													
0.00	144.66	144.66	0%	144.66	0%	144.66	0%	144.66	0%	144.66	0%	144.66	0%
0.15	204.80	162.96	20%	155.36	24%	156.78	23%	155.86	24%	143.81	30%	143.16	30%
0.30	277.20	186.40	33%	169.80	39%	173.23	38%	171.92	38%	145.04	48%	143.54	48%
0.45	339.84	211.41	38%	186.43	45%	191.31	44%	189.42	44%	145.80	57%	144.30	58%
0.60	443.25	237.45	46%	203.43	54%	210.75	52%	208.68	53%	146.52	67%	144.71	67%
0.70	507.13	256.29	49%	215.34	58%	224.30	56%	222.18	56%	146.86	71%	144.89	71%

Table 8. *Cont.*

WWR	Code	E1		F1		E2		F2		E3		F3	
	EUI	EUI	%	EUI	%	EUI	%	EUI	%	EUI	%	EUI	%
North zone													
0.00	136.09	136.09	0%	136.09	0%	136.09	0%	136.09	0%	136.09	0%	136.09	0%
0.15	165.04	146.20	11%	142.39	14%	143.02	13%	143.19	13%	135.26	18%	135.14	18%
0.30	197.30	158.13	20%	150.18	24%	151.16	23%	152.13	23%	135.77	31%	135.39	31%
0.45	227.74	169.75	25%	158.65	30%	159.73	30%	161.31	29%	135.83	40%	135.93	40%
0.60	270.60	181.41	33%	166.86	38%	168.84	38%	171.31	37%	135.76	50%	136.04	50%
0.70	294.17	188.04	36%	172.57	41%	175.01	41%	178.17	39%	135.68	54%	136.12	54%
West zone													
0.00	147.19	147.19	0%	147.19	0%	147.19	0%	147.19	0%	147.19	0%	147.19	0%
0.15	225.84	173.24	23%	163.70	28%	168.21	26%	166.40	26%	148.30	34%	147.40	35%
0.30	322.57	202.46	37%	183.45	43%	191.87	41%	188.85	41%	149.13	54%	148.45	54%
0.45	410.58	234.80	43%	203.69	50%	218.77	47%	214.79	48%	150.79	63%	148.31	64%
0.60	550.30	269.00	51%	226.21	59%	247.77	55%	242.50	56%	152.19	72%	149.29	73%
0.70	638.42	294.83	54%	241.73	62%	268.83	58%	261.77	59%	152.98	76%	149.86	77%

Table 9. District heating energy consumption (MJ/m^2-year) and percent savings compared to the GB 50189-2005 Standard for Beijing.

WWR	Code	E1		F1		E2		F2		E3		F3	
	EUI	EUI	%	EUI	%	EUI	%	EUI	%	EUI	%	EUI	%
South zone													
0.00	95.64	95.64	0%	95.64	0%	95.64	0%	95.64	0%	95.64	0%	95.64	0%
0.15	96.05	95.94	0%	95.76	0%	96.17	0%	95.78	0%	96.72	−1%	96.26	0%
0.30	96.91	95.70	1%	95.17	2%	96.35	1%	95.31	2%	97.34	0%	96.08	1%
0.45	95.07	95.79	−1%	94.80	0%	96.90	−2%	95.07	0%	98.36	−3%	96.06	−1%
0.60	93.96	96.10	−2%	94.54	−1%	97.85	−4%	94.92	−1%	99.66	−6%	96.07	−2%
0.70	94.29	96.44	−2%	94.35	0%	98.71	−5%	94.83	−1%	100.66	−7%	96.14	−2%
East zone													
0.00	97.59	97.59	0%	97.59	0%	97.59	0%	97.59	0%	97.59	0%	97.59	0%
0.15	100.19	98.78	1%	98.19	2%	99.04	1%	98.22	2%	99.48	1%	98.60	2%
0.30	102.33	98.87	3%	97.54	5%	99.58	3%	97.68	5%	100.43	2%	98.35	4%
0.45	99.08	99.18	0%	97.03	2%	100.43	−1%	97.29	2%	101.65	−3%	98.20	1%
0.60	97.15	99.68	−3%	96.62	1%	101.63	−5%	96.99	0%	103.00	−6%	98.03	−1%
0.70	97.72	100.12	−2%	96.33	1%	102.60	−5%	96.82	1%	104.01	−6%	97.99	0%

Table 9. *Cont.*

WWR	Code	E1		F1		E2		F2		E3		F3	
	EUI	EUI	%	EUI	%	EUI	%	EUI	%	EUI	%	EUI	%
North zone													
0.00	98.15	98.15	0%	98.15	0%	98.15	0%	98.15	0%	98.15	0%	98.15	0%
0.15	102.90	100.05	3%	99.14	4%	100.33	3%	99.20	4%	100.55	2%	99.42	3%
0.30	106.94	100.67	6%	98.71	8%	101.49	5%	98.89	8%	101.89	5%	99.30	7%
0.45	102.88	101.50	1%	98.37	4%	102.91	0%	98.67	4%	103.47	−1%	99.23	4%
0.60	100.27	102.43	−2%	98.07	2%	104.64	−4%	98.48	2%	105.19	−5%	99.08	1%
0.70	101.05	103.13	−2%	97.81	3%	105.99	−5%	98.33	3%	106.45	−5%	99.04	2%
West zone													
0.00	97.23	97.23	0%	97.23	0%	97.23	0%	97.23	0%	97.23	0%	97.23	0%
0.15	100.33	98.67	2%	98.04	2%	99.03	1%	98.12	2%	99.36	1%	98.42	2%
0.30	103.03	99.00	4%	97.57	5%	99.96	3%	97.79	5%	100.59	2%	98.35	5%
0.45	99.64	99.57	0%	97.21	2%	101.22	−2%	97.57	2%	102.17	−3%	98.36	1%
0.60	97.60	100.30	−3%	96.92	1%	102.83	−5%	97.41	0%	104.01	−7%	98.34	−1%
0.70	98.20	100.91	−3%	96.69	2%	104.14	−6%	97.32	1%	105.38	−7%	98.39	0%

As window area increases, thermal losses are offset by increased solar heat gains so the net effect of the window's *U*-value and SHGC on annual sensible heating loads is in proportion to these heat flows (Figure 2). The GB 50189-2005 code results in less sensible heating energy use for large-area windows because the maximum prescribed *U*-value is low and SHGC is high. The low-e window (E) and highly-insulated, low-e window (F) do not perform as well as code for WWR > 0.30 because both the *U*-value and SHGC of these windows are lower than that prescribed by code, resulting in less solar heat gains to offset heating demands. When WWR ≤ 0.30, the *U*-value of the window plays a more significant role than SHGC in affecting the zone sensible heating load. When WWR > 0.30, the opposite is true: SHGC plays a more significant role. Use of the interior or exterior shading produced very little difference in the annual load because the shades were raised when the zone was in the heating mode.

The maximum variation in zone sensible heating load between all windows modeled was small—3.8 MJ/m^2-year, which translates to a difference in annual district heating energy use of 9 MJ/m^2-year or 8% (Figure 3). For this office building, the fraction of annual heating energy used to heat up the outdoor ventilation air is significantly greater than that used to offset perimeter zone heating losses since the internal loads from people, equipment, and lights are high. Heating energy usage is therefore not a major factor when selecting windows for this cold climate—the code windows give reasonably good performance during the heating season. The lower *U*-value windows, however, will mitigate thermal discomfort due to an occupant's radiative heat losses to cold window surfaces.

On the other hand, we see that sensible cooling energy use varies significantly between the various window types modeled, where the prescribed code window performed the poorest due to its higher SHGC. Here, windows E and F with exterior shading were able to reduce sensible cooling loads by as much as 585 MJ/m^2-year (80%), resulting in performance levels that are equivalent to an opaque insulated wall for even the largest window area (WWR = 0.70).

All the variations on windows E and F improved total energy performance significantly compared to the GB 50189-2005 code. Windows E1 and F1 (unshaded) produced up to 66% savings in HVAC electricity consumption and 5% savings in district heating energy. Interior shades provided negligible benefit over unshaded windows. Exterior shades produced the least sensible cooling load and annual HVAC electricity consumption (up to 80% savings). With the same HVAC energy consumption levels for small and large area windows, the large area windows provide an added benefit of admitting more daylight. Energy savings due to daylighting were not included in this analysis.

Selecting the right window type is more consequential on the south, west, and east than on the north orientation. When a whole building facade retrofit is not economically possible, one should prioritize use of high performance window and shading systems based on this order. Overall, windows E3 and F3 provide the best performance in the cold climate. Since E3 is economically more attractive, E3 is recommended in the cold climate.

3.5. Results for a Hot Summer/Cold Winter Climate

Shanghai was chosen as the representative city for the hot summer and cold winter climate. The results are given in Tables 10–13 and Figures 4 and 5. The code is less restrictive on *U*-value and more restrictive on SHGC compared to the cold Beijing climate. Both unshaded windows E1 and F1 meet the SC and *U*-value requirements of the GB 50189-2005 code for all WWR (and all requirements for the 65% Shanghai code DGJ08-107-20 except for Window E1 for WWR between 0.50 and 0.70).

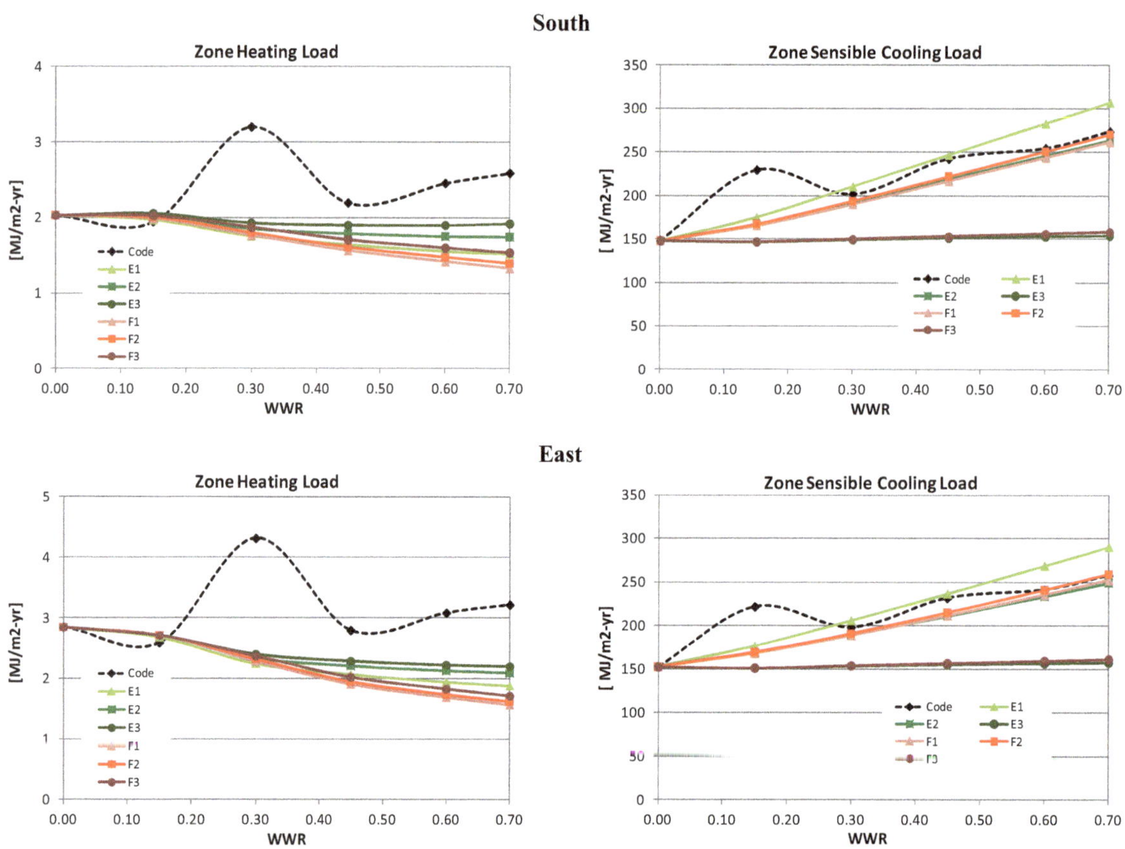

Figure 4. *Cont.*

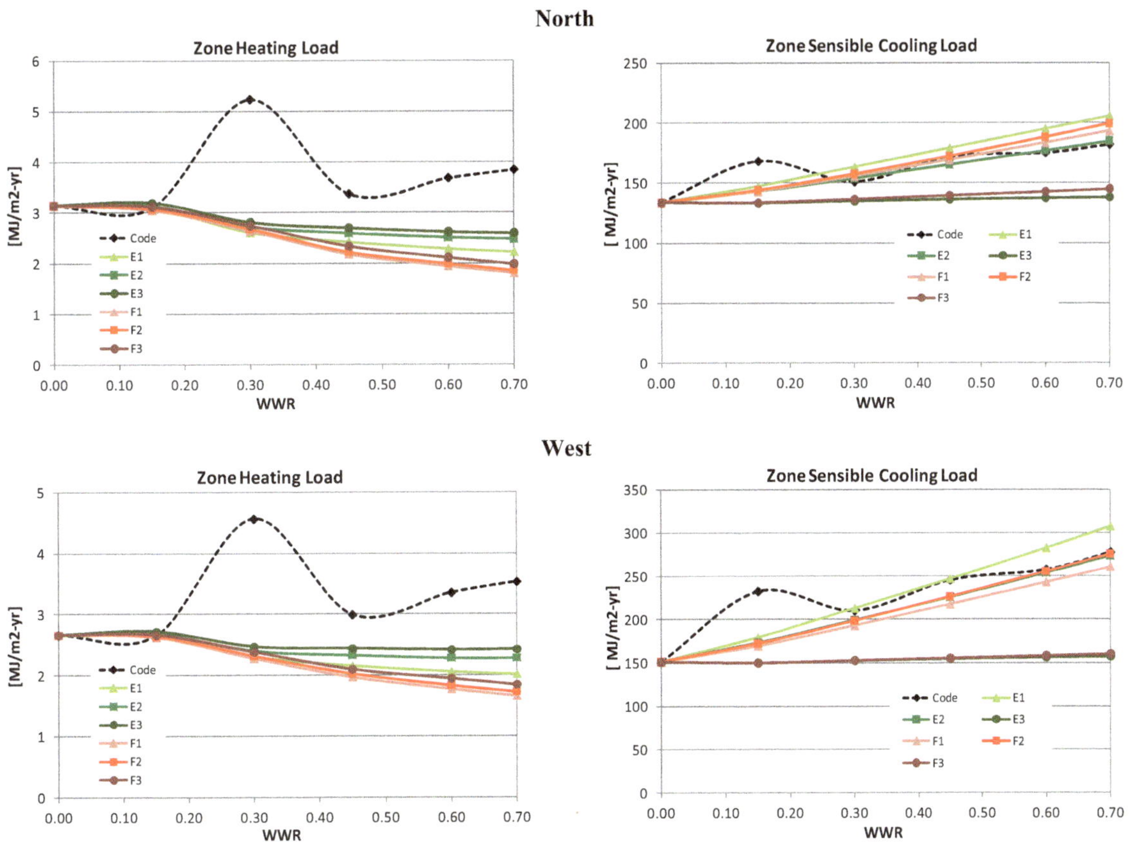

Figure 4. Heating and sensible cooling loads (MJ/m^2-year) for perimeter zones in Shanghai.

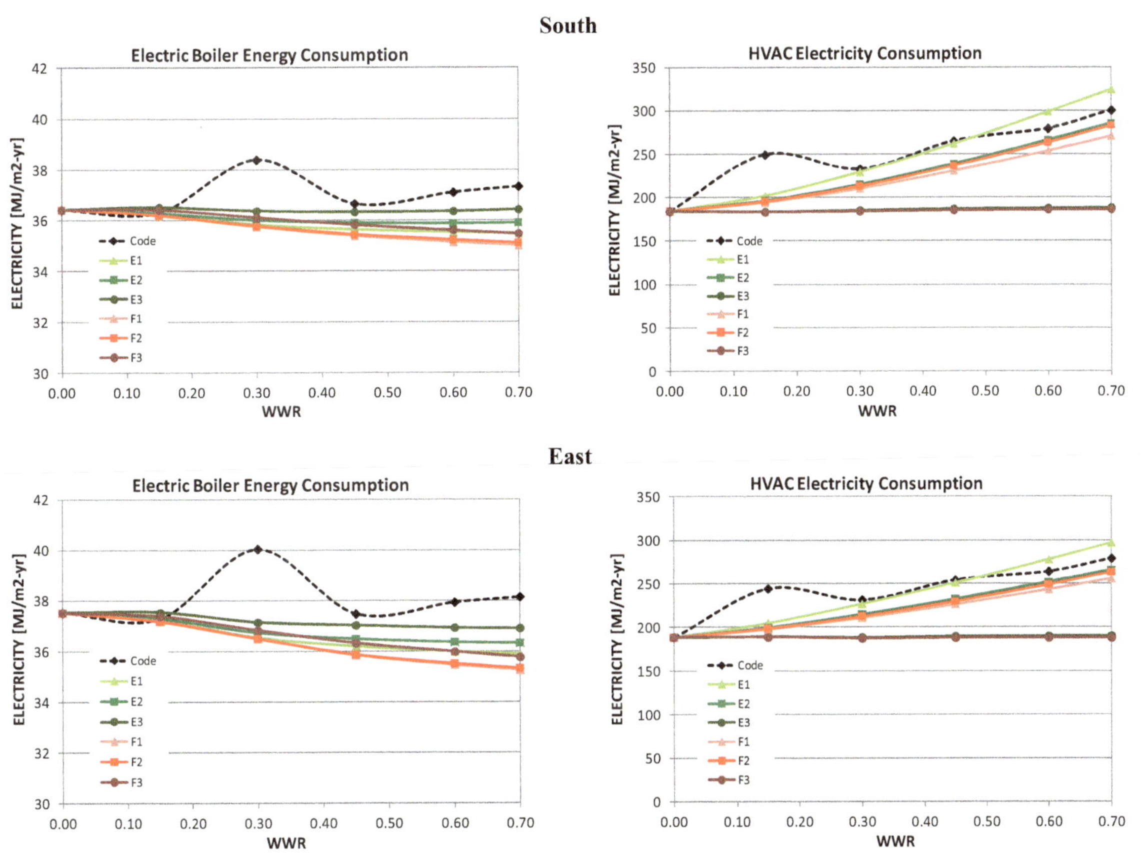

Figure 5. *Cont.*

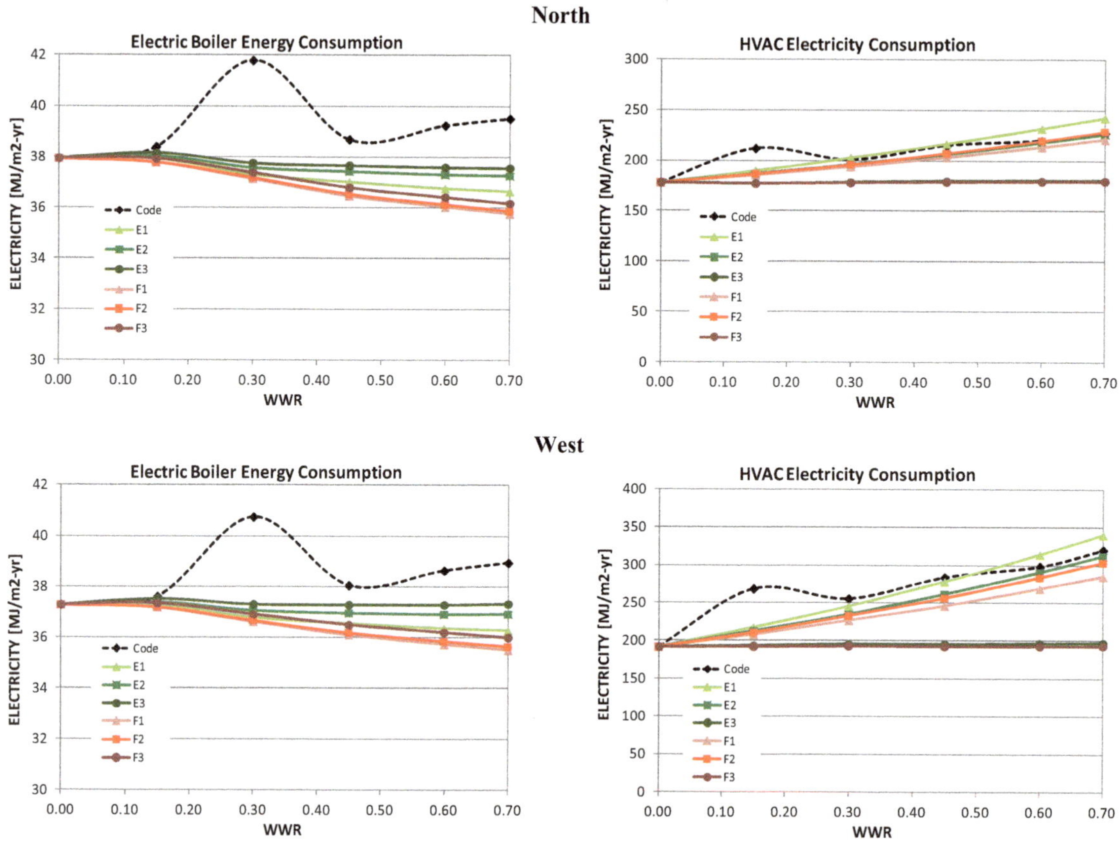

Figure 5. Disaggregated energy use (MJ/m^2-year) for perimeter zones in Shanghai.

Table 10. Zone heating load (MJ/m^2-year) and percent savings compared to the GB 50189-2005 Standard for Shanghai.

WWR	Window Type												
	Code	E1		F1		E2		F2		E3		F3	
	EUI	EUI	%	EUI	%	EUI	%	EUI	%	EUI	%	EUI	%
South zone													
0.00	2.03	2.03	0%	2.03	0%	2.03	0%	2.03	0%	2.03	0%	2.03	0%
0.15	1.94	1.96	-1%	1.98	−2%	2.01	−4%	2.00	−3%	2.05	−6%	2.02	−4%
0.30	3.20	1.75	45%	1.76	45%	1.85	42%	1.80	44%	1.93	40%	1.87	42%
0.45	2.19	1.64	25%	1.56	29%	1.78	19%	1.60	27%	1.90	14%	1.70	22%
0.60	2.45	1.56	37%	1.41	42%	1.75	29%	1.47	40%	1.90	23%	1.60	35%
0.70	2.59	1.52	41%	1.33	49%	1.74	33%	1.39	46%	1.92	26%	1.53	41%
East zone													
0.00	2.85	2.85	0%	2.85	0%	2.85	0%	2.85	0%	2.85	0%	2.85	0%
0.15	2.59	2.68	−3%	2.71	−5%	2.72	−5%	2.71	−5%	2.71	−5%	2.71	−5%
0.30	4.31	2.24	48%	2.28	47%	2.33	46%	2.32	46%	2.40	44%	2.37	45%
0.45	2.80	2.07	26%	1.90	32%	2.20	21%	1.94	31%	2.29	18%	2.02	28%
0.60	3.08	1.94	37%	1.69	45%	2.12	31%	1.73	44%	2.22	28%	1.83	41%
0.70	3.22	1.88	42%	1.56	51%	2.09	35%	1.61	50%	2.20	32%	1.71	47%

Table 10. *Cont.*

WWR	Code	E1		F1		E2		F2		E3		F3	
	EUI	EUI	%	EUI	%	EUI	%	EUI	%	EUI	%	EUI	%
North zone													
0.00	3.14	3.14	0%	3.14	0%	3.14	0%	3.14	0%	3.14	0%	3.14	0%
0.15	3.10	3.09	0%	3.05	2%	3.14	−1%	3.07	1%	3.19	−3%	3.12	0%
0.30	5.23	2.61	50%	2.64	49%	2.73	48%	2.68	49%	2.81	46%	2.75	48%
0.45	3.37	2.42	28%	2.19	35%	2.59	23%	2.23	34%	2.70	20%	2.34	30%
0.60	3.68	2.29	38%	1.94	47%	2.51	32%	1.99	46%	2.62	29%	2.12	42%
0.70	3.84	2.23	42%	1.81	53%	2.48	36%	1.86	52%	2.60	32%	1.99	48%
West zone													
0.00	2.66	2.66	0%	2.66	0%	2.66	0%	2.66	0%	2.66	0%	2.66	0%
0.15	2.66	2.63	1%	2.62	1%	2.68	−1%	2.64	1%	2.72	−2%	2.66	0%
0.30	4.56	2.31	49%	2.28	50%	2.40	47%	2.31	49%	2.47	46%	2.38	48%
0.45	2.99	2.16	28%	1.97	34%	2.33	22%	2.03	32%	2.44	18%	2.10	30%
0.60	3.35	2.06	38%	1.77	47%	2.29	32%	1.83	45%	2.42	28%	1.95	42%
0.70	3.53	2.02	43%	1.66	53%	2.28	35%	1.73	51%	2.43	31%	1.85	48%

Table 11. Zone sensible cooling load (MJ/m^2-year) and percent savings compared to the GB 50189-2005 Standard for Shanghai.

WWR	Code	E1		F1		E2		F2		E3		F3	
	EUI	EUI	%	EUI	%	EUI	%	EUI	%	EUI	%	EUI	%
South zone													
0.00	147.48	147.48	0%	147.48	0%	147.48	0%	147.48	0%	147.48	0%	147.48	0%
0.15	229.12	175.19	24%	165.20	28%	167.16	27%	167.22	27%	146.40	36%	146.32	36%
0.30	201.76	210.56	-4%	189.99	6%	192.32	5%	193.62	4%	148.90	26%	149.50	26%
0.45	242.21	246.57	-2%	216.51	11%	218.36	10%	221.71	8%	150.94	38%	152.69	37%
0.60	254.30	282.62	−11%	243.32	4%	244.98	4%	250.55	1%	152.60	40%	155.88	39%
0.70	273.54	306.52	−12%	261.45	4%	262.94	4%	270.23	1%	153.51	44%	158.07	42%
East zone													
0.00	152.22	152.22	0%	152.22	0%	152.22	0%	152.22	0%	152.22	0%	152.22	0%
0.15	221.51	176.54	20%	167.72	24%	169.04	24%	169.01	24%	151.28	32%	151.04	32%
0.30	198.22	205.98	−4%	188.47	5%	189.40	4%	190.58	4%	153.51	23%	153.81	22%
0.45	231.63	236.57	−2%	211.14	9%	210.83	9%	214.67	7%	155.29	33%	156.59	32%
0.60	241.24	268.60	−11%	235.29	2%	233.54	3%	240.75	0%	156.67	35%	159.41	34%
0.70	257.69	290.26	−13%	252.22	2%	249.13	3%	258.86	0%	157.43	39%	161.36	37%

Table 11. *Cont.*

WWR	Window Type												
	Code	E1		F1		E2		F2		E3		F3	
	EUI	EUI	%	EUI	%	EUI	%	EUI	%	EUI	%	EUI	%
North zone													
0.00	133.60	133.60	0%	133.60	0%	133.60	0%	133.60	0%	133.60	0%	133.60	0%
0.15	168.15	147.22	12%	142.72	15%	142.71	15%	143.77	14%	133.15	21%	133.46	21%
0.30	150.97	163.39	−8%	155.46	−3%	153.79	−2%	157.19	−4%	134.94	11%	136.41	10%
0.45	171.99	179.34	−4%	169.06	2%	165.29	4%	172.05	0%	136.40	21%	139.49	19%
0.60	175.07	195.38	−12%	183.35	−5%	177.09	−1%	188.07	−7%	137.56	21%	142.73	18%
0.70	182.12	206.06	−13%	193.26	−6%	185.03	−2%	199.32	−9%	138.20	24%	145.01	20%
West zone													
0.00	151.23	151.23	0%	151.23	0%	151.23	0%	151.23	0%	151.23	0%	151.23	0%
0.15	232.39	179.33	23%	169.33	27%	172.98	26%	172.24	26%	150.34	35%	149.88	36%
0.30	210.17	212.97	−1%	192.90	8%	198.85	5%	198.38	6%	152.89	27%	152.64	27%
0.45	245.37	247.02	−1%	217.56	11%	225.62	8%	226.13	8%	155.02	37%	155.44	37%
0.60	257.69	282.69	−10%	243.11	6%	253.72	2%	255.21	1%	156.64	39%	158.32	39%
0.70	277.74	307.72	−11%	260.81	6%	273.64	1%	275.48	1%	157.54	43%	160.29	42%

Table 12. HVAC electricity energy use intensity (EUI, MJ/m^2-year) and percent savings compared to the GB 50189-2005 Standard for Shanghai.

WWR	Window Type												
	Code	E1		F1		E2		F2		E3		F3	
	EUI	EUI	%	EUI	%	EUI	%	EUI	%	EUI	%	EUI	%
South zone													
0.00	184.27	184.27	0%	184.27	0%	184.27	0%	184.27	0%	184.27	0%	184.27	0%
0.15	249.32	201.88	19%	194.24	22%	196.18	21%	195.37	22%	183.53	26%	182.98	27%
0.30	232.86	229.29	2%	210.19	10%	214.95	8%	212.98	9%	185.28	20%	184.07	21%
0.45	264.93	261.92	1%	230.68	13%	238.59	10%	236.51	11%	186.95	29%	185.20	30%
0.60	279.29	298.78	−7%	253.91	9%	265.85	5%	263.43	6%	187.76	33%	186.06	33%
0.70	300.32	324.36	−8%	271.04	10%	285.23	5%	283.03	6%	188.31	37%	186.47	38%
East zone													
0.00	188.63	188.63	0%	188.63	0%	188.63	0%	188.63	0%	188.63	0%	188.63	0%
0.15	244.12	204.98	16%	197.83	19%	199.52	18%	198.62	19%	189.40	22%	188.60	23%
0.30	231.41	226.78	2%	210.86	9%	214.75	7%	212.54	8%	188.61	18%	187.21	19%
0.45	254.24	250.98	1%	226.17	11%	232.23	9%	229.61	10%	189.74	25%	187.70	26%
0.60	264.02	277.94	−5%	243.07	8%	251.71	5%	248.82	6%	190.02	28%	187.91	29%
0.70	279.03	296.90	−6%	255.62	8%	265.54	5%	263.19	6%	190.27	32%	187.87	33%

Table 12. *Cont.*

	Window Type												
WWR	Code	E1		F1		E2		F2		E3		F3	
	EUI	EUI	%	EUI	%	EUI	%	EUI	%	EUI	%	EUI	%
North zone													
0.00	177.92	177.92	0%	177.92	0%	177.92	0%	177.92	0%	177.92	0%	177.92	0%
0.15	211.83	189.64	10%	185.44	12%	186.60	12%	186.53	12%	177.30	16%	176.97	16%
0.30	200.83	202.83	−1%	193.72	4%	196.22	2%	195.96	2%	178.31	11%	177.95	11%
0.45	214.95	216.81	−1%	203.34	5%	206.28	4%	207.37	4%	179.41	17%	178.80	17%
0.60	219.80	232.06	−6%	213.55	3%	218.05	1%	219.79	0%	179.71	18%	179.50	18%
0.70	227.69	242.47	−6%	221.20	3%	226.15	1%	228.99	−1%	179.95	21%	179.81	21%
West zone													
0.00	191.50	191.50	0%	191.50	0%	191.50	0%	191.50	0%	191.50	0%	191.50	0%
0.15	268.26	217.22	19%	207.72	23%	212.26	21%	210.35	22%	193.09	28%	191.98	28%
0.30	255.32	245.36	4%	226.69	11%	234.52	8%	232.67	9%	195.48	23%	193.26	24%
0.45	283.97	278.30	2%	245.86	13%	261.44	8%	256.18	10%	195.47	31%	192.48	32%
0.60	298.61	314.38	−5%	268.93	10%	290.75	3%	283.80	5%	196.57	34%	193.13	35%
0.70	320.45	340.29	−6%	285.21	11%	312.15	3%	303.56	5%	197.35	38%	193.47	40%

Table 13. Boiler electricity consumption (MJ/m^2-year) and percent savings compared to the GB 50189-2005 Standard for Shanghai.

	Window Type												
WWR	Code	E1		F1		E2		F2		E3		F3	
	EUI	EUI	%	EUI	%	EUI	%	EUI	%	EUI	%	EUI	%
South zone													
0.00	36.41	36.41	0%	36.41	0%	36.41	0%	36.41	0%	36.41	0%	36.41	0%
0.15	36.30	36.19	0%	36.19	0%	36.26	0%	36.18	0%	36.51	−1%	36.40	0%
0.30	38.38	35.82	7%	35.74	7%	35.99	6%	35.76	7%	36.36	5%	36.11	6%
0.45	36.64	35.63	3%	35.38	3%	35.90	2%	35.43	3%	36.32	1%	35.83	2%
0.60	37.09	35.52	4%	35.14	5%	35.88	3%	35.22	5%	36.35	2%	35.61	4%
0.70	37.32	35.48	5%	35.00	6%	35.90	4%	35.10	6%	36.42	2%	35.47	5%
East zone													
0.00	37.52	37.52	0%	37.52	0%	37.52	0%	37.52	0%	37.52	0%	37.52	0%
0.15	37.28	37.21	0%	37.19	0%	37.29	0%	37.17	0%	37.53	−1%	37.39	0%
0.30	40.04	36.52	9%	36.48	9%	36.73	8%	36.49	9%	37.14	7%	36.83	8%
0.45	37.47	36.18	3%	35.85	4%	36.48	3%	35.87	4%	37.02	1%	36.32	3%
0.60	37.92	35.98	5%	35.46	7%	36.35	4%	35.51	6%	36.92	3%	35.97	5%
0.70	38.13	35.88	6%	35.25	8%	36.31	5%	35.32	7%	36.90	3%	35.76	6%

Table 13. *Cont.*

WWR	Window Type												
	Code	E1		F1		E2		F2		E3		F3	
	EUI	EUI	%	EUI	%	EUI	%	EUI	%	EUI	%	EUI	%
North zone													
0.00	37.94	37.94	0%	37.94	0%	37.94	0%	37.94	0%	37.94	0%	37.94	0%
0.15	38.39	37.93	1%	37.78	2%	38.03	1%	37.79	2%	38.16	1%	37.92	1%
0.30	41.79	37.31	11%	37.14	11%	37.56	10%	37.18	11%	37.75	10%	37.38	11%
0.45	38.69	37.01	4%	36.45	6%	37.41	3%	36.53	6%	37.66	3%	36.79	5%
0.60	39.24	36.76	6%	36.02	8%	37.29	5%	36.12	8%	37.59	4%	36.42	7%
0.70	39.51	36.65	7%	35.77	9%	37.25	6%	35.87	9%	37.56	5%	36.18	8%
West zone													
0.00	37.29	37.29	0%	37.29	0%	37.29	0%	37.29	0%	37.29	0%	37.29	0%
0.15	37.60	37.27	1%	37.18	1%	37.38	1%	37.20	1%	37.52	0%	37.34	1%
0.30	40.74	36.78	10%	36.62	10%	37.05	9%	36.68	10%	37.30	8%	36.91	9%
0.45	38.04	36.53	4%	36.10	5%	36.94	3%	36.18	5%	37.28	2%	36.49	4%
0.60	38.63	36.35	6%	35.73	8%	36.90	4%	35.85	7%	37.28	3%	36.20	6%
0.70	38.95	36.28	7%	35.52	9%	36.92	5%	35.65	8%	37.33	4%	36.01	8%

All types of windows E and F reduced the heating load and annual heating energy use with performance independent of window area. The unshaded window F1 yielded the best performance. Similar to the cold climate, the magnitude of heating energy savings resulting from the windows E and F is trivial and heating of outdoor ventilation air was the principle cause of the heating demand. Therefore, heating energy consumption is not a major consideration when selecting windows in hot summer and cold winter climates.

Exterior shaded windows (E3 and F3) produced the most significant reductions in the sensible cooling load as well as the HVAC electricity consumption compared to the 2005 code: ranging from 10% to 42% for the four different orientations as WWR increases from 0.15 to 0.70 and for some cases, resulting in less energy use than an opaque insulated wall. For the south zone with window E3 and WWR = 0.70, for example, the sensible cooling load and HVAC electricity use were reduced from 273.54 to 153.51 MJ/m^2-year (44%) and from 300.32 to 188.31 MJ/m^2-year (37%), respectively. The unshaded low-e window E1 resulted in greater HVAC electricity use than code for WWR $\geq$ 0.30 and should be avoided. Windows E2, F1, and F2 provided comparable energy performance as the code window.

In summary, the recommendation for the hot summer and cold winter climate is the same as the cold climate; *i.e.*, the low-e window with exterior shading gives the best performance in terms of energy savings and initial cost and can result in less energy use than an opaque insulated wall. With daylighting, such systems would result in ultra-low energy use in internal load dominated commercial buildings.

4. Market Context

We offer some anecdotal observations on the building construction industry in China and provide some analysis on how this market context could provide unique opportunities for broad deployment of advanced fenestration products for low energy commercial buildings.

As in the US, public buildings in China are financial instruments that are designed and built by private sector developers with the goal of achieving a short-term, profitable return on investment. In order to compete in the market, decisions are made to balance pragmatic decisions that affect the capital cost of the building against the function, aesthetics, and rentability of the end product. The GB 50189-2005 sets the floor for energy-efficient performance and the LEED or Three Star certification programs provide the upper bound for performance, the latter making the property more attractive not only for local but also foreign businesses with a vested interest in sustainable practices.

The cost of construction in China is significantly lower than in the US: about \$600–1300 US/m^2 in China *versus* \$2000–5000 US/m^2 in the U.S. (as of 2012). Full construction documents can take a few months with a dedicated team of two or three people and construction of a large building can be completed within three to six months. For a medium-size development (~60,000 m^2 of floor area), the installed cost of a curtainwall system with large-area, spectrally selective low-e IGU windows is about \$220 US/m^2 (based on conversations with an architectural/ engineering team). This cost can increase to \$285 US/m^2 if exterior shading (700 mm deep) and other curtainwall details (e.g., shadow box at spandrel, ceramic frits, transoms, caps) are added to improve the aesthetics of the façade. Simple payback periods of less than three years based on energy savings is desirable. Energy costs are similar to the US, with an average of \$0.11/kWh. In some areas, time of use rates apply with a \$0.07/kWh base and costs of \$0.13/kWh during peak periods defined by the residential load, which occur at 8:00 a.m. and 6:00 p.m.

A life cycle cost analysis showed that a simple payback of 1.5 years and an internal rate of return (IRR) of 65% was obtained if an advanced low-e window with exterior shading (Window E3) was used instead of the window prescribed by the GB 50189-2005 code in Beijing. A simple payback of 6 years and an IRR of 17% were obtained for this same scenario in Shanghai. Assumptions for this analysis were that the code and E3 windows cost \$200/m^2 and \$285/m^2 installed, respectively, heating cost \$7.1/GJ, electricity including demand cost \$0.15/kWh (flat rate), the discount rate was 6%, and the life of the technology was 30 years. Given the attractiveness of this financial investment, achieving goals of 30%–50% reductions in perimeter zone energy use below the 2005 code appear to be broadly feasible in China.

The commercial real estate developer is fully aware of the cost and limitations of maintaining reliable building operations in China. While the window E3 system does not require motorization (conscientious occupants with instructions on how to operate to the shades could achieve similar results), savings cannot be guaranteed without automation. One developer indicated that motorized systems were to be avoided since these systems require maintenance and were likely to fail in the short term. Exterior shading systems that attract pigeons or accumulate dirt or particulates from city smog were undesirable. Domestic products were preferred for practical reasons: (a) foreign products incur a 40% tariff; and (b) local Chinese manufacturers would uphold the terms of the warranty in a timely manner.

With the exception of construction cost, the business context is very similar to the US. Given the relatively low cost of construction, energy operating cost for the tenant becomes more of a critical issue for businesses that lease the space. The Three-Star Green Building certification becomes important within this context. For certification, the Ministry of Construction uses monitored data to assess energy performance relative to the energy-efficiency standard. Tenants can then use the certification for informed decision-making.

Exterior shading systems have long been known to produce the most significant reductions in HVAC energy use, particularly in internal load dominated buildings like office buildings, and have been used extensively throughout Europe. These systems are used in only a fraction of commercial buildings in the US because the codes are much less stringent and the cost of energy is lower. ASHRAE 90.1-2010 and 189.1 codes are imposing more stringent requirements that could increase use of exterior shading systems. However, in addition to cost constraints, proper implementation, code enforcement, and long-term maintenance are key concerns for operable systems. Between-pane motorized shading systems are more expensive and are subject to issues of long-term maintenance as well. Clearly, innovative products are needed to address these challenges.

Highly insulating windows have significantly greater potential for market uptake (particularly in colder climates like Beijing) since no maintenance is required and the window is likely to meet the aesthetic requirements of developers. There is significant on-going R&D worldwide to improve performance of component technologies (IGU, spacers, framing) as well as produce such systems at lower cost.

5. Discussion and Conclusions

Space heating, cooling, and lighting energy use are three of the largest energy end uses in commercial or public buildings in China and energy-efficient window and shading systems can have a significant impact on reducing these end uses in perimeter zones. To quantify these impacts, one needs to have an accurate depiction of the typical characteristics of HVAC systems and how they are operated. A prototypical large office building model originated from the DOE reference building was adapted to comply with the GB 50189-2005 Standard.

Two types of advanced window systems were evaluated by comparing zone heating and sensible cooling load as well as the HVAC energy use in the perimeter zones. A spectrally-selective low-e dual pane window and a triple pane, highly insulated window were modeled. The windows were combined with an interior or exterior roller shade that was controlled on a time step basis to minimize heating and cooling energy use. Parametric analysis of these various window types using EnergyPlus indicated that for both the cold Beijing climate, and hot summer and cold winter Shanghai climate, the spectrally-selective low-e dual pane window with an exterior roller shade and the triple pane window with an exterior roller shave provide the best performance in terms of energy savings, while the spectrally-selective low-e dual pane window with an exterior roller shade is more economically attractive due to its lower cost.

The economics of such solutions is highly dependent on the specifics of the space conditioning. If occupants are not using space conditioning either due to thermal preferences defined by adaptation to a broader range of environmental conditions or for reasons of economy given that energy costs are high, then the economics of the solution must be judged based on how the solution improves thermal comfort and habitability of indoor spaces. If a developer is selecting energy-efficiency measures based on a short return on investment and low capital cost, then advanced window solutions may be made cost-effective by including not only projected energy use reductions into the calculation but also potential capital reductions due to downsizing of HVAC capacity and air distribution systems and increased comfort and amenity provided by such solutions. Even so, with the low costs for labor and materials in China, anecdotal data suggests that the simple payback periods for even the more aggressive energy efficiency measures will be less than 5–7 years.

This study illustrates that the goal of achieving 30%–50% lower perimeter zone energy use compared to the current GB 50189-2005 code can be straightforward with unshaded high performance low-e windows that are commercially available today. To meet even more aggressive energy-efficiency and greenhouse gas emission reduction goals, low-e windows with operable exterior shading can be used to achieve up to 50%–80% reductions in perimeter zone energy use within a simple payback of 2–6 years in internal load dominated commercial buildings.

Acknowledgments

This study was conducted within the USA/China Clean Energy Resource Center—Building Energy Efficiency program. We would like to thank our LBNL colleagues, Walter Fred Buhl, Christian Kohler, Kaustubh Phalak, and Liping Wang for their assistance on this project.

This work was supported by the Assistant Secretary for Energy Efficiency and Renewable Energy, Office of Building Technology, State and Community Programs, Office of Building Research and Standards of the U.S. Department of Energy under Contract No. DE-AC02-05CH11231.

Author Contributions

Eleanor S. Lee and Xiufeng Pang were the primary authors; Andrew McNeil, Sabine Hoffmann, and Anothai Thanachareonkit contributed towards the simulations and analysis; Zhengrong Li and Yong Ding provided technical and market insights related to commercial building fenestration systems in China.

Appendix A.

Assumptions for the reference model (1983 base case) and GB 50189–2005 model.

Table A1. Common building information for both reference building model and the GB 50189 code compliant building model.

Building Characteristic	Description
Total Floor Area	43,000 m^2, 73 m (L) × 49 m (W) × 48 m (H)
Number of Floors	12
Floor to Ceiling Height	2.7 m
Floor to Floor Height	4 m
Roof type	Built-up flat roof, insulation entirely above deck
Exterior Wall Type	Mass wall

Table A1. *Cont.*

Building Characteristic	Description
Wall to Skin Ratio	0.77
Roof Construction Type	IEAD
HVAC System Type	4-pipe fan coil unit with central DOAS
Heating Type	Various depending on climates
Cooling Type	two water cooled chillers
Indoor design temperature	20 °C (winter), 25 °C (summer)
Ventilation	30 m^3/h·person

Table A2. Characteristics of the reference model per climate zone.

Characteristic	Harbin	Beijing	Shanghai	Guangzhou
U-value for envelope (W/m^2·K)	1.28	1.7	2	2.35
U-value for roof (W/m^2·K)	0.77	1.26	1.5	1.55
U-value for window (W/m^2·K)	3.26	6.4	6.4	6.4
SC	0.8	0.8	0.8	0.8
Boiler efficiency	0.55	0.55	0.55	0.55
Centrifugal chiller efficiency	4.2	4.2	4.2	4.2
Lighting (W/m^2)	25	25	25	25

Table A3. Characteristics of the GB 50189–2005 code compliant building model.

Element	Value
Chiller efficiency COP	5.13
Gas boiler efficiency	0.89
Fan total efficiency (combined motor and fan efficiency)	0.49

Table A4. Interior loads for the GB 50189–2005 code-compliant building model.

Zone	Lighting (W/m^2)	Interior Equipment (W/m^2)	Occupant Density (m^2/person)
Core zone (standard office)	11	20	4
Exterior zone (luxury office)	18	13	8

Table A5. Schedules for the GB 50189–2005 code-compliant building model.

Heating Set Point												
Time (h)	**1**	**2**	**3**	**4**	**5**	**6**	**7**	**8**	**9**	**10**	**11**	**12**
Weekday (°C)	12	12	12	12	12	12	18	20	20	20	20	20
Weekend (°C)	12	12	12	12	12	12	12	12	12	12	12	12
Time (h)	**13**	**14**	**15**	**16**	**17**	**18**	**19**	**20**	**21**	**22**	**23**	**24**
Weekday (°C)	20	20	20	20	20	20	12	12	12	12	12	12
Weekend (°C)	12	12	12	12	12	12	12	12	12	12	12	12

Table A5. *Cont.*

Cooling Set Point												
Time (h)	**1**	**2**	**3**	**4**	**5**	**6**	**7**	**8**	**9**	**10**	**11**	**12**
Weekday (°C)	37	37	37	37	37	37	28	26	26	26	26	26
Weekend (°C)	37	37	37	37	37	37	37	37	37	37	37	37
Time (h)	**13**	**14**	**15**	**16**	**17**	**18**	**19**	**20**	**21**	**22**	**23**	**24**
Weekday (°C)	26	26	26	26	26	26	37	37	37	37	37	37
Weekend (°C)	37	37	37	37	37	37	37	37	37	37	37	37
Lighting												
Time (h)	**1**	**2**	**3**	**4**	**5**	**6**	**7**	**8**	**9**	**10**	**11**	**12**
% of design value during weekday	0	0	0	0	0	0	10	50	95	95	95	80
% of design value during weekend	0	0	0	0	0	0	0	0	0	0	0	0
Time (h)	**13**	**14**	**15**	**16**	**17**	**18**	**19**	**20**	**21**	**22**	**23**	**24**
% of design value during weekday	80	95	95	95	95	30	30	0	0	0	0	0
% of design value during weekend	0	0	0	0	0	0	0	0	0	0	0	0
Equipment												
Time (h)	**1**	**2**	**3**	**4**	**5**	**6**	**7**	**8**	**9**	**10**	**11**	**12**
% of design value during weekday	0	0	0	0	0	0	10	50	95	95	95	50
% of design value during weekend	0	0	0	0	0	0	0	0	0	0	0	0
Time (h)	**13**	**14**	**15**	**16**	**17**	**18**	**19**	**20**	**21**	**22**	**23**	**24**
% of design value during weekday	50	95	95	95	95	30	30	0	0	0	0	0
% of design value during weekend	0	0	0	0	0	0	0	0	0	0	0	0
Occupancy												
Time (h)	**1**	**2**	**3**	**4**	**5**	**6**	**7**	**8**	**9**	**10**	**11**	**12**
% of design value during weekday	0	0	0	0	0	0	10	50	95	95	95	80
% of design value during weekend	0	0	0	0	0	0	0	0	0	0	0	0
Time (h)	**13**	**14**	**15**	**16**	**17**	**18**	**19**	**20**	**21**	**22**	**23**	**24**
% of design value during weekday	80	95	95	95	95	30	30	0	0	0	0	0
% of design value during weekend	0	0	0	0	0	0	0	0	0	0	0	0

Conflicts of Interest

The authors declare no conflict of interest.

References

1. Li, B.; Yao, R. Urbanisation and its impact on building energy consumption and efficiency in China. *Renew. Energy* **2009**, *34*, 1994–1998.
2. Xiao, H.; Wei, Q.; Jiang, Y. The reality and statistical distribution of energy consumption in office buildings in China. *Energy Build.* **2012**, *50*, 259–265.
3. Bressand, F.; Zhou, N.; Lin, J. Energy Use in Commercial Building in China: Current Situation and Future Scenarios. In Proceedings of the ECEEE 2007 Summer Study, La Colle sur Loup, France, 4–9 June 2007.

4. Fridley, D.G.; Zheng, N.; Zhou, N. *Estimating Total Energy Consumption and Emissions of China's Commercial and Office Buildings*; Technical Report LBNL-248E; Lawrence Berkeley National Laboratory: Berkeley, CA, USA, 2008.
5. People's Republic of China National Standard. *Design Standard for Energy Efficiency of Public Buildings*; GB 50189–2005; Chinese Construction Industrial Publish House: Beijing, China, 2005. (in Chinese)
6. Zhou, N.; Fridley, D.; McNeil, M.; Zheng, N.; Ke, J.; Levine, M. *China's Energy and Emissions Outlook to 2050*; Technical Report LBNL-4472E; Lawrence Berkeley National Laboratory: Berkeley, CA, USA, 2011.
7. Hong, T. A close look at the China design standard for energy efficiency of public buildings. *Energy Build.* **2009**, *41*, 426–435.
8. *Standard 90.1–2004 Energy Standard for Buildings Except Low-Rise Residential Buildings*; American Society of Heating, Refrigerating, Air-Conditioning Engineers, Inc.: Atlanta, GA, USA, 2007.
9. *The International Glazing Database*, Version 23.0; Lawrence Berkeley National Laboratory: Berkeley, CA, USA. Available online: http://windowoptics.lbl.gov/data/igdb (accessed on 15 April 2012).
10. *Window 6.2/Therm 6.2 Research Version User Manual*; LBNL-941; Mitchell, R., Kohler, C., Klems, J., Rubin, M., Arasteh, D., Huizenga, C., Yu, T., Curcija, D., Eds.; Lawrence Berkeley National Laboratory: Berkeley, CA, USA, 2008.
11. Klems, J.H. A new method for predicting the solar heat gain of complex fenestration systems: I. Overview and derivation of the matrix layer calculation. *ASHRAE Trans.* **1994**, *100*, 1065–1072.
12. McNeil, A.; Lee, E.S. A validation of the Radiance three-phase simulation method for modeling annual daylight performance of optically complex fenestration systems. *J. Build. Perform. Simul.* **2012**, *6*, 24–37.
13. Lam, J.C.; Wan, K.K.W.; Tsang, C.L.; Yang, L. Building energy efficiency in different climates. *Energy Convers. Manag.* **2008**, *49*, 2354–2366.
14. Yang, L.; Lam, J.C.; Tsang, C.L. Energy performance of building envelopes in different climate zones in China. *Appl. Energy* **2008**, *85*, 800–817.
15. Johnson, R.; Sullivan, R.; Nozaki, S.; Selkowitz, S.; Conner, C.; Arasteh, D. *Building Envelope Thermal and Daylighting Analysis in Support of Recommendations to Upgrade ASHRAE/IES Standard 90: Vol 1, Parametric Study, Analysis, and Results*; Technical Report LBNL-16770; Lawrence Berkeley National Laboratory: Berkeley, CA, USA, 1983.
16. Energy Information Administration. *2003 Commercial Buildings Energy Consumption Survey*; Energy Information Administration: Washington, DC, USA, 2005. Available online: http://www.eia.gov/consumption/commercial/ (accessed on 20 April 2012).
17. Deru, M.; Field, K.; Studer, D.; Benne, K.; Griffith, B.; Torcellini, P.; Halverson, M.; Winiarski, D.; Liu, B.; Rosenberg, M.; *et al. U.S. Department of Energy Commercial Reference Building Models of the National Building Stock*; Technical Report NREL/TP-5500–46861; National Renewable Energy Laboratory: Golden, CO, USA, 2011.

18. *EnergyPlus Energy Simulation Software*; U.S. Department of Energy, Energy Efficiency and Renewable Energy: Washington DC, USA. Available online: http://apps1.eere.energy.gov/buildings/energyplus/ (accessed on 15 April 2012).
19. Zhou, N.; Lin, J. The reality and future scenarios of commercial building energy consumption in China. *Energy Build.* **2008**, *40*, 2121–2127.
20. Zhao, T.; Zhang, J.; Sun, D. Experimental study on a duty ratio fuzzy control method for fan-coil units. *Build. Environ.* **2011**, *46*, 527–534.

13

Simulating Performance Risk for Lighting Retrofit Decisions

Jia Hu *, Eric Shen † and Yun Gu †

Philips Research North America, Briarcliff Manor, NY 10510, USA;
E-Mails: e.shen@philips.com (E.S.); yun.gu@philips.com (Y.G.)

† These authors contributed equally to this work.

* Author to whom correspondence should be addressed; E-Mail: jia.hu@philips.com

Academic Editor: Richard Mistrick

Abstract: In building retrofit projects, dynamic simulations are performed to simulate building performance. Uncertainty may negatively affect model calibration and predicted lighting energy savings, which increases the chance of default on performance-based contracts. Therefore, the aim of this paper is to develop a simulation-based method that can analyze lighting performance risk in lighting retrofit decisions. The model uses a surrogate model, which is constructed by adaptively selecting sample points and generating approximation surfaces with fast computing time. The surrogate model is a replacement of the computation intensive process. A statistical method is developed to generate extreme weather profile based on the 20-year historical weather data. A stochastic occupancy model was created using actual occupancy data to generate realistic occupancy patterns. Energy usage of lighting, and heating, ventilation, and air conditioning (HVAC) is simulated using EnergyPlus. The method can evaluate the influence of different risk factors (e.g., variation of luminaire input wattage, varying weather conditions) on lighting and HVAC energy consumption and lighting electricity demand. Probability distributions are generated to quantify the risk values. A case study was conducted to demonstrate and validate the methods. The surrogate model is a good solution for quantifying the risk factors and probability distribution of the building performance.

Keywords: risk; lighting; retrofit; simulation; EnergyPlus; surrogate model

---

1. Introduction

In a retrofit decision process, professionals perform energy audits and construct dynamic simulation models to benchmark the performance of existing buildings and predict the effect of retrofit interventions [1]. Retrofit decisions can be evaluated using multi-objective optimization [2,3]. However, deterministic models do not provide insights into underperforming risks associated with each retrofit intervention. A more reliable approach is performing sensitivity analysis to quantify the influence of different risk factors on building operation performance [4,5]. To do this, a distribution is chosen for a number of inputs and numerous models are generated to simulate and determine statistics of important outputs. For instance, sensitivity analysis for thermal parameter ranking can help to choose the envelope material type.

A number of researchers have studied uncertainty and sensitivity analysis in building design and retrofit decisions. For example, Eisenhower *et al.* [6] used EnergyPlus to perform uncertainty analysis with one thousand uncertain parameters (almost all numeric parameters in the EnergyPlus input file were included, except those parameters such as architectural parameters and text-based parameters), and quantified which intermediate processes were contributing the most to this uncertainty using a decomposition-based approach. Shen *et al.* [7,8] studied lighting control strategies using different parameters such as climate zones, blind type and window-to-wall ratios. The combination of only a few parameters can result in a large number of building models. The number of simulations significantly increases when more parameters are evaluated. Heo *et al.* [1] used Bayesian calibration of normative energy models for energy model calibration in retrofit projects. However, the normative energy model is essentially lightweight formulations of heat balance equations, resulting in limitations compared to the advanced energy modeling available in programs such as EnergyPlus.

Two main problems occur when using uncertainty and sensitivity analysis to evaluate a building retrofit decision. The first problem is the long simulation times, especially when large numbers of simulations are needed. Sensitivity analysis requires performing energy and lighting simulations using advanced tools such as EnergyPlus [9], Radiance, and Daysim [10,11]. The current workflow of sensitivity analysis [4] shows that a large number of simulations may have to be performed [11]. In addition, when the variation of input parameters becomes significant, this results in a large number of simulations requiring a long time to complete the analysis. Therefore, two solutions were proposed to solve this problem. The first solution is to develop new models that run faster. For example, Hu and Olbina [12] developed an analytical model to simulate the day-lighting performance of blinds with different blind reflectance and glazing transmittances. The model can quickly predict the interior luminance without repeatedly simulating all the possible parameters. A neural network model was also developed to predict the luminance and building energy by reducing the simulation complexity [13,14]. An expert system can provide realistic results using prior knowledge from lighting designers and simulation data [15,16]. The second solution is to reduce the number of simulations. Instead of sampling parameters in the whole space, the samples can be reduced by intelligently selecting optimum positions. For example, a surrogate model uses adaptive sampling to select sample points with maximization of errors. The surrogate model treats the computationally expensive simulation code as a data generating black-box, only taking into account its input-output behavior and constructing approximation models. The approximation surface of the surrogate model can be constructed using a small sample size of simulation data. Once the surrogate model is developed, the input variables can be directly fed into the surrogate model to generate the output values (*i.e.*, energy consumption).

The second problem is identifying critical risk factors and sampling these factors efficiently. Some risk factors tend to be ignored or simulated inappropriately because these parameters are not scalar and distribution functions are not clearly defined. For example, the sky model or weather conditions affect the indoor daylight levels, directly affecting daylight sensor readings or indirectly affecting human control behavior. The sky model has been recognized as the key factor in lighting and energy simulation. The sky model is generated from a weather file whose data are measured from a weather station. A weather file contains many meteorological variables where radiation and temperature are typically the most important factors. Currently, most methods focus on temperature only [17]. In lighting simulation, other parameters such as global and solar radiation and sky clearness have more weight than temperature. Therefore, the weather data should be further analyzed when used in lighting retrofit projects. Another essential parameter for lighting and energy simulation is the occupancy pattern, which affects occupant control behavior and internal heat generation. When occupancy based control is used for lighting, occupancy will directly influence the lighting on/off status [12,18]. The third parameter is lighting control strategy. Different control strategies significantly affect building energy, especially lighting energy consumption [8]. For example, significant savings can be achieved in lighting energy consumption using integrated closed-loop blind and lighting control strategy.

Therefore, based on the problems identified in the analysis of lighting retrofit projects, the aim of this research is to develop a surrogate model to quantify the influence of risk factors (*i.e.*, weather condition, occupancy pattern, and lighting control strategy) on building performance (*i.e.*, HVAC/lighting energy savings, and lighting demand).

2. Model Development

2.1. Model Structure

A surrogate model is used to predict energy consumption and lighting electricity demand. The surrogate model uses adaptive sampling to reduce the difference between the surrogate model and simulation model [19]. For this sampling technique, the sampling position in current iteration is determined by the previous iteration. For example, in Figure 1, in the first iteration, the predicted value at point A has a large deviation from the simulation model even though the sample points (white dots) have a small error. In the next iteration, a new sample point is added at position *A*, having a maximum error (*i.e.*, the largest vertical distance between the curve of the simulation model and the surrogate model). In this way, new sample points are calculated sequentially.

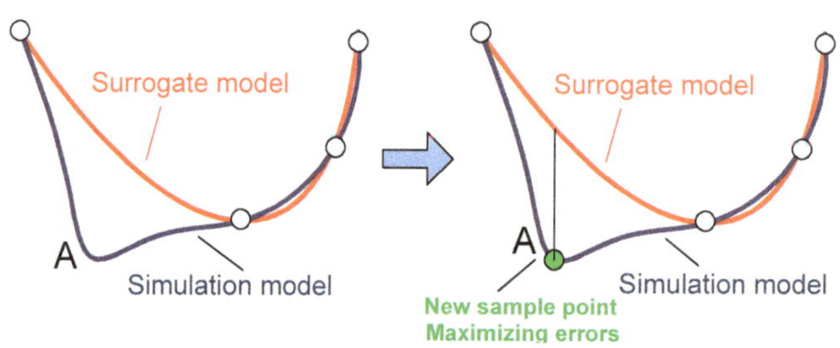

Figure 1. Adaptive sampling technique.

The model structure is shown in Figure 2. The input is a building model containing building thermal, location, and geometry data. The first step is sampling input variables (*i.e.*, risk factors) using Latin hypercube sampling method. The input variables include occupancy level, weather, control strategy and luminaire type. Other parameters such as envelope thermal properties and surface reflectances are obtained from design drawings or on-site auditing.

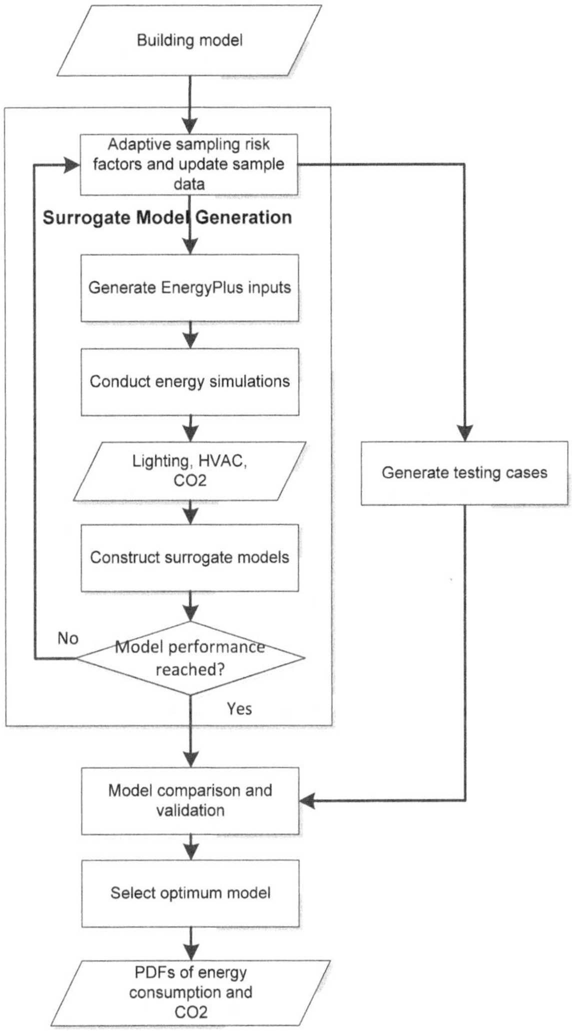

Figure 2. Model structure.

In the next steps, shown in Figure 2, energy simulation is performed using EnergyPlus. The input file of EnergyPlus is automatically generated by incorporating risk factors into an input file template. The output of the energy simulation includes energy consumption, electricity demand, *etc*. Then, surrogate models can be built by using methods such as polynomial response surfaces, Kriging, support vector machines and neural networks. A preliminary test using a DOE reference office building located in Chicago was performed which showed that the neural network model generally provides better results than the other approaches, which is consistent with a previous study on energy consumption prediction using neural networks [20].

In the next two steps, the surrogate model is validated and the optimum one is selected. Mean square error (MSE) is a metric for measuring the model performance. The dataset is divided into training and testing datasets for validation. After the surrogate model is constructed, the input variables are sampled using the Monte Carlo method. A large number of combinations of the input variables are generated and fed into the surrogate model for energy consumption prediction. Then, probability distributions of lighting and heating, ventilation, and air conditioning (HVAC) energy consumption can be generated.

2.2. Risk Factor Sampling

The risk factors considered in this model include occupancy level, weather, luminaire input wattage, and lighting control strategy. The use case is to quantify the performance risks when a fluorescent luminaire is replaced by another more energy efficient luminaire and advanced lighting control system. This section will introduce how to sample the risk factors: occupancy level, lighting control strategies, and weather conditions.

2.2.1. Occupancy Level

Occupancy patterns can be simulated (e.g., using occupancy model) or measured (e.g., onsite survey and audit). A stochastic occupancy model was developed using actual office data [8]. Occupancy data was monitored and logged every minute for nine offices over 42 weekdays and 18 weekend days. The model can be used to simulate the occupancy status for every occupant in the building. Each individual occupancy state is represented as a Markov Chain where the state transition probabilities are derived from the actual occupancy data. Figure 3 shows occupancy patterns (total occupancy) with three dynamic occupancy levels (low, medium, and high) calculated from the dynamic occupancy model. The occupants arrive around 7:00–9:00 a.m. and leave around 5:00–6:00 p.m.

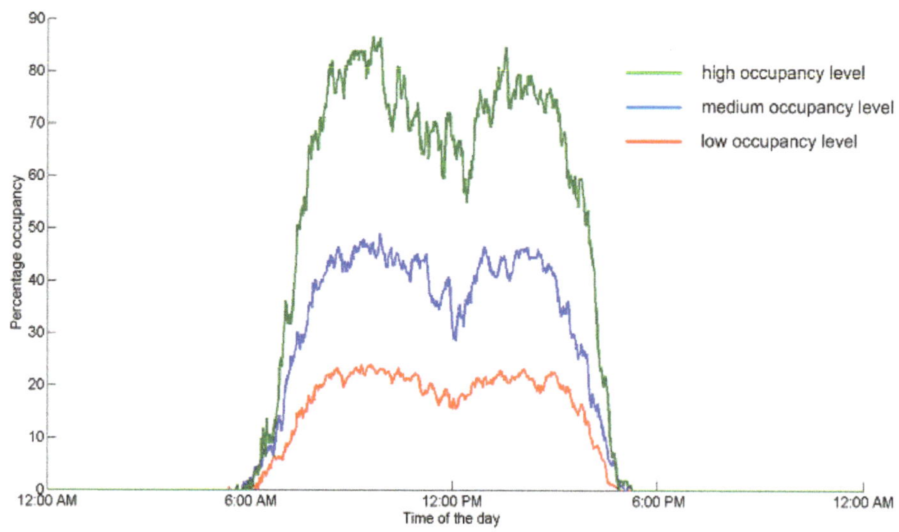

Figure 3. An example of occupancy patterns according to three peak occupancy levels

2.2.2. Lighting Control Strategies

Three lighting control strategies are evaluated: manual control (or fixed schedule based control); dynamic occupancy based control, or occupancy + day-lighting control.

- Manual Control: The lights are turned on in compliance with a pre-defined schedule. There is no automation in this process. A retrofitted building often uses this control strategy.

- Occupancy Control: This control strategy controls lighting on/off status based on the presence of occupants. In building simulation, a dynamic occupancy model can generate occupancy patterns that are used to simulate the occupant presence.
- Occupancy and Day-Lighting Control: This control strategy uses occupancy sensors to detect occupant presence and uses daylight sensors to measure daylight. Lights in perimeter zones dim in response to the daylight sensor readings. It is expected that more energy savings can be achieved while still maintaining desired luminance levels.

2.2.3. Weather Conditions

Sky clearness, solar radiation, and temperature affect HVAC and lighting energy consumption, especially when daylight control is used. In order to evaluate risks caused by variations in weather conditions, the meteorological data needs to be sampled. Previous research only sampled temperature in order to create a new weather file for energy simulation. This is inappropriate because sampling temperature alone fails to evaluate risks that influence lighting energy consumption. We developed a method to generate dynamic weather conditions relevant to lighting simulation using statistical analysis of historical weather data. This method is based on the Sandia method [21], which is an empirical approach that selects individual months from different years of the period of record to formulate a TMY weather file that is used in deterministic building simulations [22]. The weather files generated for our analysis represents extreme historical weather conditions based on probability analysis. These weather files are different from TMY weather files typically used for EnergyPlus simulations. Figure 4 shows the main steps of how to generate the dynamic weather files. The major steps are described as follows.

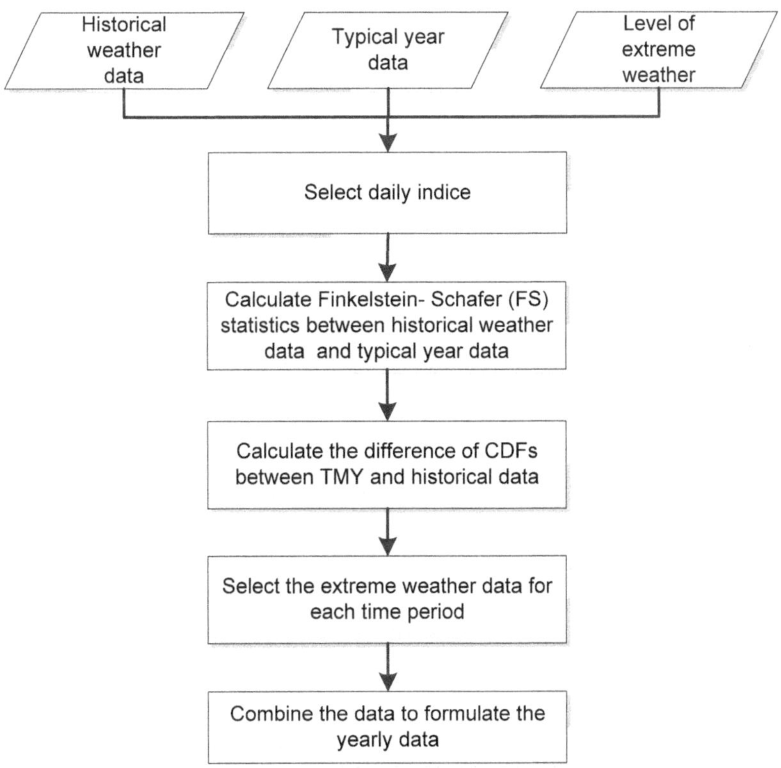

Figure 4. Flow chart of generating extreme weather data from historic weather data.

In the first step, the model input is defined. The historical weather data can be retrieved from the NREL website, which provides 20 years of actual weather data. TMY weather data are generated by processing historical weather data using the Sandia method. Data sources used in creating TMY data include the Typical Meteorological Year (TMY2, TMY3), and International Weather for Energy Calculation (IWEC) datasets. The input to the model is the level of extreme weather, where the level can be categorized into: (1) cloudy; (2) partially cloudy; and (3) clear. A Gaussian distribution is used for each level.

In the next step, the daily index such as clearness and solar radiation can be selected, depending on the simulation purpose. For example, when evaluating lighting energy, we use C = total radiation/diffuse radiation. Then a cloudy day would be represented by a small value of C. In order to generate a yearly weather file, statistical differences between historical data and typical data were quantified according to the Sandia method by calculating Finkelstein-Schafer (*FS*) statistics for relevant weather data. *FS* is defined in Equation (1).

$$FS = (1/n)\sum_{i=1}^{n} \delta_i \tag{1}$$

where δ_i = absolute difference between the long-term cumulative probability functions (CDF) and the candidate month CDF at x_i, and n = the number of daily readings in a month.

In the next step, the differences between CDFs of typical year and CDFs of other years are calculated. The calculation can be done using histograms, similar to calculating the area between the two CDFs. Based on the calculated statistics, extreme weather is selected based on the CDF difference. For example, assume the weather data in January of 1995 has the largest CDF difference from the CDF of TMY data. Then this month data is selected for January when building the extreme cloudy year data. Note that different time periods may be used. For example, the time period may be weekly instead of monthly. Figure 5 shows an example of CDFs for global solar radiation. The whole year extreme weather data are created by combining the data calculated from each time period (e.g., month, week).

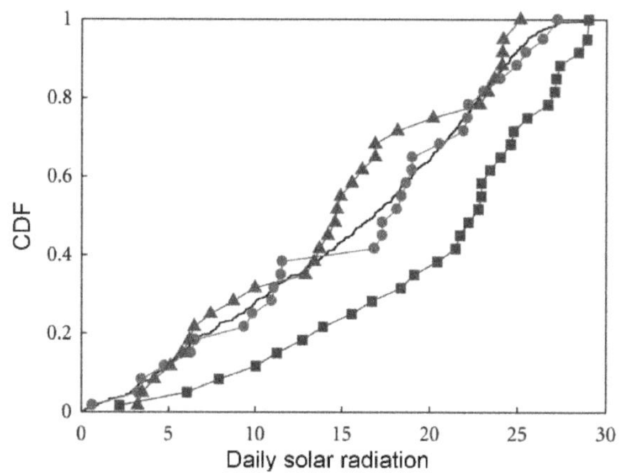

Figure 1. Cumulative Probability Functions (CDFs) of solar radiation in a month.

3. Case Study

In this case study, a lighting system retrofit is performed to reduce the power density from 13.83 W/m^2 (existing luminaire) to 11.52 W/m^2 (retrofit luminaire) and to add advanced lighting control strategies.

3.1. Model Implementation

The surrogate model is generated by using the SUrrogate MOdeling (SUMO) Toolbox [23], which is a Matlab toolbox that automatically builds accurate surrogate models (or response surface models) from a given data source (e.g., simulation code) within accuracy and time constraints. The toolbox minimizes the number of data points by selecting the best sample position adaptively and automatically, requiring no user input besides some initial configuration. In this case study, we used neural network model as the training model. The model was implemented as shown in Figure 6.

- Configuration File: This file defines the parameters of the surrogate model. For example, parameter of neural networks, and data sampling method (Latin hypercube sampling method) are used in this case study.
- Matlab Script/SUMO Tool: The tool calculates the objective function (*i.e.*, lighting and HVAC energy consumption) from the output of EnergyPlus simulation. In the meantime, the tool calls the data generator to generate the sample for the next round simulation.
- Weather File Generator: Implemented in Python, it can dynamically produce the weather file by combing monthly actual weather data, given the probability from Matlab script.
- Occupancy Pattern Generator: Implemented in Matlab, it uses the stochastic occupancy model developed from collected office data [8].
- EnergyPlus Input File Generator: Implemented in Python, it reads risk factors, the reference building model template, and the control configuration file. The control configuration file defines the control strategy readable by EnergyPlus. The reference building model is modified from U.S. DOE reference model.
- EnergyPlus: In each iteration of the surrogate modeling process, the complete input file is fed into EnergyPlus. Parallel simulation uses a total of eight threads. The output file is processed by the SUMO tool.

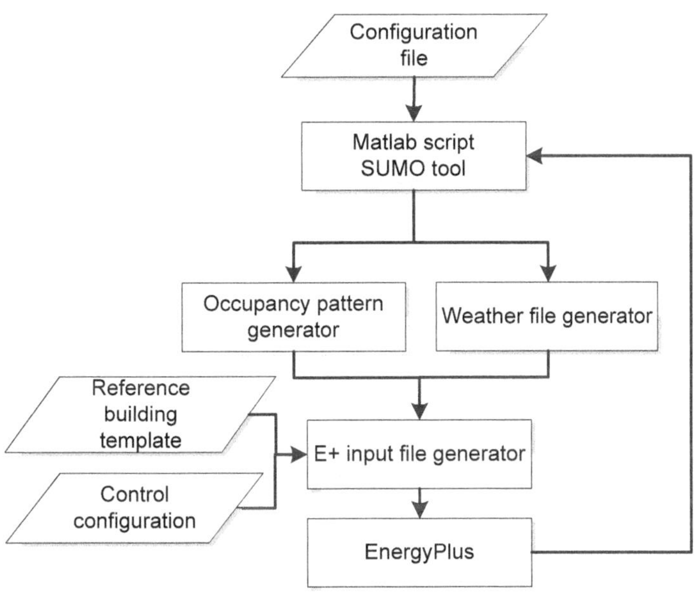

Figure 6. Structure of the implementation method.

3.2. Building Model

The US DOE commercial reference model for medium office building (post 1980 building) is selected as the retrofit building for this case study [24]. The building is located in Chicago, IL and has three floors. Each floor has four perimeter zones and a core zone (see Figure 7). The main building parameters are:

- Envelope: The envelope thermal properties comply with ASHRAE Standard 90.1-1989 [25]. The exterior wall is set to steel frame walls. The window-to-wall ratio is set to 33.0%. The glass U-Factor is set to 3.354 W/m^2-K and SHGC (Solar Heat Gain Coefficient) is set to 0.385.
- HVAC: Packaged multizone variable air volume system with plenum zones, gas furnace, and electric reheat. The fan efficiency is 0.59.
- Lighting Control: Manual control is used for the existing luminaire A. The luminaires are fully turned on when the first occupant arrives. They remain fully on until 6:30 p.m. or when the last occupant leaves the office, whichever is earlier. During the night (e.g., no occupants), a minimum of 10% of the existing luminaires are turned on. For retrofit luminaires automatic lighting control strategies are used: occupancy control, and occupancy + day-lighting control. The control information is stored in the control configuration file. During the night (e.g., no occupants), a minimum of 10% of the retrofit luminaires are turned on.

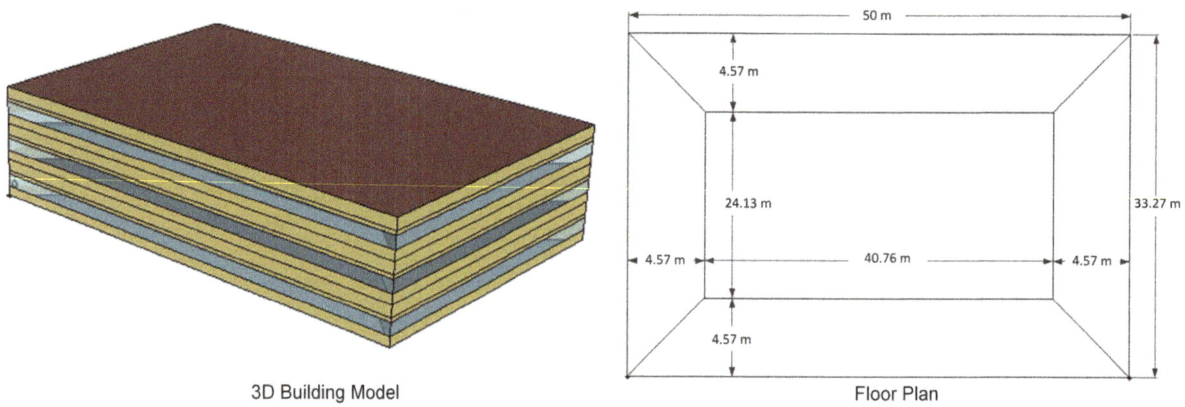

Figure 7. Reference building model and floor plan.

3.3. Risk Factor Sampling

Four risk factors are simulated and sampled using the following methods:

- Lighting Control Strategy: Two control strategies are evaluated: occupancy control and occupancy + daylight control. Because the number of control strategies is limited (two in this case study), a surrogate model is developed for each type of control strategy and evaluated separately.
- Weather Condition: Three levels of weather and sky condition are generated: overcast (<30%), medium (30%–70%), and clear (>70%). A Gaussian distribution is used for each level with the standard deviation set to 20%. For example, when the sampled probability is 25%, then the sky condition is set to overcast. In each sample datum, the weather file generator will generate the corresponding weather file based on the probability value.

- Luminaire Input Wattage: The mean wattage is set to 11.52 W/m^2 for retrofit luminaire and 13.83 W/m^2 for the existing luminaire. The input wattage complies with Gaussian distribution with a standard deviation that is set to 5% of the mean value (*i.e.*, standard deviation is set to 0.576 W/m^2 for the retrofit luminaire and 0.69 W/m^2 for the existing luminaire).
- Occupancy Level: The occupancy level is set to medium (e.g., mean peak occupancy level = 50% with a standard deviation of 10%), low (e.g., mean peak occupancy level = 30% with a standard deviation of 10%), or high (e.g., mean peak occupancy level = 80% with a standard deviation of 10%). Gaussian distribution is applied to each level. For example, if the occupancy level is set to medium (50%), a number of occupancy patterns are sampled using Gaussian distribution.

3.4. Results

Surrogate models (neural network models) were developed to predict the lighting and HVAC energy consumption, and electricity demand. The surrogate model serves as a replacement for computationally expensive simulation and provides a fast calculation of the risk factors sampled using Monte Carlo simulations.

3.4.1. Lighting Energy Consumption

The performance of the neural network model is shown in Figure 8. The target value in Y-axis denotes the simulated value from EnergyPlus. The output value in *X*-axis denotes the predicted value using surrogate model. The testing and training results indicate that very good performance has been achieved in the model. When the occupancy based control is used, RMSE (root mean square error) is 21.6 kWh, and CV (the coefficient of variation) is less than 0.1%. When the occupancy + daylight control is used, RMSE is 144.3 kWh, and CV is 0.16%. Both of the test results show very good performance with an R value close to one. This result is consistent with the previous study showing that lighting energy consumption has a linear relationship with many building parameters [12].

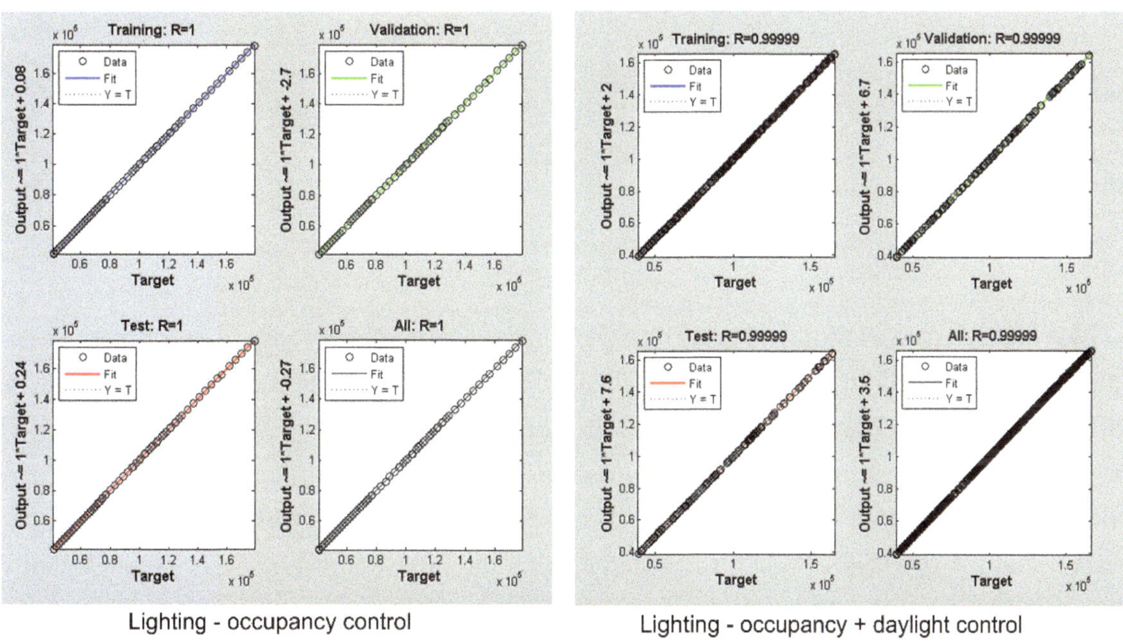

Figure 8. Neural network training and testing results for lighting energy consumption.

To evaluate the risk of reducing lighting power density by replacing an existing luminaire with a retrofit luminaire, one of the requirements is to calculate the distribution of lighting energy savings. In current retrofit decisions, only a fixed value is calculated without knowing the probability, which results in risk. Users can predict the probability distribution of lighting energy savings by setting different scenarios. For example, users can evaluate the risk level when medium sky condition happens in a medium occupied building. In Figure 9, the distribution of the difference between the energy consumption of the two luminaires (A and B) is plotted. The risk scenario is set so that the average occupancy level is about 50% with a standard deviation of 10%, and the weather occurrence is M (medium) level. Different results are shown in the two control strategy types. In the left plot in Figure 9, when the occupancy based control is used, replacing an existing luminaire with a retrofit luminaire with advanced lighting controls results in energy savings ranging from about 240 MWh to 140 MWh. Similarly, in the right plot in Figure 9, the energy savings ranges from 240 MWh to 170 MWh. It is observed that using an occupancy + daylight control has less risk for the lighting retrofit project. The guaranteed minimum energy savings using an occupancy + daylight control (170 MWh) is higher than the minimum savings (140 MWh) using occupancy control. When making retrofit decisions, the cost of adding additional control system should be taken into account.

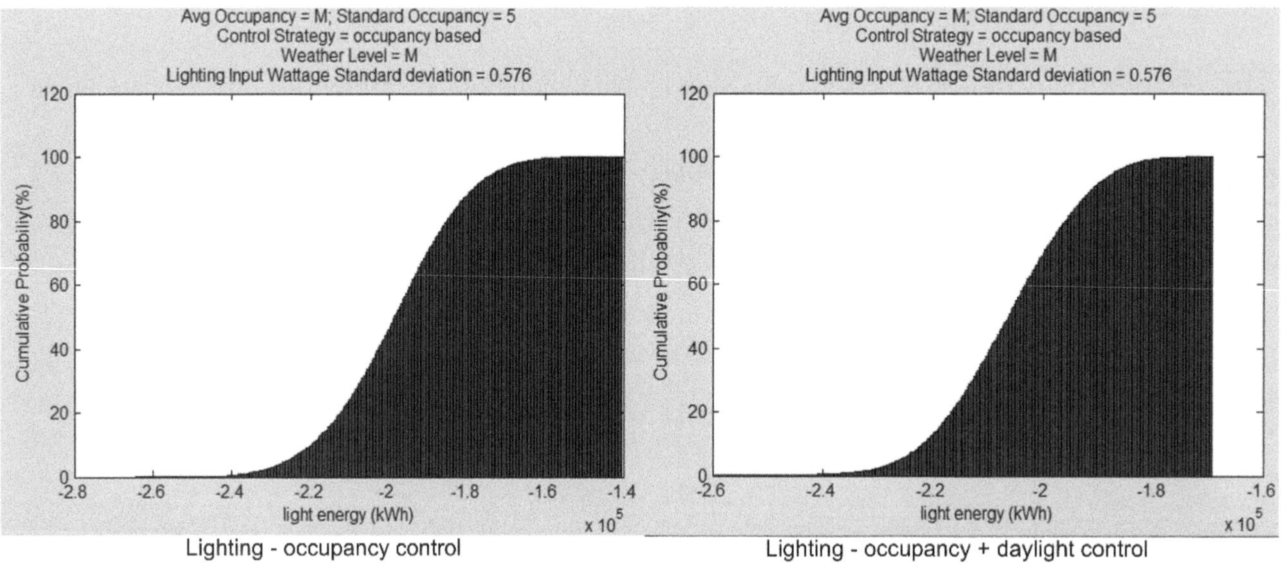

Figure 9. CDFs of lighting energy difference between retrofit and existing lighting system.

3.4.2. HVAC Energy Consumption

Internal heat emitted from lighting affects HVAC usage, and thus HVAC energy consumption. Neural network model is also used to predict HVAC energy consumption, which has been studied in the literature. In general, the model achieves good performance with $R = 0.97–0.98$ (see Figure 10). There are a number of factors affecting HVAC energy usage and the prediction accuracy is lower than the accuracy of the lighting energy usage prediction. When the occupancy control strategy is used, RMSE is 1777.5 kWh and CV is 0.54%. When the occupancy + daylight control strategy is used, RMSE is 2693.6 kWh and CV is 0.82%. The error increases when using a complicated lighting control strategy (*i.e.*, occupancy + daylight control).

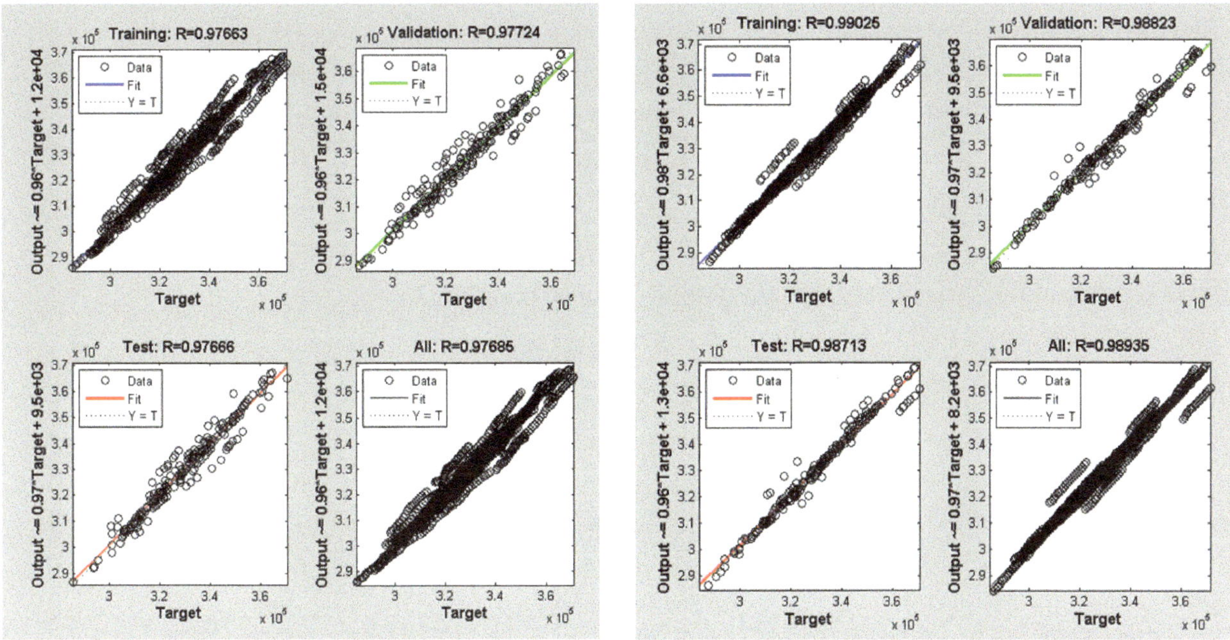

HVAC - occupancy control

HVAC - occupancy + daylight control

Figure 10. Neural network training and testing results for HVAC energy consumption.

The probability of the difference between HVAC energy consumption when using the retrofit luminaire with advanced lighting controls and the existing luminaire is plotted (see Figure 11). The occupancy level and weather condition are set to medium. When the occupancy based control is used, the energy consumption increase changed from 15 MWh to 60 MWh. When the occupancy + daylight control is used, the energy consumption increase changed from −100 MWh to 150 MWh, which indicates a larger changing range than the occupancy based control. This is opposite from the case for lighting energy savings. In addition, the CDF curve using occupancy + daylight control (right plot in Figure 11) is smoother than the curve using occupancy control, which indicates a higher standard deviation for HVAC energy savings. The reason that using occupancy + daylight control has a wider range is because the daylight level affects the lighting levels and thus results in fluctuating internal heat gains.

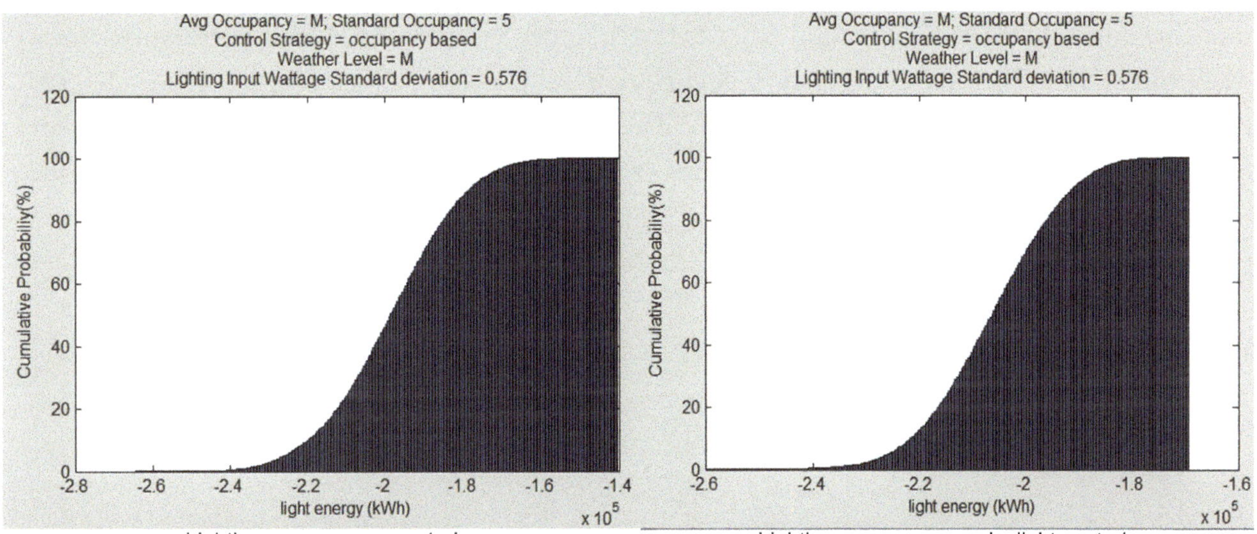

Lighting - occupancy control

Lighting - occupancy + daylight control

Figure 11. CDFs of HVAC energy difference between retrofit and existing lighting system.

3.4.3. Electricity Demand

An advantage of using the retrofit luminaires with advanced lighting controls is to reduce the electricity peak demand. The electricity demand for the retrofit luminaires with advanced lighting controls is predicted using neural network with good performance (see Figure 12). When the occupancy control is used, RMSE is 3.9 kW and CV is 0.88%. When the occupancy + daylight control is used, RMSE is 7.8 kW and CV is 1.7%. The prediction is reasonably accurate.

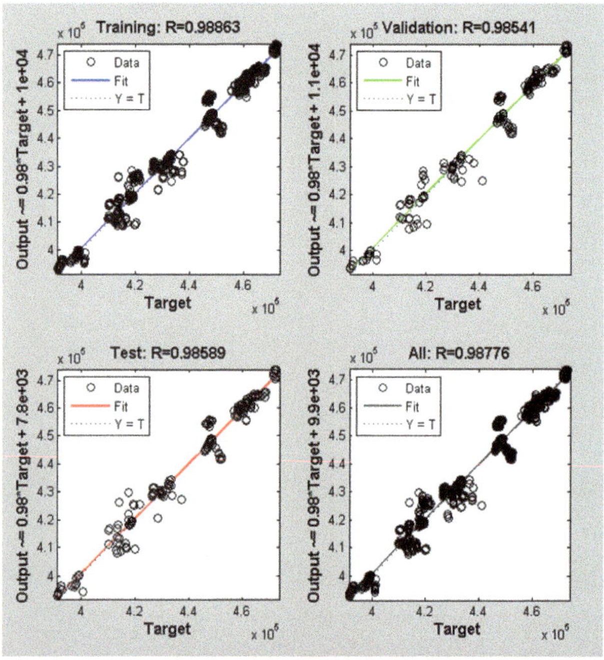

Demand - occupancy control

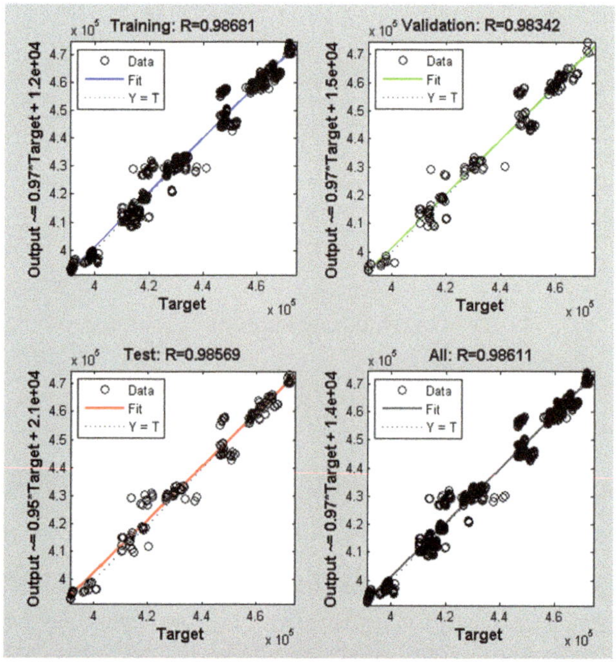

Demand - occupancy + daylight control

Figure 12. Neural network training and testing results for lighting electric demand.

The difference of the whole building electricity demand using retrofit luminaires with advanced lighting controls and existing luminaires is shown in Figure 13. The two scenarios have a similar pattern. One possible reason is that electricity demand only represents the peak value and thus is less sensitive to luminaire dimming. When the occupancy control is used, the demand difference changes from −100 kW to 70 kW. When the occupancy + daylight control is used, the demand difference changes from −90 kW to 55 kW. The range changes from negative value to positive value, which implies replacing existing luminaires with retrofit luminaires does not always reduce the electricity demand. One possible reason is that HVAC electricity peak value increases while lighting electricity demand decreases. The mean value is expected to reduce the electricity demand.

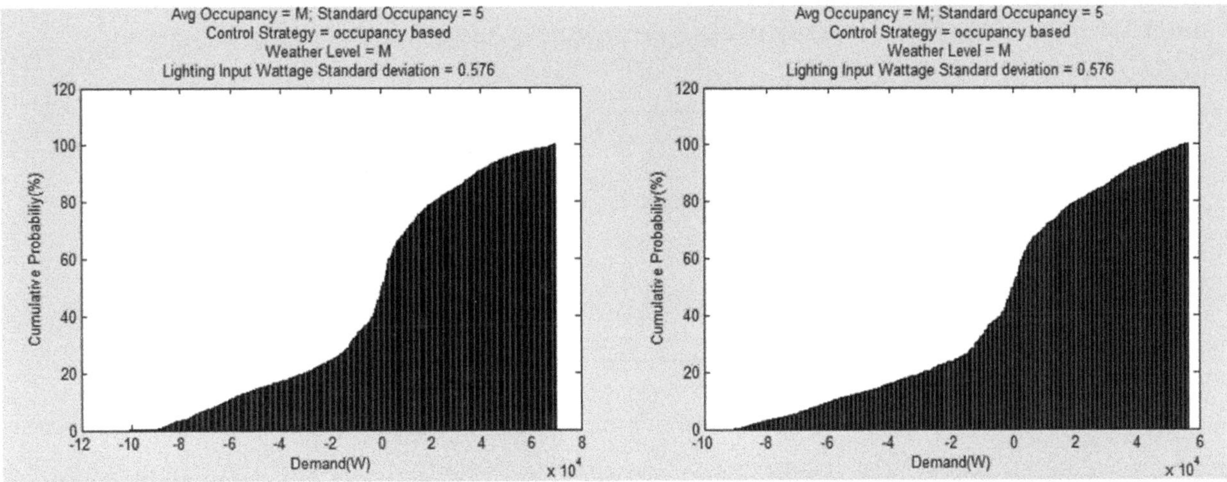

Demand - occupancy control

Demand - occupancy + daylight control

Figure 13. CDFs of lighting demand difference between retrofit and existing lighting system.

3.5. Discussion

In the previous section, the probability distribution plots of lighting and HVAC energy savings and demand difference are presented. Using this information, the mean and standard deviation of these values can be calculated to estimate the risk levels. The expected values and standard deviation of lighting, HVAC, and demand savings are listed in Table 1.

- Lighting Energy Savings: Lighting has the largest energy savings. Both of the two control strategies have small standard deviation, indicating a lower risk.
- HVAC Energy Savings: HVAC energy savings is about 15% of the lighting energy savings. When the occupancy based control is used, the standard deviation is small. However, when the occupancy + daylight control is used, the standard deviation becomes much larger. This is caused by the dimming control of retrofit luminaire with a daylight control system. Thus, the risk level for HVAC energy savings becomes higher when the daylight control is used.
- Electricity Demand: The mean values for both control strategies are similar because the peak value of electricity is less sensitive to the luminaire dimming, and the peak value does not change too much even when using day-lighting control. The standard deviation is much larger. As stated in Section 3.4.3, using the retrofit luminaire with advanced lighting controls may not always reduce the electricity demand because the electricity demand of HVAC may increase. Another reason is that using an existing luminaire may generate more internal heat in winter, and help reduce the HVAC electricity demand. In a whole year, using a retrofit luminaire with advanced lighting controls can still help reduce the electricity demand by about 9 kW on average.

Table 1. Mean and standard deviation of lighting energy savings, HVAC energy savings and demand savings.

Control	Lighting (MWh)		HVAC (MWh)		Demand (kW)	
	Occupancy	Occupancy + Daylight	Occupancy	Occupancy + Daylight	Occupancy	Occupancy + Daylight
Mean	201.5	208.0	−30.6	−31.2	9.3	9.4
Standard deviation	16.9	12.3	4.1	39.6	30.8	33.6

The probability plots in Section 3.4 are based on the scenario that occupancy and weather control are set to medium. In some buildings or locations, the occupancy levels may change significantly during different months or seasons, or the weather is more variable. Thus the expected value should be calculated from the distribution of the different occupancy levels and weather conditions. The expected value can be calculated using Equation (2). The surrogate model calculates the expected value of the joint distribution.

$$E = \sum_{c}\sum_{j} P_c P_w \left(S_L(c,w) + S_H(c,w)\right) \tag{2}$$

where E = expected value of energy savings; P_i^c = probability of the occupancy level c; P_j^w = probability of the weather condition w; $S_L(c,w)$ = expected value of the lighting energy savings under occupancy level = c and weather condition = w; and $S_H(c,w)$ = expected value of HVAC energy savings under occupancy level = c and weather condition = w.

Equation (3) shows the runtime ratio between the runtime using the surrogate model and the runtime using a simulation approach. In the equation, the total runtime for the surrogate model based approach includes the training time T_{train} and calculation time T_{cal}: (a) In the training time, the training data is generated from simulation. The size of the data samples are generally small; (b) In the calculation time, the surrogate model reads a large number of sampled points to generate a distribution of the energy savings or demand level. Generally, the value of m is much smaller than n ($m \ll n$). For example, in the case study, n is more than one million while m is less than one thousand. The total runtime for simulation is easy to compute. In the case study, the surrogate model takes less than one second to calculate the results. However, simulations using EnergyPlus may take more than 1.5 min. The simulation time is significantly reduced.

$$R = \frac{T_{cal} + T_{train}}{t_s \times n} = \frac{t_m \times n + t_s \times m}{t_s \times n} \approx \frac{t_m}{t_s} \tag{3}$$

where R = Ratio between the runtime using surrogate model and the runtime using simulated based approach; T_{cal} = The total calculation time for surrogate model to generate the probability distribution of the energy savings; T_{train} = The total training time for training the surrogate model using simulated data; t_m = Runtime using surrogate model to generate a single result; t_s = Runtime for a single simulation; n = The number of sampled inputs for surrogate model; m = The number of sampled inputs for training surrogate model.

4. Conclusions

Dynamic simulation can be performed for energy efficient building retrofits in order to predict the effect of retrofit interventions. The large number of simulations done in this process results in long simulation times, and underestimates or overestimates the risk factors because the risk factor space is not correctly sampled. In this research, we developed a method that uses surrogate models as an alternative to current methods that are usually done using physics based simulation tools. The surrogate model uses adaptive sampling to reduce the sample space while maintaining prediction accuracy. The model can predict performance without repeatedly sampling the risk parameters. The probability distributions of the risk factors are evaluated by using Monte Carlo simulation coupled with the surrogate model.

This research targets four main risk factors during a lighting retrofit decision: luminaire input wattage, weather or sky condition, occupancy pattern, and lighting control strategy. Weather distribution data is sampled using historical data and the Sandia method. Occupancy patterns are generated based on a stochastic occupancy model. Two luminaire types (existing luminaire and retrofit luminaire with advanced lighting controls) are evaluated with an input wattage distribution. To test the feasibility and accuracy of this model, a case study was performed using a standard commercial reference office building. The neural network model selected from the surrogate modeling demonstrates good performance. The model can accurately predict lighting and HVAC energy consumption and electricity demand. The probability distribution of retrofit performance is evaluated. The lighting control strategy affects the magnitude of building performance. The lighting energy consumption and HVAC energy consumption show different trends for the two control strategies. The four risk factors affect the distribution function of the building performance and are quantified to provide feedback for decision-makers.

There are some limitations in this research. For example, the surrogate models were validated during the training process. However, the energy savings results (e.g., Figure 9) calculated using surrogate model are not validated and compared with the results using EnergyPlus simulation because it is practically impossible to simulate all the cases (a few million samples) using EnergyPlus. Future work includes adding a ranking function to quantify under which conditions which factor has a higher risk than others, and exploring the financial analysis of the retrofit decision.

Acknowledgements

The authors gratefully acknowledge the support and funding from Philips Research North America.

Author Contributions

Jia Hu and Eric Shen conceived and designed the research methods and experiments. Jia Hu performed main experiments and data analysis. Eric Shen and Yun Gu actively contributed to experiment setup and data analysis.

Conflicts of Interest

The authors declare no conflict of interest.

References

1. Heo, Y.; Choudhary, R.; Augenbroe, G.A. Calibration of building energy models for retrofit analysis under uncertainty. *Energy Build.* **2012**, *47*, 550–560.
2. Asadi, E.; da Silva, M.G.; Antunes, C.H.; Dias, L. Multi-objective optimization for building retrofit strategies: A model and an application. *Energy Build.* **2012**, *44*, 81.
3. Murray, S.N.; Walsh, B.P.; Kelliher, D.; O'Sullivan, D.T.J. Multi-variable optimization of thermal energy efficiency retrofitting of buildings using static modelling and genetic algorithms—A case study. *Build. Environ.* **2014**, *75*, 98–107.
4. Tian, W. A review of sensitivity analysis methods in building energy analysis. *Renew. Sustain. Energy Rev.* **2013**, *20*, 411–419.
5. Lam, J.C.; Wan, K.K.W.; Yang, L. Sensitivity analysis and energy conservation measures implications. *Energy Convers. Manag.* **2008**, *49*, 3170–3177.
6. Eisenhower, B.; O'Neill, Z.; Fonoberov, V.A.; Mezić, I. Uncertainty and sensitivity decomposition of building energy models. *J. Build. Perform. Simul.* **2012**, *5*, 171–184.
7. Shen, E.; Hu, J.; Patel, M. Comparison of Energy and Visual Comfort Performance of Independent and Integrated Lighting and Daylight Controls Strategies. In *ACEEE Summer Study on Energy Efficiency in Industry*; American Council for an Energy-Efficient Economy: New York, NY, USA, 2013.
8. Shen, E.; Hu, J.; Patel, M. Energy and visual comfort analysis of lighting and daylight control strategies. *Build. Environ.* **2014**, *78*, 155–170.
9. Olbina, S.; Hu, J. Daylighting and thermal performance of automated split-controlled blinds. *Build. Environ.* **2012**, *56*, 127.
10. Reinhart, C.F.; Herkel, S. The simulation of annual daylight illuminance distributions—A state-of-the-art comparison of six RADIANCE-based methods. *Energy Build.* **2000**, *32*, 167–187.
11. Hu, J.; Olbina, S. Radiance-based Model for Optimal Selection of Window Systems. In Proceedings of the ASCE 2012 International Conference on Computing in Civil Engineering, Clearwater Beach, FL, USA, 17 June 2012.
12. Hu, J.; Olbina, S. Simulation-based model for integrated daylighting system design. *J. Comput. Civil Eng.* **2013**, *28*, doi:10.1061/(ASCE)CP.1943-5487.0000336.
13. Hu, J.; Olbina, S. Illuminance-based slat angle selection model for automated control of split blinds. *Build. Environ.* **2011**, *46*, 786–796.
14. Magnier, L.; Haghighat, F. Multiobjective optimization of building design using TRNSYS simulations, genetic algorithm, and artificial neural network. *Build. Environ.* **2010**, *45*, 739–746.
15. Hu, J.; Olbina, S. An Expert System Based on OpenStudio Platform for Evaluation of Daylighting System Design. In Proceedings of the ASCE International Workshop on Computing in Civil Engineering, Los Angeles, CA, USA, 23–25 June 2013; pp. 186–193.
16. Gagne, J.M.L.; Andersen, M.; Norford, L.K. An interactive expert system for daylighting design exploration. *Build. Environ.* **2011**, *46*, 2351–2364.
17. Jentsch, M.F.; Bahaj, A.S.; James, P.A. Climate change future proofing of buildings—Generation and assessment of building simulation weather files. *Energy Build.* **2008**, *40*, 2148–2168.

18. Yun, G.Y.; Kim, H.; Kim, J.T. Effects of occupancy and lighting use patterns on lighting energy consumption. *Energy Build.* **2012**, *46*, 152–158.
19. Forrester, A.I.J.; Keane, A.J. Recent advances in surrogate-based optimization. *Prog. Aerosp. Sci.* **2009**, *45*, 50–79.
20. Neto, A.H.; Fiorelli, F.A.S. Comparison between detailed model simulation and artificial neural network for forecasting building energy consumption. *Energy Build.* **2008**, *40*, 2169–2176.
21. Wilcox, S.; Marion, W. *Users Manual for TMY3 Data Sets*; Technical Report NREL/TP-581-43156; National Renewable Energy Laboratory: Golden, CO, USA, 2008.
22. Ebrahimpour, A.; Maerefat, M. A method for generation of typical meteorological year. *Energy Convers. Manag.* **2010**, *51*, 410–417.
23. SUMO Lab SUrrogate MOdeling (SUMO) Toolbox. Available online: http://www.sumo.intec.ugent.be/SUMO (accessed on 1 August 2013).
24. U.S. Department of Energy Commericial Reference Buildings. Available online: http://energy.gov/eere/buildings/commercial-reference-buildings (accessed on 1 September 2013).
25. American Society of Heating, Refrigerating and Air Conditioning, Inc (ASHREA). *Energy Standard for New Buildings except Low-Rise Residential Buildings*; ANSI/ASHRAE/IESNA Standard 90.1-1989; ASHREA: Atlanta, GA, USA, 1989.

14

Renewable Substitutability Index: Maximizing Renewable Resource Use in Buildings

Ravi S. Srinivasan [1,*], Daniel E. Campbell [2,†] and Wei Wang [1,†]

[1] M.E. Rinker, Sr. School of Construction Management, University of Florida, Gainesville, FL 32611, USA; E-Mail: swingingbach@gmail.com

[2] U.S. Environmental Protection Agency, Office of Research and Development, National Health and Environmental Effects Research Laboratory, Atlantic Ecology Division, Narragansett, RI 02882, USA; E-Mail: campbell.dan@epa.gov

† These authors contributed equally to this work.

* Author to whom correspondence should be addressed; E-Mail: sravi@ufl.edu

Academic Editor: Elma Durmisevic

Abstract: In order to achieve a material and energy balance in buildings that is sustainable in the long run, there is an urgent need to assess the renewable and non-renewable resources used in the manufacturing process and to progressively replace non-renewable resources with renewables. Such progressive disinvestment in the non-renewable resources that may be substituted with renewable resources is referred to as "Renewable Substitutability" and if implemented, this process will lead to a paradigm shift in the way building materials are manufactured. This paper discusses the development of a Renewable Substitutability Index (RSI) that is designed to maximize the use of renewable resources in a building and quantifies the substitution process using solar emergy (*i.e.*, the solar equivalent joules required for any item). The RSI of a building or a building component, *i.e.*, floor or wall systems, *etc.*, is the ratio of the renewable resources used during construction, including replacement and maintenance, to the building's maximum renewable emergy potential. RSI values range between 0 and 1.0. A higher RSI achieves a low-energy building strategy promoting a higher order of sustainability by optimizing the use of renewables over a building's lifetime from formation-extraction-manufacturing to maintenance, operation, demolition, and recycle.

Keywords: Renewable Substitutability Index (RSI); emergy analysis; Renewable Emergy Balance; environmental building design

1. Introduction

Environmental evaluation of buildings has played an essential role in conserving resource use and improving overall performance. Approaches such as Life Cycle Assessment (LCA) and other ecological accounting models offer environmental decision-making solutions. While on one hand, LCA methods explicitly include manufacturing and maintenance energies or costs, they exclude much of the environmental work provided by the biosphere in the formation and concentration of resources, particularly raw materials used in the manufacture of building materials. With respect to inventory data and aggregation methods, LCA primarily focuses on emissions, and direct and indirect consumption of resources through every stage of a product's life cycle. This involves a cradle-to-grave approach to evaluating the product or process, since its environmental impact is assessed from the point of obtaining the raw materials for product development through production, distribution, and its final disposal or recycling [1,2].

In the building sector, embodied energy of the building materials and energy consumption and the associated emissions over the period of use were evaluated using the LCA method [3]. While almost all of the embodied energy is accumulated during the construction phase of the building, energy used for operating the building, *i.e.*, electricity, chilled water, water, steam, wastewater treatment, *etc.*, constitutes the operational phase of the building. A study by Sartori and Hestnes [4] revealed a linear relation between operating and total energy use across sixty cases in a survey of nine countries. A similar study by Ramesh *et al.* [5] showed that most of the building lifecycle energy is expended during its operational phase. In [6], a methodological approach was proposed to design and evaluate eco-design of buildings. Besides concluding that the operational phase had the highest contribution over the entire lifecycle of the three buildings studied in this paper, Asdrubali *et al.* implemented energy optimizations to reduce operational energy use. Nevertheless, the use of LCA for buildings is not as ubiquitous as in the manufacturing industry. Moreover, issues such as setting system boundaries, procedural responsibility, allocation of inputs used (resource consumption), undesired outputs (emissions), and heavy dependency on data collection expenses and data quality, in some cases impose restrictions on the application of LCA. Besides, most of the ecosystem services that are required for dispersing the emissions and absorbing their impact are not specified in the LCA methods and the links between LCA and natural resource depletion are largely missing [7]. Moreover, lacking a rigorous thermodynamic framework in LCA leads to the incompleteness and inaccuracy of ecosystem analysis and in certain situations may even violate thermodynamic laws [8].

On the other hand, emergy analysis, as an ecological accounting model offers environmental decision-making solutions, which include ecological considerations related to material and energy flows during a building's entire lifetime. Emergy was developed to systematically include all the work required for the preparation and delivery of a resource or service [9]. In emergy analysis, first, a system diagram of the process or system under consideration is constructed. This is followed by accounting for all inputs and outflows over the lifespan of the material or product, including an inventory of essential processes, storages, and flows [10]. Emergy analysis uses thermodynamic principles and offers a "total environmental analysis" that includes all environmental energies involved in the system under investigation. A few research efforts have focused on assessing built environments using emergy analysis. They are: evaluation of recycling and reuse of building materials [11]; emergy associated with the operation of a building [12]; building manufacturing, maintenance, and development of Em-building indices [13]; emergy evaluation of a green façade [14]; comparison of life cycle methodologies [15]; emergy-energy assessment of buildings [16]. Among these, an energy and emergy based cost-benefit evaluation of building envelopes relative to geographical location and climate [17] focused on the building envelope. Also, an emergy-based method to assess the Renewable Emergy Balance (REB) of a building to guide movement towards the highest order of sustainability was developed and demonstrated using a case study building [18].

The central aspect of a REB is the estimation of the maximum renewable emergy potential of the building. This quantity is a limit, also a moving target, which improves as the technology improves to integrate and/or generate more renewable resources. Moreover, this limit sets a benchmark that is required to achieve a higher level of sustainability in buildings.

In order to strive for material and energy balance in buildings, there is an urgent need to assess the renewable and non-renewable resources used in the manufacturing process and to progressively replace non-renewable resources with renewables. Such progressive disinvestment in non-renewable resources that may be substituted with renewable resources is referred to as Renewable Substitutability (RS), and if implemented, will lead to a paradigm shift in the way building materials are manufactured [18]. For this paper, emergy analysis was chosen as the tool to assess the resources used in building and to develop the Renewable Substitutability Index (RSI). This paper develops a Renewable Substitutability Index to maximize renewable resource use in buildings using emergy analysis as the means of assessment. A case study of an existing building in a university campus setting was used to demonstrate the application of RSI to attain maximum renewable resource investment. This paper is organized as follows: while Section 2 discusses the concept of RSI and its significance, Section 3 uses a case study building in a university campus setting to apply the concept of RSI and to discover how this index can be used for optimal decision-making during material replacement.

2. Renewable Substitutability Index (RSI) of a Building

A building is supported by a variety of inputs and outputs, some purchased and others provided free (*i.e.*, without explicit cost) by the environment. Inputs include embodied energies of all kinds and qualities, from the sun shining on the window to the ancient sunlight expended in the formation of the fossil fuels used to prepare the glass through which the sun is shining. By translating all inputs into emergy with a common energy unit, the solar emjoule (sej), meaningful comparisons can be established between renewable and non-renewable resources, and between natural and manmade energies, including refining the assessment of real costs over time. The quasi-sustainability principle as discussed by Holling [19] and Odum [20,21] can be extended to buildings, *i.e.*, related to renew-non-renew substitution [18]. Theoretical possibilities of using non-renewables to take advantage of renewable resources have been discussed by Bastianoni *et al.* [22].

In other words, as emergy is accounted for in a particular system, the renewability and non-renewability of materials are appropriately identified. This, then, is used to develop Renewable Substitutability, followed by, calculating RSI values for decision-making. For example, the electric energy from coal used in a cement manufacturing unit might be replaced with electricity generated from renewable resources such as solar, kinetic energy of wind, water or geothermal heat. Thus, the use of non-renewable resources to improve the system's capacity to exploit renewable resources will aid the development of a quasi-sustainable solution, over the long run. Such non-renewable resources may be replaced with renewable resources progressively. Figure 1 shows a simplified systems diagram.

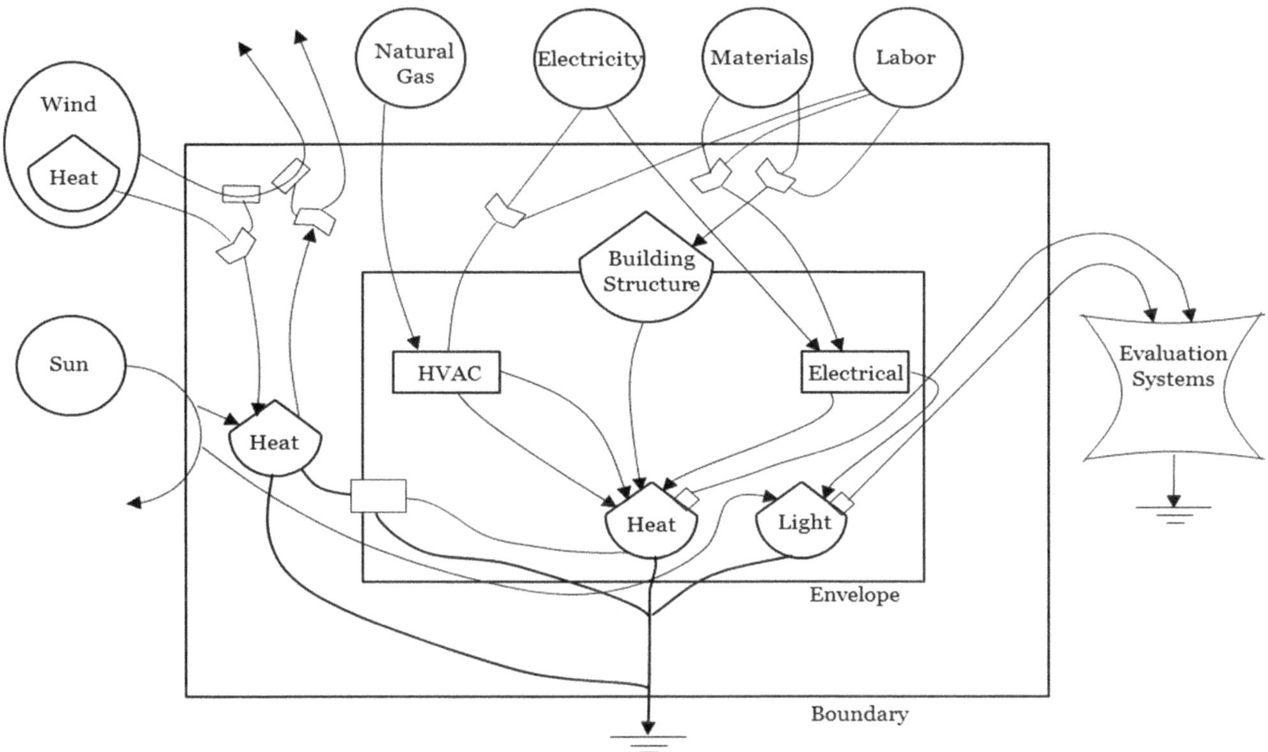

Figure 1. Typical systems diagram of a building. Only building structure was used for this study.

Using emergy analysis and through the identification of the Renewable Substitutability of all non-renewable resources, the emergy content may be split into non-renewables with Renewable Substitutability (RS) and unsubstitutable Non-Renewables (NR), *i.e.*, non-renewable resources that cannot be substituted with renewables using the best available technology. Detailed steps and explanation of developing RS and NR for building materials is discussed in Srinivasan *et al.* [18]; Table A.1 in the paper of Srinivasan *et al.* [18] lists the RS and NR for a subset of building materials. Non-renewable resources with Renewable Substitutability have, at least, a potential to be replaced by Renewables (R) and this should be exploited to attain more sustainable buildings. Thus, the Renewable Substitutability Index in a building's useful life is defined as the ratio of the Renewables used in construction, including replacement and maintenance, to its maximum renewable emergy potential. The maximum renewable potential is sum of R and RS.

$$\text{Renewable Substitutability Index} = \sum R_n (R_n + RS_n)^{-1} \quad (1)$$

where R_n and $(R_n + RS_n)$ are the amount of renewable materials used and the maximum renewable potential remaining of *n* materials.

A building structure may be described as a collection of element-level (e.g., brick, wood flooring, *etc.*) or component-level (e.g., building envelope, HVAC, *etc.*) parts. RSI values range between 0 and 1.0. Thus, by making progressive improvements, over the useful life of a building, all non-renewable resources with Renewable Substitutability can be replaced by renewables, and the RSI of the building will approach 1.0, at which point it will achieve a Renewable Emergy Balance. The central aspect of RSI is that it can be used to evaluate improvement strategies during the operation and maintenance phases of existing structures. RSI can also be used to compare element-level building data such as brick, concrete, steel, aluminum, ceramic, wood, *etc.*

Figure 1 compares element-level data; the *X*- and *Y*- axes represent Renewables (R) and maximum renewable emergy potential (R + RS) respectively. The highest, or optimal, RSI is achieved if the elemental data falls on the bisector, *i.e.*, when R equals (R + RS), *i.e.*, RSI equals 1.0. For example, in Figure 2 (refer to Section 3 for RSI calculation), concrete (RSI, 0.91) falls closer to the bisector when compared to brick (RSI, 0.89). In other words, concrete has a higher renewable emergy potential compared to brick. Similarly, aluminum (RSI, 0.99) has the highest renewable emergy potential compared to concrete and brick. A similar analysis can be performed at a component-level, for example, groundwork, building frame, external wrapping (sidewalls and facades), internal walls, pavements and coverings, doors and windows, sheet-metal works, and drainage systems, Figure 3. Compared to other building components, "pavements and coverings" possess high RSI indicating that the element-level materials of the components (*i.e.*, carpet, linoleum, red brick, rubber, enamel and vinyl acrylic latex) hold high renewable emergy potential and the most desirable renewable resource recovery rate.

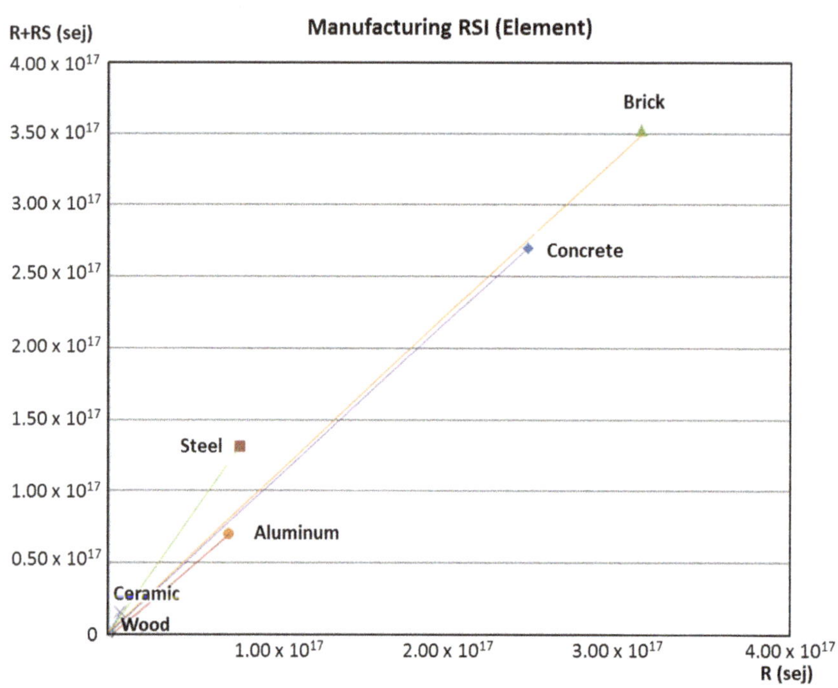

Figure 2. Renewable Substitutability Index (RSI): Element-level.

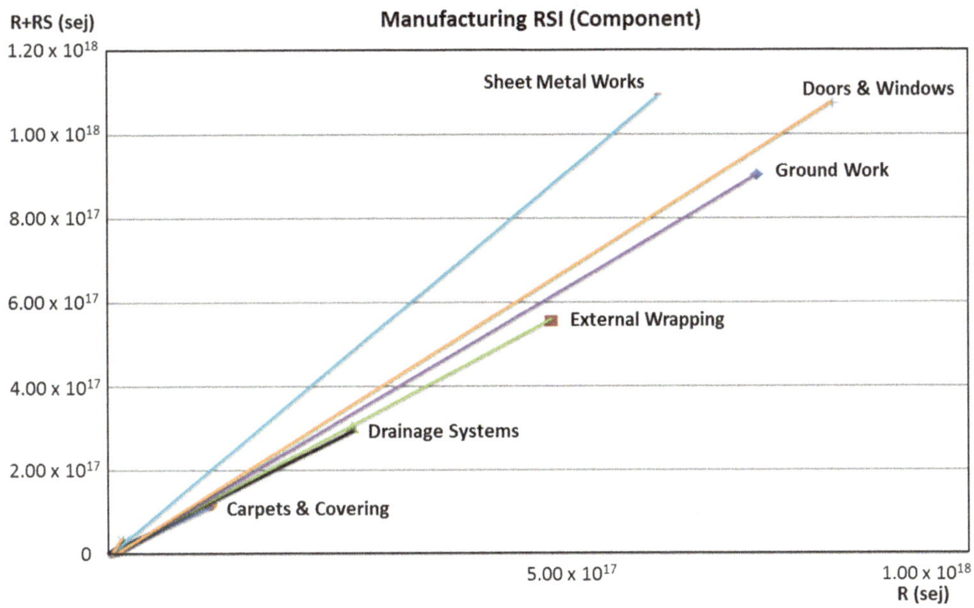

Figure 3. RSI: Component-level.

As the RSI of particular element- or component- level data and the associated design decision that affects the building's overall performance tends to move towards 1.0, it significantly reduces the total non-renewable energy use of a building over its entire useful life through progressive disinvestment of all substitutable non-renewable resources in a building, thereby, achieving a low- non-renewable energy building with a more sustainable status. Using this index, improvement strategies for existing buildings can be evaluated from a holistic perspective in order to maximize renewable resource use. The following section discusses a case study of an existing facility in a university campus setting that is used to demonstrate the application of RSI to strive for the maximum renewable resource investment.

3. RSI of an Existing Building: Case Study

Completed in January 2003, Rinker Hall is a three-story, 47,300-square-foot building that houses the School of Construction Management at the University of Florida, Figure 4. As the first gold standard LEED™ building in the State of Florida, Rinker Hall sets a high bar for design excellence and it incorporates a range of green building features to provide a productive, healthy environment for students, staff, and faculty [23]. Located on the site of a former parking lot, Rinker Hall includes a mix of classrooms, construction labs, teaching labs, faculty and administrative offices, and student facilities. The building was designed to maximize natural light with lighting controls such as motion detectors and low-flow plumbing fixtures and waterless urinals were installed to promote efficiency of water use. Building materials were reviewed for proximity in manufacturing, sustainable harvesting, renewable-resource content, *i.e.*, by using reused bricks from demolition, recycled cellulose insulation material, local and regional assemblies, certified wood, and renewable flooring material. During the operation phase, the building's energy is supplied by three separate energy sources: electricity, chilled water, and steam.

Figure 4. (**Left**) Exterior view of Rinker Hall; (**Right**) Typical classroom (Photo credits: Croxton Collaborative Architects).

The facility's Renewable Substitutability was first evaluated by studying the building construction, operation, and maintenance phases. Using this data, RSI values, for both elemental- and component-levels, were calculated. These represented each of the retrofit options. These options were plotted to identify optimal solutions for progressive replacement of non-renewable resources with renewables. Technical documents such as construction plans, finish schedules, and commissioning reports were used to generate an inventory of inputs. The inventory included a list of materials used and their mass. For this paper, a few assumptions were used, namely: (i) land use was assumed to be a renewable resource because the site was previously a parking lot and not virgin land; (ii) demolition, site pavement, and human services were excluded; (iii) for materials whose specific emergy values were unavailable, RS was assumed to be 0 (zero), *i.e.*, NR is 1.0; (iv) for materials whose total emergy was unavailable, R was assumed to be 0 (zero) such that the total emergy is (RS + NR); and (v) the RS of operational energy is assumed as 10.6% based on DOE (2010) data, *i.e.*, total electricity generation in the United States included 10.6% renewable generation.

3.1. RSI Assessment Methodology

The RSI assessment methodology is comprised of four steps. While the first three steps estimate the emergy content during the building's manufacturing, operation, and maintenance phases, the final step calculates the RSI based on the data from the previous steps. The resultant RSI values can be used to evaluate improvement strategies from a holistic perspective in order to maximize renewable resource content of the existing facility.

3.1.1. Step 1: Emergy Analysis of the Building Manufacturing Process

In this first step, the emergy content of the building materials used during the construction process were quantified. Material emergy is the product of the material quantity derived from contract documents (mass) and specific emergy (sej/kg). In order to identify the Renewable Substitutability portions of materials, the specific emergy was split into three portions namely R, RS, and NR. For this study, the materials were aggregated into ten major categories and each category was further separated into various units for clarity, Table 1. In the case of Rinker Hall building, future replacement and maintenance efforts may be targeted based on the percentage of Renewable Substitutability of the components. For example, the roofs have 71.1% of RS potential.

Table 1. Emergy analysis of building manufacturing (RS—Renewable Substitutability; NR—Non Renewables; R—Renewables; El—Elemental; Cp—Component).

Item	Specification	Unit	Specific Emergy (sej/kg)				RSI		% of RS
			RS	NR	Total (R + RS + NR)	R	R/(R + RS) El.	R/(R + RS) Cp.	
Solar irradiation	Irradiation on building yard	J	0	0	1.00	1.00	1.00	1.00	0.00
Land use (soil erosion)	Soil Organic matter (3% of 1 m depth vol.)	J	1.24×10^{5}	0	1.24×10^{5}	0	0	0	1.91
Groundwork								0.84	0.50
Foundation	Concrete	kg	3.47×10^{10}	1.41×10^{12}	1.81×10^{12}	$3.65 \times 10^{1'}$	0.91		
Footing	Steel	kg	1.17×10^{12}	4.18×10^{12}	6.97×10^{12}	1.62×10^{12}	0.58		
Lean concrete	Concrete	kg	3.47×10^{10}	1.41×10^{12}	1.81×10^{12}	$3.65 \times 10^{1'}$	0.91		
Lean concrete	Steel	kg	1.17×10^{12}	4.18×10^{12}	6.97×10^{12}	1.62×10^{12}	0.58		
Building frame								0.84	0.68
Bearing wall	Concrete	kg	3.47×10^{10}	1.41×10^{12}	1.81×10^{12}	3.65×10^{11}	0.91		
Beams and columns	Concrete	kg	3.47×10^{10}	1.41×10^{12}	1.81×10^{12}	$3.65 \times 10^{1'}$	0.91		
Exterior structure framing	Steel	kg	1.17×10^{12}	4.18×10^{12}	6.97×10^{12}	1.62×10^{12}	0.58		
Gypsum dry construction Framing	Steel	kg	1.17×10^{12}	4.18×10^{12}	6.97×10^{12}	1.62×10^{12}	0.58		
Overhangs	Concrete	kg	3.47×10^{10}	1.41×10^{12}	1.81×10^{12}	3.65×10^{11}	0.91		
Stairs	Concrete	kg	3.47×10^{10}	1.41×10^{12}	1.81×10^{12}	3.65×10^{11}	0.91		
Elevator Box	Concrete	kg	3.47×10^{10}	1.41×10^{12}	1.81×10^{12}	$3.65 \times 10^{1'}$	0.91		
Elevator Box	Steel	kg	1.17×10^{12}	4.18×10^{12}	6.97×10^{12}	1.62×10^{12}	0.58		
External wrapping (side walls + facades)								0.92	0.14
Side wall (16" thick)	CMU	kg	0	1.00	1	0	0		
Side walls thermal insulation	fiberglass	kg	0	1.00	1.00	0	0		
Binder (Side wall)	Mortar	kg	0	1.00	3.31×10^{12}	3.31×10^{12}	1		
Precast side wall coping	Concrete	kg	3.47×10^{10}	1.41×10^{12}	1.81×10^{12}	3.65×10^{11}	0.91		

Table 1. *Cont.*

Item	Specification	Unit	Specific Emergy (sej/kg)				RSI		% of RS
			RS	NR	Total (R + RS + NR)	R	R/(R + RS)		
							El.	Cp.	
Precast screen wall coping	Concrete	kg	3.47×10^{10}	1.41×10^{12}	1.81×10^{12}	3.65×10^{11}	0.91		
Façades (External skin)	CMU	kg	0	1.00	1.00		0		
Façades	Brick	kg	1.89×10^{11}	2.03×10^{12}	3.68×10^{12}	1.46×10^{12}	0.89		
Binder	Mortar	kg	0	1.00	3.31×10^{12}	3.31×10^{12}	1.00		
Thermal insulation	Extruded polystyrene	kg	0	1.00	1.00	0	0		
Thermal insulation	Fiberglass	kg	0	1.00	1.00	0	0		
Filling	Concrete	kg	3.47×10^{10}	1.41×10^{12}	1.81×10^{12}	3.65×10^{11}	0.91		
Floors								0.60	23.2
Floor (20 cm thick)	Concrete	kg	3.47×10^{10}	1.41×10^{12}	1.81×10^{12}	3.65×10^{11}	0.91		
Floor (3-3/8" & 7 3/8" thick)	Red Brick	kg	1.89×10^{11}	2.03×10^{12}	3.68×10^{12}	1.46×10^{12}	0.89		
Insulation	High-density polyethylene	kg	0	1.00	8.85×10^{12}	8.85×10^{12}	1.00		
Vapor barrier	Concrete	kg	3.47×10^{10}	1.41×10^{12}	1.81×10^{12}	3.65×10^{11}	0.91		
Tile	Ceramic	kg	1.03×10^{12}	2.03×10^{12}	3.68×10^{12}	6.20×10^{11}	0.38		
Floor deck	Steel	kg	1.17×10^{12}	4.18×10^{12}	6.97×10^{12}	1.62×10^{12}	0.58		
Roof								0.58	71.1
Roof	Gypsum	kg	0	1.00	1.00	0	0		
Roof (Shingle-ply)	Built-up 3 ply & gravel	kg	1.16×10^{12}	4.76×10^{11}	1.64×10^{12}	0	0		
Roof blocking	Wood	kg	7.98×10^{11}	8.70×10^{10}	2.40×10^{12}	1.52×10^{12}	0.65		
Electro welding net	Steel	kg	1.17×10^{12}	4.18×10^{12}	6.97×10^{12}	1.62×10^{12}	0.58		
Metal roof deck	Steel	kg	1.17×10^{12}	4.18×10^{12}	6.97×10^{12}	1.62×10^{12}	0.58		
Internal walls								0.60	0.04
Wall	Gypsum	kg	0	1.00	1.00	0	0		
Wall	CMU	kg	0	1.00	1.00	0	0		
Tape and finish	Fiberglass	kg	0	1.00	1.00	0	0		
Binder	Mortar	kg	0	1.00	3.31×10^{12}	3.31×10^{12}	1.00		
Tile	Ceramic	kg	1.03×10^{12}	2.03×10^{12}	3.68×10^{12}	6.20×10^{11}	0.38		
Pavements and coverings								0.99	0.01
Carpeting	Carpet	kg	0	1.00	1.00	0	0		
Resilient flooring	Linoleum	kg	0	1.00	1.00	0	0		
Ext. pavement	Red brick	kg	1.89×10^{11}	2.03×10^{12}	3.68×10^{12}	1.46×10^{12}	0.89		

Table 1. *Cont.*

Item	Specification	Unit	Specific Emergy (sej/kg)				RSI		% of RS
			RS	NR	Total (R + RS + NR)	R	R/(R + RS)		
							El.	Cp.	
Cove base	Rubber	kg	0	1.00	7.22×10^{12}	7.22×10^{12}	1.00		
Tread and riser	Rubber	kg	0	1.00	7.22×10^{12}	7.22×10^{12}	1.00		
Tile	Rubber	kg	0	1.00	7.22×10^{12}	7.22×10^{12}	1.00		
Walls and ceilings	Enamel	kg	0	1.00	1.00	0	0		
Walls and ceilings	Vinyl acrylic latex	kg	5.29×10^{9}	6.32×10^{12}	6.33×10^{12}	0\	0		
Doors and windows								0.78	0.78
Door	Hollow metal	kg	0	1.00	1.00	0	0		
Door	Glass	kg	6.22×10^{12}	1.65×10^{12}	7.87×10^{12}	0	0		
Door	Wood	kg	7.98×10^{11}	8.70×10^{10}	2.40×10^{12}	1.52×10^{12}	0.65		
Door frame	Hollow metal	kg	0	1.00	1.00	0	0		
Door frame	Wood	kg	7.98×10^{11}	8.70×10^{10}	2.40×10^{12}	1.52×10^{12}	0.65		
Door frame	Aluminum	kg	4.70×10^{10}	1.27×10^{13}	2.13×10^{13}	8.55×10^{12}	0.99		
Window	Glass	kg	6.22×10^{12}	1.6×10^{12}	7.87×10^{12}	0	0		
Int. window casing frame	Wood	kg	7.98×10^{11}	8.70×10^{10}	2.40×10^{12}	1.52×10^{12}	0.65		
Ext. window casing frame	Aluminum	kg	4.70×10^{10}	1.27×10^{13}	2.13×10^{12}	8.55×10^{12}	0.99		
Int. window casings	Wood	kg	7.98×10^{11}	8.70×10^{10}	2.40×10^{12}	1.52×10^{12}	0.65		
Ext. window casings	Aluminum	kg	4.70×10^{10}	1.27×10^{13}	2.13×10^{13}	8.55×10^{12}	0.99		
Sheet-metal works								0.58	1.54
Tube	Steel	kg	1.17×10^{12}	4.18×10^{12}	6.97×10^{12}	1.62×10^{12}	0.58		
Structural tees	Steel	kg	1.17×10^{12}	4.18×10^{12}	6.97×10^{12}	1.62×10^{12}	0.58		
Channel	Steel	kg	1.17×10^{12}	4.18×10^{12}	6.97×10^{12}	1.62×10^{12}	0.58		
Angles	Steel	kg	1.17×10^{12}	4.18×10^{12}	6.97×10^{12}	1.62×10^{12}	0.58		
Plates, bent and flat	Steel	kg	1.17×10^{12}	4.18×10^{12}	6.97×10^{12}	1.62×10^{12}	0.58		
Drainage system								0.94	0.06
Standard PVC pipe	PVC	kg	0	1.00	9.86×10^{12}	9.86×10^{12}	1.00		
Junction box	Concrete	kg	3.47×10^{10}	1.41×10^{12}	1.81×10^{12}	3.65×10^{11}	0.91		
Reinforced concrete pipe	Concrete	kg	3.47×10^{10}	1.41×10^{12}	1.81×10^{12}	3.65×10^{11}	0.91		
Ductile iron pipe	Iron	kg	0	1.00	6.97×10^{12}	6.97×10^{12}	1.00		
Covering	Concrete	kg	3.47×10^{10}	1.41×10^{12}	1.81×10^{12}	3.65×10^{11}	0.91		

3.1.2. Step 2: Emergy Analysis of the Building Operation Process

In this step, energy use during building operation was assessed. Energy use data was obtained from the university's Physical Plant Division. For this building, steam is used primarily for space heating. Owing to an insignificant quantity of steam used, it was not included. Emergy for the operational phase is a product of the actual usage data of energy sources and their respective transformities. For this study, year 2010 energy data was used. Using the above data, total emergy flows including the Renewable Substitutability portions of the operational energy were calculated, Table 2.

Table 2. Emergy analysis of building operation.

Operational Energy Use	Unit (MBtu)	Emergy (sej)	Renewable Resource (R)	Renewable Substitutability Index (RSI)
Electrical	1446.5	2.21×10^{12}	3.35×10^{10}	0.11
Chilled Water (Cooling)	3864	5.91×10^{12}	8.94×10^{10}	0.11
Total	**5310.5**	$\mathbf{8.12 \times 10^{12}}$	$\mathbf{1.23 \times 10^{11}}$	**0.11**

3.1.3. Step 3: Emergy Analysis of the Building Maintenance Process

In this step, the depreciation of the stored emergy associated with building elements such as windows and doors was evaluated. For this paper, replacement of glass based on the building maintenance schedule was studied. Assuming a 100-year useful life for the building, the conventional float glass is replaced with traditional recycled float glass product every 30 years. As the Renewable Substitutability of the replacement product is larger than the ones that will be replaced, the overall Renewable Substitutability of the building is slated to improve, Table 3.

Table 3. Emergy analysis of building maintenance (using windows as an example).

Building Element	Lifetime (years)	Specific Emergy (sej/kg)		
		Non-Renewables (NR)	Total	Renewable (R)
Window	30	1.65×10^{12}	7.87×10^{12}	0
		1.04×10^{12}	7.69×10^{12}	0

3.1.4. Step 4: RSI Calculation

In this final step, the renewable emergy potential at the component-level and of the overall building is assessed to determine the respective RSI values, Table 4. It is to be noted that the RSI of a material remained constant relative to its specific emergy irrespective of its mass quantity. Alternatively put, a material retains its renewable emergy potential, which is derived based on its manufacturing process and does not alter when used in varied forms. This offers an opportunity to seek materials that have both higher Renewable Substitutability and are renewable (*i.e.*, RSI → 1.0), refer Table 1.

Table 4. Emergy analysis of energy efficient measures.

Energy Efficient Measures (EEM)	Energy Use	Improvement (Saving)		Emergy	Improvement Emergy (Saving)	Renewable Resource (R)
	MBtu	MBtu	%	R (sej)	R (sej)	10.6% Renew (sej)
EEM1: *Change lighting power density—ASHRAE 90.1 space-by-space application*	5214.1	96.4	0.018	1.14×10^{12}	2.11×10^{10}	1.21×10^{11}
EEM2: *Replace inefficient refrigerator—with Energy Star performer*	5221.7	88.8	0.017	1.14×10^{12}	1.94×10^{10}	1.21×10^{12}
EEM3: *Modify temperature setpoint to 76F—SHRAE 90.1 as long as 80% of occupants are comfortable*	5219.4	91.1	0.017	1.14×10^{12}	1.99×10^{10}	1.21×10^{11}
EEM4: *Install 20KW PV system—total area 2000 ft^2 on the roof*	5221.1	89.4	0.017	1.14×10^{12}	1.95×10^{10}	1.21×10^{11}
EEM5: *Shorten Hours of Operation to 7 a.m.-9 p.m.—modify California Alternative Calculation Method (ACM) Manual from 6 a.m. to 10 p.m.*	4633.7	676.8	0.127	1.01×10^{12}	1.48×10^{11}	1.07×10^{11}
EEM6: *Improve Energy Recovery Ventilation (ERV)—install ERV to handle additional outside air and install premium motor*	5130.5	180.0	0.034	1.12×10^{12}	3.93×10^{10}	1.19×10^{11}
EEM7: *Reduce cooling capacity—specifically during weekends*	4690.1	620.4	0.117	1.02×10^{12}	1.35×10^{11}	1.09×10^{11}
EEM8: *Repair lighting controls—classroom lights do not remain on during the night*	5167.0	143.5	0.027	1.13×10^{12}	3.13×10^{10}	1.20×10^{11}
EEM9: *Install Variable Frequency Displays (VFD)—on both heating hot water pumps*	5230.8	79.7	0.015	1.14×10^{12}	1.74×10^{10}	1.21×10^{11}
EEM10: *Provide higher temperature delta—in chilled water supply and return lines*	5138.1	172.4	0.032	1.12×10^{12}	3.76×10^{10}	1.19×10^{11}
EEM11: *Install shut-off timers—in classroom to prevent 24 h idle load*	3630.5	233.5	0.060	7.93×10^{11}	5.10×10^{10}	8.40×10^{10}
EEM12: *Repair lighting controls—classroom lights do not remain on during the night*	5208.8	101.7	0.019	1.14×10^{12}	2.22×10^{10}	1.21×10^{11}

3.2. Selection of Improvement Scenarios

RSI assessment methodology was used to calculate the RSI values, for both element- and component-levels. Using this data, several scenarios for improvement were evaluated from a holistic perspective in order to maximize renewable resource use. For this paper, the operational and maintenance phases were assessed as discussed below.

3.2.1. Operational Phase

To reduce energy use in building operation, Energy Efficiency Measures (EEMs) are typically considered. Through evaluations of Renewable Substitutability potential of energy sources, an optimum solution that maximizes renewable resource use can be identified. Take, for example, the nine EEMs that were developed based on detailed energy audits and onsite building assessment [24].

Using the energy breakdown of the electricity and chilled water systems (steam was not included owing to insignificant quantities), each of the EEMs' equivalent emergy flows including their renewable resource use were assessed, Table 4. In this study, 10.6% renewable electricity generation was the only renewable operational source for the building. Evaluating the RSI of all EEMs shows that EEM5 approaches 1.0 rapidly when compared with other EEMs, Figure 5. By calculating the ratio of the operational renewable energy use to total energy use, an RSI of 0.11 was calculated and remained a constant for all EEMs. However, if the renewable mix changes, these EEMs will not co-locate in the bisector, rather they will spread across the plot depicting their emergy qualities.

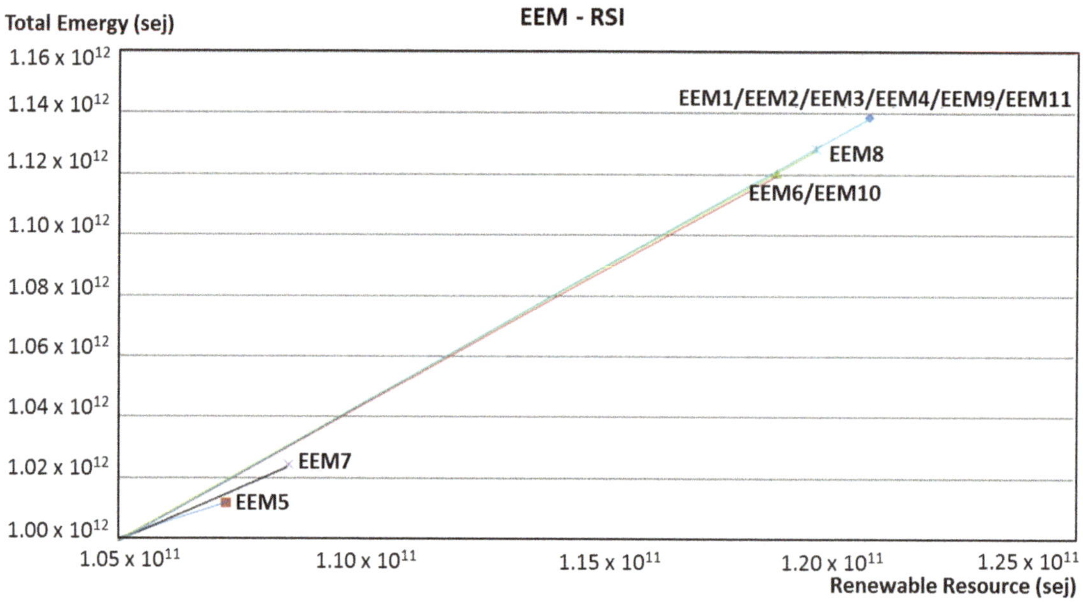

Figure 5. RSI of energy efficient measures.

3.2.2. Maintenance Phase

By realizing the Renewable Substitutability of replacement during the maintenance phase, optimal solutions can be identified. Take, for example, the scenario that includes two replacement glazing, one with 10% renewables and another, 15%, Table 5, Figure 6. The second alternative possesses a higher RSI and is recommended for use as replacement glass.

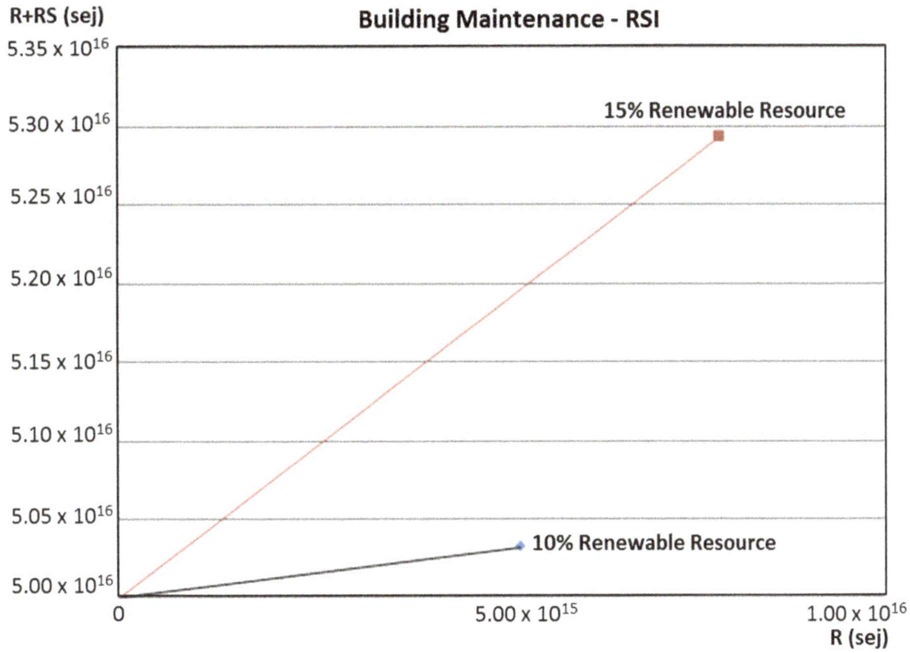

Figure 6. Emergy analysis of 10% and 15% renewable resource of replacement glass.

Table 5. Emergy analysis of renewable substitutability of replacement glass (using 10% and 15% of the total emergy required for manufacture as renewable resource).

Item	Emergy (sej)			%	
	RS	NR	Total	RS	NR
Current Glazing Systems	4.22×10^{16}	1.12×10^{16}	5.33×10^{16}	79%	21%
Replacement Glazing	4.51×10^{16}	7.05×10^{15}	5.22×10^{16}	86%	14%
w/Renewable resources (10%)	4.96×10^{16}	2.53×10^{15}	5.22×10^{16}	95%	5%
w/Renewable resources (15%)	5.19×10^{16}	2.79×10^{15}	5.22×10^{16}	99%	1%

4. Conclusions

This paper discussed the development of a Renewable Substitutability Index to maximize the incorporation of renewable resources in a building using emergy analysis. The central aspect of RSI is that it can be used to evaluate improvement strategies during the operation and maintenance phases of existing structures. A case study of an existing building in a university campus setting was used to demonstrate the application of RSI to attain maximum renewable resource investment. Using emergy (which accounts for energy qualities), the EEMs were assessed both for operational improvement and improvements in the maintenance phase. One of the limiting factors for the use of emergy evaluation in building assessments is the lack of a comprehensive database of building materials' specific emergy values with accurate data on renewable resource use. The configuration of raw materials, as well as, both the renewable and non-renewable portions (e.g., the renewable fraction of electricity) may differ regionally, as compared to national data. This may pose issues when calculating operational energy use and associated quantities of renewable resources. In addition, calculating the Renewable Substitutability of energy sources requires additional effort to provide location-specific values. For this purpose, authors Srinivasan, Campbell, and their team are currently developing an emergy database for building components. Over sixty building components were identified and evaluated for their material compositions in terms of weights, which then are used with transformities to reconstruct the emergy content of a unit weight of the building component.

Conflicts of Interest

Although EPA contributed to this article, the research presented was not performed by or funded by EPA and was not subject to EPA's quality system requirements. Consequently, the views, interpretations, and conclusions expressed in this article are solely those of the authors and do not necessarily reflect or represent EPA's views or policies.

References

1. Guinee, J.B.; Heijungs, R.; Udo de Haes, H.A.; Huppes, G. Quantitative life cycle assessment of products: 2. Classification, valuation and improvement analysis. *J. Clean. Prod.* **1993**, *1*, 81–91.
2. Madu, C.N. *Handbook of Environmentally Conscious Manufacturing*; Kluwer Academic Publishers: Boston, MA, USA, 2001.
3. Zabalza Bribián, I.; Aranda Usón, A.; Scarpellini, S. Life cycle assessment in buildings: State-of-the-art and simplified LCA methodology as a complement for building certification. *Build. Environ.* **2009**, *44*, 2510–2520.
4. Sartori, I.; Hestes, A.G. Energy use in the life cycle of conventional and low-energy buildings: A review article. *Energy Build.* **2007**, *39*, 249–257.
5. Ramesh, T.; Prakash, R.; Shukla, K.K. Life cycle energy analysis of buildings: An overview. *Energy Build.* **2010**, *42*, 1592–1600.
6. Asdrubali, F.; Baldassarri, C.; Fthenakis, V. Life cycle analysis in the construction sector: Guiding the optimization of conventional Italian buildings. *Energy Build.* **2013**, *64*, 73–89.
7. Zhang, Y.; Singh, S.; Bakshi, B.R. Accounting for ecosystem services in life cycle assessment, part I: A critical review. *Environ. Sci. Technol.* **2010**, *7*, 2232–2242.
8. Hau, J.L. Toward Environmentally Conscious Process Systems Engineering via Joint Thermodynamic Accounting of Industrial and Ecological Systems. Ph.D. Thesis, The Ohio State University, Columbus, OH, USA, May 2005.
9. Odum, H.T. *Environmental Accounting: Emergy and Environmental Decision Making*; Wiley: New York, NY, USA, 1996.
10. Brown, M.T.; Williams, S.B.; Tilley, D.R.; Ulgiati, S. Emergy Synthesis: Theory and Applications of the Emergy Methodology. In Proceedings of the First Biennial Emergy Analysis Research Conference, Gainesville, FL, USA, 2–4 September, 1999.
11. Buranakarn, V. Evaluation of Recycling and Reuse of Building Materials Using the Emergy Analysis Method. PhD Thesis, University of Florida, Gainesville, FL, USA, 1998.
12. Meillaud, F.; Gay, J.B.; Brown, M.T. Evaluation of a building using the emergy method. *Sol. Energy* **2005**, *79*, 204–212.
13. Pulselli, R.M.; Simoncini, F.M.; Pulselli, S.; Bastianoni, S. Emergy analysis of building manufacturing, maintenance and use: Em-building indices to evaluate housing sustainability. *Energy Build.* **2007**, *39*, 620–628.

14. Price, J.; Tilley, D.R. Emergy Evaluation of Green Façade. In Proceedings of the Sixth Biennial Emergy Research Conference, Gainesville, FL, USA, 14–16 January 2010.
15. Srinivasan, R.S.; Ingwersen, W.; Trucco, C.; Ries, R.; Campbell, D. Comparing energy-based indicators used in life cycle assessment tools for buildings. *Build. Environ.* **2014**, *79*, 138–151.
16. Yi, H.; Srinivasan, R.S.; Braham, W.W. An integrated energy-emergy approach to building form optimization: Use of energyplus, emergy analysis, and taguchi-regression method. *Build. Environ.* **2015**, *84*, 89–104.
17. Pulselli, R.M.; Simoncini, F.M.; Marchettini, N. Energy and emergy based cost benefit evaluation of building envelopes relative to geographical location and climate. *Build. Environ.* **2009**, *44*, 920–928.
18. Srinivasan, R.S.; Braham, W.W.; Campbell, D.E.; Curcija, C.D. Re(de)fining net zero energy: Renewable emergy balance in environmental building design. *Build. Environ.* **2012**, *47*, 300–315.
19. Holling, C.S. The Resilience of Terrestrial Ecosystems: Local Surprise and Global Change. In *Sustainable Development in the Biosphere*; Clark, W.M., Munn, R.E., Eds.; Oxford University Press: Oxford, UK, 1986; pp. 292–320.
20. Odum, H.T. *Systems Ecology*; Wiley: New York, NY, USA, 1983.
21. Odum, H.T. *Ecological and General Systems*; Reprint of Systems Ecology; John Wiley: Hoboken, NJ, USA, 1983; University Press of Colorado: Niwot, CO, USA, 1994.
22. Bastianoni, S.; Pulselli, R.M.; Pulselli, F.M. Models of withdrawing renewable and non-renewable resources based on Odum's energy systems theory and Daly's quasi-sustainability principle. *Ecol. Model.* **2009**, *220*, 1926–1930.
23. Rinker Hall, University of Florida. *AIA/COTE Top Ten Green Projects*; Rinker Hall, University of Florida: Gainesville, FL, USA, 2004. Available online: http://www2.aiatopten.org/hpb/overview.cfm?projectid=286 (accessed on 14 January 2014).
24. Stephany, D. The Value of Building Energy Audit Accuracy for Defining Effective Energy Efficient Measures. Master Thesis, University of Florida, Gainesville, FL, USA, August 2011.

Permissions

All chapters in this book were first published in Buildings, by MDPI; hereby published with permission under the Creative Commons Attribution License or equivalent. Every chapter published in this book has been scrutinized by our experts. Their significance has been extensively debated. The topics covered herein carry significant findings which will fuel the growth of the discipline. They may even be implemented as practical applications or may be referred to as a beginning point for another development.

The contributors of this book come from diverse backgrounds, making this book a truly international effort. This book will bring forth new frontiers with its revolutionizing research information and detailed analysis of the nascent developments around the world.

We would like to thank all the contributing authors for lending their expertise to make the book truly unique. They have played a crucial role in the development of this book. Without their invaluable contributions this book wouldn't have been possible. They have made vital efforts to compile up to date information on the varied aspects of this subject to make this book a valuable addition to the collection of many professionals and students.

This book was conceptualized with the vision of imparting up-to-date information and advanced data in this field. To ensure the same, a matchless editorial board was set up. Every individual on the board went through rigorous rounds of assessment to prove their worth. After which they invested a large part of their time researching and compiling the most relevant data for our readers.

The editorial board has been involved in producing this book since its inception. They have spent rigorous hours researching and exploring the diverse topics which have resulted in the successful publishing of this book. They have passed on their knowledge of decades through this book. To expedite this challenging task, the publisher supported the team at every step. A small team of assistant editors was also appointed to further simplify the editing procedure and attain best results for the readers.

Apart from the editorial board, the designing team has also invested a significant amount of their time in understanding the subject and creating the most relevant covers. They scrutinized every image to scout for the most suitable representation of the subject and create an appropriate cover for the book.

The publishing team has been an ardent support to the editorial, designing and production team. Their endless efforts to recruit the best for this project, has resulted in the accomplishment of this book. They are a veteran in the field of academics and their pool of knowledge is as vast as their experience in printing. Their expertise and guidance has proved useful at every step. Their uncompromising quality standards have made this book an exceptional effort. Their encouragement from time to time has been an inspiration for everyone.

The publisher and the editorial board hope that this book will prove to be a valuable piece of knowledge for researchers, students, practitioners and scholars across the globe.

List of Contributors

Ronald Rovers
Sustainable Building Support and Consultancy, Waalre 5581 HH, The Netherlands
Research Institute Built Environment of Tomorrow (RiBuilT), Zuyd University of Applied Sciences, Heerlen 6419 DJ, The Netherlands

Jianchao Zhang
Department of Electrical and Electronic Engineering, Auckland University of Technology, Auckland 1010, New Zealand

Boon-Chong Seet
Department of Electrical and Electronic Engineering, Auckland University of Technology, Auckland 1010, New Zealand

Tek Tjing Lie
Department of Electrical and Electronic Engineering, Auckland University of Technology, Auckland 1010, New Zealand

Mikael Boulic
School of Engineering and Advanced Technology, Massey University, Manawatu Campus, Palmerston North 4474, New Zealand

Robyn Anne Phipps
School of Engineering and Advanced Technology, Massey University, Albany Campus, Auckland 0632, New Zealand

Malcolm Cunningham
Building Research Association of New Zealand, Porirua 5381, New Zealand

Donald John Cleland
School of Engineering and Advanced Technology, Massey University, Manawatu Campus, Palmerston North 4474, New Zealand

Pär Fjällström
IVL Swedish Environmental Research Institute, P.O. Box 210 60, SE-100 31 Stockholm, Sweden

Keiko Abe
Institute of Environmental Biology, JDC Corporation, Kanagawa 243-03, Japan

Philippa Howden-Chapman
Department of Public Health, University of Otago, Wellington 6021, New Zealand

Natalie Allen
School of Architecture and Planning, The University of Auckland, 26 Symonds St, Auckland 1142, New Zealand

Xabat Oregi
Energy and Environment Division, Tecnalia Research and Innovation, Tecnalia, Azpeitia 20730, Spain

Patxi Hernandez
Energy and Environment Division, Tecnalia Research and Innovation, Tecnalia, Azpeitia 20730, Spain

Cristina Gazulla
UNESCO Chair in Life Cycle and Climate Change (ESCI – Pompeu Fabra University), Barcelona 08003, Spain

Marina Isasa
UNESCO Chair in Life Cycle and Climate Change (ESCI – Pompeu Fabra University), Barcelona 08003, Spain

Hideki Takebayashi
Department of Architecture, Graduate School of Engineering, Kobe University, 1-1 Rokkodai, Nada, Kobe 657-8501, Japan

Chihiro Yamada
Department of Architecture, Graduate School of Engineering, Kobe University, 1-1 Rokkodai, Nada, Kobe 657-8501, Japan

Paola Leardini
School of Architecture and Planning, The University of Auckland, 26 Symonds Street, 1010 Auckland, New Zealand

Manfredo Manfredini
School of Architecture and Planning, The University of Auckland, 26 Symonds Street, 1010 Auckland, New Zealand

Wajiha Shahzad
School of Engineering and Advanced Technology, Massey University, P/Bag 102904, Auckland 0745, New Zealand

Jasper Mbachu
School of Engineering and Advanced Technology, Massey University, P/Bag 102904, Auckland 0745, New Zealand

Niluka Domingo
School of Engineering and Advanced Technology, Massey University, P/Bag 102904, Auckland 0745, New Zealand

Javier González
Faculty of Architecture, Design and Planning, The University of Sydney, 148 City Road, Sydney 2006, Australia

Francesco Fiorito
Faculty of Architecture, Design and Planning, The University of Sydney, 148 City Road, Sydney 2006, Australia

Honnie Aguilar Leinartas
Department of Civil, Architectural and Environmental Engineering, Illinois Institute of Technology, Chicago, IL 60616, USA

Brent Stephens
Department of Civil, Architectural and Environmental Engineering, Illinois Institute of Technology, Chicago, IL 60616, USA

Shaan Cory
Centre for Building Performance Research, Victoria University of Wellington, P.O. Box 600, Wellington 6140, New Zealand

Michael Donn
Centre for Building Performance Research, Victoria University of Wellington, P.O. Box 600, Wellington 6140, New Zealand

Andrew Pollard
Building Research Association of New Zealand (BRANZ) Ltd, Private Bag 50908, Porirua 5240, New Zealand

Eleanor S. Lee
Building Technology and Urban Systems Division, Energy Technologies Area, Lawrence BerkeleyNational Laboratory, Mailstop 90-3111, 1 Cyclotron Road, Berkeley, CA 94720, USA

Xiufeng Pang
Building Technology and Urban Systems Division, Energy Technologies Area, Lawrence BerkeleyNational Laboratory, Mailstop 90-3111, 1 Cyclotron Road, Berkeley, CA 94720, USA

Andrew McNeil
Building Technology and Urban Systems Division, Energy Technologies Area, Lawrence BerkeleyNational Laboratory, Mailstop 90-3111, 1 Cyclotron Road, Berkeley, CA 94720, USA

Sabine Hoffmann
Building Technology and Urban Systems Division, Energy Technologies Area, Lawrence BerkeleyNational Laboratory, Mailstop 90-3111, 1 Cyclotron Road, Berkeley, CA 94720, USA

Anothai Thanachareonkit
Building Technology and Urban Systems Division, Energy Technologies Area, Lawrence BerkeleyNational Laboratory, Mailstop 90-3111, 1 Cyclotron Road, Berkeley, CA 94720, USA

Zhengrong Li
College of Mechanical Engineering, Tongji University, 1239 Siping Road, Shanghai 200092,

Yong Ding
Urban Construction and Environment Engineering, Chongqing University, 83 Shabei Street, Shapingba, Chongqing 400045, China

Jia Hu
Philips Research North America, Briarcliff Manor, NY 10510, USA

Eric Shen
Philips Research North America, Briarcliff Manor, NY 10510, USA

Yun Gu
Philips Research North America, Briarcliff Manor, NY 10510, USA

Ravi S. Srinivasan
M.E. Rinker, Sr. School of Construction Management, University of Florida, Gainesville, FL 32611, USA

Daniel E. Campbell
U.S. Environmental Protection Agency, Office of Research and Development, National Health and Environmental Effects Research Laboratory, Atlantic Ecology Division, Narragansett, RI 02882, USA

Wei Wang
M.E. Rinker, Sr. School of Construction Management, University of Florida, Gainesville, FL 32611, USA